Computational and Data-Driven Modeling of Combustion in Reciprocating Engines or Gas Turbines

Computational and Data-Driven Modeling of Combustion in Reciprocating Engines or Gas Turbines

Editors

Maria Cristina Cameretti
Roberta De Robbio

MDPI

Basel • Beijing • Wuhan • Barcelona • Belgrade • Novi Sad • Cluj • Manchester

Editors

Maria Cristina Cameretti	Roberta De Robbio
University of Naples	University of Naples
"Federico II"	"Federico II"
Naples	Naples
Italy	Italy

Editorial Office
MDPI AG
Grosspeteranlage 5
4052 Basel, Switzerland

This is a reprint of articles from the Special Issue published online in the open access journal *Energies* (ISSN 1996-1073) (available at: https://www.mdpi.com/journal/energies/special_issues/CDDM).

For citation purposes, cite each article independently as indicated on the article page online and as indicated below:

Lastname, A.A.; Lastname, B.B. Article Title. *Journal Name* **Year**, *Volume Number*, Page Range.

ISBN 978-3-7258-2573-8 (Hbk)
ISBN 978-3-7258-2574-5 (PDF)
doi.org/10.3390/books978-3-7258-2574-5

Contents

About the Editors

Maria Cristina Cameretti

Maria Cristina Cameretti (Ph.D) is a Full Professor at the Department of Industrial Engineering of the University of Naples "Federico II". She is the author of numerous papers in peer-reviewed scientific journals and conference proceedings. She has gained solid experience in CFD computation through 3D codes. Her main research activities are focused on simulation methodologies for internal combustion engines and gas turbines, cogeneration plants with gas turbines, and the use of innovative configurations or non-conventional fuels. She also has specific experience with three-dimensional computation codes as regards modeling injection and combustion phenomena in combustion chambers. Numerical investigation was also extended to ICE–electric hybrid powertrains and hybrid thermal/electric configurations for turboprop engines. Currently, she is in charge of the following courses: "Innovative Energy Systems" (level II degree in Electrical Engineering), "Power Plants with Gas Turbine", and "Modelling and Optimization of Internal Combustion Engines" (level II degree in Mechanical Engineering).

Roberta De Robbio

Roberta De Robbio (Ph.D) is currently an Assistant Professor at the Department of Industrial Engineering of the University of Naples "Federico II". The main topic of her research activity is combustion analysis and modeling by means of CFD tools. To this end, she conducted parallel studies on two different types of thermal plants: a micro gas turbine fueled with bio-fuels and hydrogen, and diesel engines supplied with natural gas, hydrogen and ammonia in dual-fuel mode. In the latter case, De Robbio was involved in experimental campaigns, specifically regarding an optically accessible single cylinder research engine. She teaches "Hydraulic and pneumatic drives" in the level I degree program in Mechatronics Engineering and "Thermo Fluid-dynamics of Machines" in the level II degree program in Mechanical Engineering for the Energy and the Environment. Further, she received the ATI "Alfano e Caputo" Award for the best PhD Thesis in 2020 and the Travel Award from the Energies Journal in 2023.

energies

Editorial

Computational and Data-Driven Modeling of Combustion in Reciprocating Engines or Gas Turbines

Maria Cristina Cameretti and Roberta De Robbio *

Department of Industrial Engineering, University of Naples Federico II, Via Claudio 21, 80125 Naples, Italy
* Correspondence: roberta.derobbio@unina.it

1. Introduction

The targets set by the Paris Agreement to limit greenhouse gas emissions and global warming aim to significantly reduce the levels of pollutants emitted in the atmosphere from all sectors, including transportation and land use energy production [1].

To this aim, new technologies are progressively entering the market as represented by electrification [2,3] or solutions to avoid the enhancement of climate-altering species, like carbon capture and storage [4,5].

Nevertheless, the well-established internal combustion energy systems, i.e., reciprocating engines and gas turbines, still represent a key in the short and medium term; hence, reducing all kinds of emissions and optimizing their operation is mandatory [6–9]. As a matter of fact, having a deep knowledge of the several processes occurring in the reactive domain or predicting the performance of the overall system can help to prevent the formation of harmful species [10], to study the feasibility of alternative primary energy sources and to optimize their combustion development [11].

To this aim, computational modeling represents a powerful tool, making it possible to investigate a wide range of operating conditions, even the extreme ones, avoiding risk of damage and reducing general costs, i.e., facility setup, maintenance, and supplies. Numerical modeling includes several approaches that can study the same problem on different levels. For example, one-dimensional and machine learning can analyze and predict the entire system performance [12,13], while CFD focuses in detail on the processes occurring inside a fluid domain [14]. Of course, the models must be validated before considering them reliable. Experimental investigations can provide the data to understand the behavior of the thermal energy system and, thus, provide the base for the building of the model itself. On the other hand, it is known that experimental activities and numerical computations are strictly linked.

The present Special Issue includes nine research articles and a review on the role of micro gas turbines (MGTs) in the distributed energy production. The nine articles' applications cover reciprocating internal combustion engines (ICEs) and gas turbines (GTs), making use of all the above-mentioned different modeling approaches, demonstrating a complete contribution to each field of the topic.

Citation: Cameretti, M.C.; De Robbio, R. Computational and Data-Driven Modeling of Combustion in Reciprocating Engines or Gas Turbines. *Energies* **2024**, *17*, 3863. https://doi.org/10.3390/en17163863

Received: 27 May 2024
Accepted: 12 June 2024
Published: 6 August 2024

2. Modeling of Reciprocating Internal Combustion Engines

Clean mobility is surely the most urgent topic. ICEs still represent the predominant energy conversion system for all sizes and kinds of land [15,16] and sea applications [17,18].

Diesel engines have been progressively banned in several countries due to the high emissions of particulate matter (PM) and nitrogen oxides (NO_x). However, it is well known that they feature levels of power, weight, efficiency, and reliability that are compatible with off-road and marine applications. For this reason, researchers are continuing to study new methodologies and strategies to reduce the above-mentioned pollutants in this type of engine.

To this aim, the Dual Fuel (DF) combustion concept is a promising solution. It consists of the partial substitution of diesel fuel with a cleaner fuel, such as natural gas. However, the high emissions of unburned hydrocarbons (UHCs) have led to attempts to improve the combustion characteristics with the addition of hydrogen [19,20].

Cameretti et al. (Contribution 1) examined the overall combustion development in a single-cylinder research engine compression ignition engine via CFD simulations of hydrogen/methane blends as primary fuels. Starting from experimental measurements to validate the full methane and full hydrogen cases, a progressive enhancement of hydrogen energy share is numerically analyzed. The results demonstrated that the presence of hydrogen leads not only to reduced emissions of carbon-based species but also to a strong increase in temperature and consequently NO_x.

Abdelhameed and Tashima (Contribution 2) studied two important techniques for a reduction in NO_x in marine engines: exhaust gas recirculation (EGR) and water-in-fuel emulsion. Both cause a decrease in thermal efficiency. Therefore, their combination is modeled with a synergetic approach that involved CFD and 0D/1D tools on the basis of experimental results to minimize NO_x while maintaining an adequate level of thermal efficiency.

Falai and Misul (Contribution 3) used machine learning to develop virtual sensors for monitoring NO_x emissions in diesel engines. The virtual sensor is calibrated via the Extreme Gradient Boosting (XGBoost) algorithm on the basis of collected data from a test bench at steady-state conditions, claiming a prediction accuracy of almost 98%. Instead, the accuracy reduces to 85% in dynamic conditions as demonstrated in an on-road driving mission.

Doppalapudi et al. (Contribution 4) numerically investigated the effects of the shape of the combustion chamber on injection profiles and in-cylinder distributions of temperature, velocity, and turbulent motions (swirl and tumble ratio), with the aim to assess the most improved configuration in terms of combustion characteristics. They evaluated five chamber shapes and suggested avoiding wide bowls that may induce uneven combustion across the cylinder. Instead, better results can be obtained with split bowls.

Spark ignition (SI) engines, as leading powertrain technologies, are being investigated and improved as well. As widely discussed by Sforza et al. (Contribution 5), a possible solution to minimize thermal losses, while reducing to zero soot particles and drastically reducing NO_x, is to operate the engine with homogeneous ultra-lean premixed mixtures (CH_4-air). Nevertheless, the progressive shift towards increasingly high values of the air-to-fuel ratios raises doubts in the applicability of the most commonly used numerical models for SI engines. To this aim, the authors of the paper evaluated two different flamelet-based models, typically used for premixed mixtures: the Coherent Flame Model and the Flame Area Model. The results were compared with experiments carried out on a Darmstadt TSF burner. They observed that, near the burner exit, the flamelet assumption is still valid and the models accurately predict the combustion behavior. In contrast, in the typical conditions of homogeneous ultra-lean mixtures, the models are no more capable of providing reliable results. Thus, the authors suggested modifications and extensions to the flamelet-based models to improve CFD simulations.

Following the path of using carbon-free fuel in order to drastically reduce CO_2 emissions, the paper by D'Antuono et al. (Contribution 6) concerned the study of an ammonia/hydrogen blend, 85% and 15%, respectively, by volume, in a light-duty SI engine using a 1D predictive model. Ammonia features low combustion characteristics that limit its operating capacity as pure fuel; however, with respect to hydrogen, it is easier to store on vehicles. To overcome such an issue, the authors demonstrated that the addition of H_2 to the air–fuel mixture allows us to improve the flammability of NH_3, extending the range of operation of the engine.

3. Modeling of Gas Turbines

Since gas turbines of all sizes are employed in the aviation industry and power generation, researchers are committed to reducing the emissions from this type of internal combustion engine, which is also the object of this Special Issue, as well as complying with the regulations set for these sectors [21,22].

Castiglione et al. (Contribution 7) evidenced that the most viable option to achieve a low-emission aero-propulsion system is the utilization of alternative bio and synthetic fuels, the so-called sustainable aviation fuels (SAFs), instead of fossil fuels. When supplied with new fuels, the engine might undergo architectural modifications; consequently, advanced control systems are mandatory to ensure a safe and efficient operation. Therefore, the authors developed linear models for direct fuel control of a turboshaft aircraft engine obtained using two approaches: Small Perturbation and System Identification. Moreover, the linear models included the component degradation, demonstrating a reliable link with performance, especially when the Small Perturbation technique is adopted.

Clean bioenergy generation in gas turbines is the focus of the paper by Lu at al. (Contribution 8). The authors proposed an optimized methodology for the developed pressurized Chemical Looping Combustors via a zero-dimensional model validated by experimental data. Through such technology, combustion occurs in two separated reactors (air and fuel). The modeling of the air reactor can be found in a previous paper [23], while in the present Special Issue, a metal oxide is reduced in the fuel reactor, providing the oxygen for the oxidation of the fuel, and then regenerated in the air reactor. At the fuel reactor outlet, pure carbon dioxide can be obtained by condensing water vapor. In this way, the CO_2 can be easily compressed and stored.

As already observed for ICEs, hydrogen and ammonia are interesting alternative fuels for gas turbines as well, as demonstrated by Giacomazzi et al. (Contribution 9). However, to use such alternative fuels and simultaneously reduce NO_x emissions, it is necessary to operate a reassessment of the combustor. Indeed, the flame structure of NH_3 could favor NO formation. Thus, the authors investigated the micro-mixing combustion system of a non-premixed jet flame of $NH_3/H_2/N_2$ blend in air crossflow via Direct Numerical Simulation (DNS). Such a simple configuration provided evidence of the turbulent mixing and its effects on the flame structure.

Finally, this Special Issue includes a wide and detailed review on the use of micro gas turbines [24]. MGTs are gaining more attention in the current energy production scenario because it is shifting from a centralized structure to a distributed one with the necessity to exploit local renewable resources. In this regard, MGTs are compact and highly flexible, being able to operate with different fuels and at different load levels [25]. The review illustrates in detail the combustion of several liquid and gaseous fuels, also obtained from biomasses considered to emit net zero CO_2. In addition, due to the reduced size and the typical high temperature at the exhaust of gas turbines, this particular power plant can be perfectly integrated with other energy systems with a low environmental impact, for example, solar fields, ORC, and fuel cells, and it can be utilized in CHP and CCHP assets.

Conflicts of Interest: The authors declare no conflicts of interest.

List of Contributions:

1. Cameretti, M.C.; De Robbio, R.; Mancaruso, E.; Palomba, M. CFD Study of Dual Fuel Combustion in a Research Diesel Engine Fueled by Hydrogen. *Energies* **2022**, *15*, 5521. https://doi.org/10.3390/en15155521.
2. Abdelhameed, E.; Tashima, H. EGR and Emulsified Fuel Combination Effects on the Combustion, Performance, and NOx Emissions in Marine Diesel Engines. *Energies* **2022**, *16*, 336. https://doi.org/10.3390/en16010336.
3. Falai, A.; Misul, D.A. Data-Driven Model for Real-Time Estimation of NOx in a Heavy-Duty Diesel Engine. *Energies* **2023**, *16*, 2125. https://doi.org/10.3390/en16052125.
4. Doppalapudi, A.T.; Azad, A.K.; Khan, M.M.K. Analysis of Improved In-Cylinder Combustion Characteristics with Chamber Modifications of the Diesel Engine. *Energies* **2023**, *16*, 2586. https://doi.org/10.3390/en16062586.
5. Sforza, L.; Abdelwahid, S.; Lucchini, T.; Onorati, A. Ultra-Lean Premixed Turbulent Combustion: Challenges of RANS Modelling. *Energies* **2022**, *15*, 5947. https://doi.org/10.3390/en15165947.

6. D'Antuono, G.; Lanni, D.; Galloni, E.; Fontana, G. Numerical Modeling and Simulation of a Spark-Ignition Engine Fueled with Ammonia-Hydrogen Blends. *Energies* **2023**, *16*, 2543. https://doi.org/10.3390/en16062543.
7. Castiglione, T.; Perrone, D.; Strafella, L.; Ficarella, A.; Bova, S. Linear Model of a Turboshaft Aero-Engine Including Components Degradation for Control-Oriented Applications. *Energies* **2023**, *16*, 2634. https://doi.org/10.3390/en16062634.
8. Lu, W.; Bartocci, P.; Abad, A.; Bischi, A.; Yang, H.; Cabello, A.; de Las Obras Loscertales, M.; Zampilli, M.; Fantozzi, F. Dimensioning Air Reactor and Fuel Reactor of a Pressurized CLC Plant to Be Coupled to a Gas Turbine: Part 2, the Fuel Reactor. *Energies* **2023**, *16*, 3850. https://doi.org/10.3390/en16093850.
9. Giacomazzi, E.; Cecere, D.; Cimini, M.; Carpenella, S. Direct Numerical Simulation of a Reacting Turbulent Hydrogen/Ammonia/Nitrogen Jet in an Air Crossflow at 5 Bar. *Energies* **2023**, *16*, 7704. https://doi.org/10.3390/en16237704.

References

1. Paris Agreement to the United Nations Framework Convention on Climate Change, December 12, 2015, T.I.A.S. No. 16-1104. Available online: https://www.state.gov/16-1104/ (accessed on 11 May 2024).
2. Yildiz, A.F.; Sahinler, R.; Ozturk, C. Data-Driven Modelling of a Plug-In Hybrid Electric Vehicle. In Proceedings of the 2023 14th International Conference on Electrical and Electronics Engineering (ELECO), Bursa, Turkiye, 30 November–2 December 2023; pp. 1–5. [CrossRef]
3. Camacho, M.d.l.N.; Jurburg, D.; Tanco, M. Hydrogen Fuel Cell Heavy-Duty Trucks: Review of Main Research Topics. *Int. J. Hydrog. Energy* **2022**, *47*, 29505–29525. [CrossRef]
4. Li, X.; Zhou, R.; Yang, H.; Liang, Z.; Yao, Y.; Yu, Z.; Du, M.; Lou, D.; Li, K. Properties of Carbonic Anhydrase-Containing Active Coatings for CO_2 Capture. *Processes* **2024**, *12*, 810. [CrossRef]
5. Shervani, S.; Tansug, L.P.; Tezel, F.H. Microporous Adsorbent-Based Mixed Matrix Membranes for CO_2/N_2 Separation. *Energies* **2024**, *17*, 1927. [CrossRef]
6. Godiño, J.A.V.; García, M.T.; Aguilar, F.J.J.-E. Experimental Investigation and Modelling of Biodiesel Combustion in Engines with Late Direct Injection Strategy. *Energy Rep.* **2022**, *8*, 7476–7487. [CrossRef]
7. Napolitano, P.; Di Domenico, D.; Di Maio, D.; Guido, C.; Golini, S. Ultra-Fine Particle Emissions Characterization and Reduction Technologies in a NG Heavy Duty Engine. *Atmosphere* **2022**, *13*, 1919. [CrossRef]
8. Zucareli de Souza, T.A.; Frez, G.V.; Pinto, G.; Costa, R.; Roque, L.F.A.; Coronado, C.J.R.; Vidigal, L.P.V. *CFD Analysis of Different Biogas Upgrading Levels for Dual-Fuel Operation in Diesel Engines*; SAE Technical Paper 2023-24-0055; SAE International: Capri, Naples, Italy, 2023. [CrossRef]
9. Barone, G.; Buonomano, A.; Del Papa, G.; Maka, R.; Palombo, A. How to Achieve Energy Efficiency and Sustainability of Large Ships: A New Tool to Optimize the Operation of on-Board Diesel Generators. *Energy* **2023**, *282*, 128288. [CrossRef]
10. Breer, B.; Rajagopalan, H.; Godbold, C.; Johnson, H.; Emerson, B.; Acharya, V.; Sun, W.; Noble, D.; Lieuwen, T. Numerical Investigation of NOx Production from Premixed Hydrogen/Methane Fuel Blends. *Combust. Flame* **2023**, *255*, 112920. [CrossRef]
11. Zhu, J.; Wang, Z.; Li, R.; Liu, S.; Hua, Y. Experimental Study and Prediction Model of Combustion Stability and Combustion Mode Variation of Burning Methanol/Biodiesel Blends for Diesel Engines. *Fuel* **2023**, *335*, 127038. [CrossRef]
12. Galloni, E.; Lanni, D.; Fontana, G.; D'Antuono, G.; Stabile, S. Performance Estimation of a Downsized SI Engine Running with Hydrogen. *Energies* **2022**, *15*, 4744. [CrossRef]
13. Shetty, S.; Shetty, R.P.; Shrinivasa Rao, B.R. Experimental Investigation and Artificial Neural Network Modelling of Combustion Pressure Parameters of Dual Spark Plug SI Engine. *Mater. Today Proc.* **2023**, *92*, 189–201. [CrossRef]
14. Füzesi, D.; Wang, S.; Józsa, V.; Chong, C.T. Ammonia-Methane Combustion in a Swirl Burner: Experimental Analysis and Numerical Modeling with Flamelet Generated Manifold Model. *Fuel* **2023**, *341*, 127403. [CrossRef]
15. Perrone, D.; Castiglione, T.; Falbo, L.; Bova, S.; D'Epiro, C. *1-D Numerical Model of a Spark Ignition Engine Fueled with Methanol for Off-Grid Charging Stations*; SAE Technical Paper 2023-24-0098; SAE International: Capri, Naples, Italy, 2023. [CrossRef]
16. Ramognino, F.; Sforza, L.; D'Errico, G.; Gomez-Soriano, J.; Onorati, A.; Novella, R. *CFD Modelling of Hydrogen-Fueled SI Engines for Light-Duty Applications*; SAE Technical Paper 2023-24-0017; SAE International: Capri, Naples, Italy, 2023. [CrossRef]
17. Ichikawa, Y.; Niki, Y.; Takasaki, K.; Kobayashi, H.; Miyanagi, A. Experimental Study of Combustion Process of NH3 Stratified Spray Using Imaging Methods for NH3 Fueled Large Two-Stroke Marine Engine. *Appl. Energy Combust. Sci.* **2023**, *13*, 100119. [CrossRef]
18. Nyongesa, A.J.; Kim, J.K.; Lee, W.-J. Investigation on the Combustion of Ammonia Using Direct High/Medium-Pressure-Otto Injection Approach in a Diesel Two-Stroke Marine Slow Speed Engine. *J. Energy Inst.* **2024**, *114*, 101641. [CrossRef]
19. Sehili, Y.; Loubar, K.; Tarabet, L.; Cerdoun, M.; Lacroix, C. Development of Predictive Model for Hydrogen-Natural Gas/Diesel Dual Fuel Engine. *Energies* **2023**, *16*, 6943. [CrossRef]
20. De Robbio, R.; Cameretti, M.C.; Mancaruso, E.; Tuccillo, R.; Vaglieco, B.M. *CFD Analysis of Different Methane/Hydrogen Blends in a CI Engine Operating in Dual Fuel Mode*; SAE Technical Paper 2022-01-1056; SAE International: Krakow, Poland, 2022. [CrossRef]

21. Inoue, C.; Higaki, S.; Fujii, H.; Ishikawa, N.; Ueda, Y. On-Site Direct Imaging of Carbon Dioxide Emissions from Aircraft during Landing and Take-off Cycle. *Environ. Adv.* **2023**, *13*, 100432. [CrossRef]
22. Carusotto, S.; Goel, P.; Baratta, M.; Misul, D.A.; Salvadori, S.; Cardile, F.; Forno, L.; Toppino, M.; Valsania, M. Combustion Characterization in a Diffusive Gas Turbine Burner for Hydrogen-Compliant Applications. *Energies* **2022**, *15*, 4117. [CrossRef]
23. Bartocci, P.; Abad, A.; Bischi, A.; Wang, L.; Cabello, A.; de Las Obras Loscertales, M.; Zampilli, M.; Yang, H.; Fantozzi, F. Dimensioning Air Reactor and Fuel Reactor of a Pressurized Chemical Looping Combustor to Be Coupled to a Gas Turbine: Part 1, the Air Reactor. *Energies* **2023**, *16*, 2102. [CrossRef]
24. De Robbio, R. Micro Gas Turbine Role in Distributed Generation with Renewable Energy Sources. *Energies* **2023**, *16*, 704. [CrossRef]
25. Di Nardo, A.; Bo, A.; Calchetti, G.; Giacomazzi, E.; Messina, G. Study on the Fuel Flexibility of a Microgas Turbine Combustor Burning Different Mixtures of H_2, CH_4, and CO_2. *J. Eng. Gas Turbine Power* **2020**, *142*, 061001. [CrossRef]

energies MDPI

Article

CFD Study of Dual Fuel Combustion in a Research Diesel Engine Fueled by Hydrogen

Maria Cristina Cameretti [1], Roberta De Robbio [1,*], Ezio Mancaruso [2] and Marco Palomba [1]

[1] Department of Industrial Engineering, University of Naples Federico II, Via Claudio 21, 80125 Naples, Italy; mc.cameretti@unina.it (M.C.C.); marco.palomba3@studenti.unina.it (M.P.)
[2] STEMS—CNR, Viale G. Marconi 4, 80125 Naples, Italy; ezio.mancaruso@stems.cnr.it
* Correspondence: roberta.derobbio@unina.it

Abstract: Superior fuel economy, higher torque and durability have led to the diesel engine being widely used in a variety of fields of application, such as road transport, agricultural vehicles, earth moving machines and marine propulsion, as well as fixed installations for electrical power generation. However, diesel engines are plagued by high emissions of nitrogen oxides (NO_x), particulate matter (PM) and carbon dioxide when conventional fuel is used. One possible solution is to use low-carbon gaseous fuel alongside diesel fuel by operating in a dual-fuel (DF) configuration, as this system provides a low implementation cost alternative for the improvement of combustion efficiency in the conventional diesel engine. An initial step in this direction involved the replacement of diesel fuel with natural gas. However, the consequent high levels of unburned hydrocarbons produced due to non-optimized engines led to a shift to carbon-free fuels, such as hydrogen. Hydrogen can be injected into the intake manifold, where it premixes with air, then the addition of a small amount of diesel fuel, auto-igniting easily, provides multiple ignition sources for the gas. To evaluate the efficiency and pollutant emissions in dual-fuel diesel-hydrogen combustion, a numerical CFD analysis was conducted and validated with the aid of experimental measurements on a research engine acquired at the test bench. The process of ignition of diesel fuel and flame propagation through a premixed air-hydrogen charge was represented the Autoignition-Induced Flame Propagation model included ANSYS-Forte software. Because of the inefficient operating conditions associated with the combustion, the methodology was significantly improved by evaluating the laminar flame speed as a function of pressure, temperature and equivalence ratio using Chemkin-Pro software. A numerical comparison was carried out among full hydrogen, full methane and different hydrogen-methane mixtures with the same energy input in each case. The use of full hydrogen was characterized by enhanced combustion, higher thermal efficiency and lower carbon emissions. However, the higher temperatures that occurred for hydrogen combustion led to higher NO_x emissions.

Keywords: diesel engine; dual fuel; hydrogen; combustion modeling; CFD

Citation: Cameretti, M.C.; De Robbio, R.; Mancaruso, E.; Palomba, M. CFD Study of Dual Fuel Combustion in a Research Diesel Engine Fueled by Hydrogen. *Energies* **2022**, *15*, 5521. https://doi.org/10.3390/en15155521

Academic Editor: Constantine D. Rakopoulos

Received: 8 July 2022
Accepted: 25 July 2022
Published: 29 July 2022

1. Introduction

The diesel engine is used in a variety of fields of application, such as road transport, agricultural vehicles, earth-moving machines, rail, marine propulsion, and fixed installations for electrical power generation, due to its superior fuel economy, high torque and durability [1]. However, high nitrogen oxide (NO_x) and particulate matter (PM) emissions have led to the development of low-temperature combustion (LTC) strategies, such as homogeneous charge compression ignition (HCCI), premixed charge compression ignition (PCCI), and reactivity controlled compression ignition (RCCI) which emit less NO_x and PM because of lower flame temperature and lean combustion, respectively [2].

Dual fuel (DF) combustion is a potential alternative to these combustion technologies as it requires minimal engine modifications. DF combustion involves the use of two fuels: the first with a high octane number (primary fuel), which is injected in the intake manifold, and the second with a lower octane number (secondary fuel, usually diesel fuel), which is injected just before the top dead center (TDC) to act as a source of ignition for the propagation of multiple flame fronts which burn the air-primary fuel mixture [3].

In recent years, research has focused on the study of DF combustion with natural gas (NG) used as the primary fuel [4–6]. The low C/H ratio of NG results in less CO_2 emissions compared to an engine that works in full diesel mode. Moreover, the lean NG/air lean mixture leads to a decrease in the maximum temperatures and, consequently, of NO_x emissions. At the same time, large quantities of carbon monoxide (CO) and unburned hydrocarbons (UHC) are produced since the equivalence ratio falls outside flammability limits, preventing efficient propagation of the flame front [7,8].

The use of gaseous hydrogen as a primary fuel provides more significant advantages in terms of emissions since hydrogen is a carbon-free fuel. Furthermore, the high low heating value and laminar flame speed result in improved thermal efficiency and a faster combustion process [9]. However, the higher adiabatic flame temperature of hydrogen compared to other fossil fuels results in large emissions of nitrogen oxides [10].

The use of hydrogen causes a decrease in volumetric efficiency as a result of the low density of hydrogen, which leads to the displacement of greater volumes of air with increasing hydrogen [11], and a strong tendency to knocking (autoignition of hydrogen before the ignition of the diesel fuel) that occurs particularly with large hydrogen energy contributions, and with high loads and compression ratios [12]. One of the biggest problems in the use of hydrogen is the high production cost. Therefore, researchers have focused on the use of hydrogen–methane mixtures as a primary fuel in DF engines to combine the advantages of the two fuels. Experimental studies [13–15] have demonstrated that blends with a higher quantity of hydrogen lead to a major peak in pressure, enhanced diesel fuel combustion (due to an advanced start to the combustion), and reduced combustion duration of the gaseous fuel phase (due to the rapid combustion speed and wider combustion limits of hydrogen). Furthermore, the use of large hydrogen proportions has been shown to be effective in lowering carbon emissions, while NO_x emissions sharply increase.

The study of combustion as the fuel used varies is of fundamental importance. A correct understanding of combustion phenomena can be achieved using a numerical three-dimensional approach. CFD 3D techniques are widely used since they provide an accurate description of the chemical and physical processes which occur in the combustion chamber and a detailed description of the flow field. Interest in this topic is demonstrated by various reports available in the scientific literature for the use of both DF methane [16–19] and hydrogen [20,21]. However, the main challenge in numerical simulations concerns the combustion model, since it needs to take into account both the combustion and the interaction of the primary with the secondary fuel. Most of the studies available use a reduced kinetic mechanism which involves a diesel surrogate (typically n-heptane or n-dodecane), methane and hydrogen [22–25].

In alignment with current research, the authors sought to investigate the modeling of the combustion of a dual fuel diesel engine, with a particular emphasis on the propagation of the flame front and evaluation of the laminar flame speed (LFS). To simulate the combustion process, a kinetic mechanism was created by merging GRIMECH 3.0 with a detailed scheme proposed by Ra and Reitz [26] for n-dodecane, used as a surrogate for diesel fuel. The mechanism, containing 124 species and 660 reactions, coupled with an autoignition-induced flame propagation model, was implemented in ANSYS Forte® code. The model was shown to be reliable for describing dual fuel combustion in a previous paper [27]. Using the GRIMECH 3.0 mechanism in ANSYS Chemkin-Pro, tables with values for laminar flame speed as a function of pressure, temperature and equivalence ratio were obtained. The accuracy of this approach was verified by comparing the trend in the laminar flame speed derived from tables with the LFS obtained using well-established correlations. The approach described represents a valuable method for modeling dual-fuel combustion phenomena.

Since the authors have focused extensively on DF operation with methane in a single cylinder research engine in previous experimental and numerical work [28–31], this paper mainly seeks to highlight differences in the use of hydrogen as a primary fuel compared to methane for the same research engine. In particular, an analysis was conducted by comparing the experimental results with methane and hydrogen and then by comparing the numerical 3D CFD results obtained for different test cases with hydrogen, methane and hydrogen-methane mixtures, with the same total energy input, but varying the energy provided by hydrogen (i.e., 0, 25, 50, 75 and 100%). In the simulations performed, the focus was on the development of the combustion process and evaluation of chemical species trends, including pollutant emissions, by testing the goodness of fit of the combustion model used.

2. Experimental Test Cases

The experiments were performed on a research single-cylinder diesel engine in dual fuel mode as described in previous papers [28–31].

The engine (Table 1) was equipped with a common rail injection system and the cylinder head of a real four-cylinder diesel engine. Valve timings were fixed for all tests; their values are listed in Table 2. Diesel injection was achieved with the use of a seven-hole-injector, and was controlled by an electronic control unit (ECU). The methane or hydrogen gaseous fuel was injected in the intake manifold using a commercial gas PFI electro-injector with a maximum pressure of injection of 5 bar. The port fuel injection was controlled by a delay unit synchronized with the engine shaft encoder. The injection system specifications are summarized in Table 3.

Table 1. Engine specifications.

Engine Type	Stroke [mm]	Bore [mm]	Cylinder Volume [cm^3]	Bowl [cm^3]	Compression Ratio
4-stroke single-cylinder 4 valves	92	85	522	19.7	16.5:1

Table 2. Opening and closing valve timing.

EVO	EVC	IVO	IVC
116° ATDC	340° BTDC	344° ATDC	132° BTDC

Table 3. Injection system specifications.

Diesel Injection System	Number of Holes	Cone Angle Axis [deg]	Hole Diameter [mm]	H_2/CH_4 Injection System	Holes Number of H_2/CH_4 Injector	Maximum PFI Pressure [bar]
Common Rail	7	148	0.141	PFI	1	5

A piezoelectric pressure sensor installed in the cylinder head, in place of a preheating glow plug, alongside a multichannel acquisition system, enabled measurement of the pressure cycles. The experimental in-cylinder pressure cycles were taken as benchmarks to represent the average cycle over 200 cycles, assuming the uncertainty of the measurements to be equal to ±0.5 bar.

The test cases involved dual-fuel operation, varying the premixed fuel (natural gas or hydrogen); the engine operating conditions are listed in Tables 4 and 5.

The initiation of the pilot and main injections (SOI) was dependent on the current signal, so a mechanical delay (about 300 μs) was needed for opening of the nozzle. Experimental measurements of the fuel and air mass flow rates taken enabled calculation of the thermal energy provided by the fuels, the equivalence ratio and premixed ratio (*RP*).

$$RP = \frac{(m_p LHV_p)}{(m_p LHV_p + m_d LHV_d)} \times 100 \tag{1}$$

where m_p and m_d are the mass flow rate of the premixed fuel (methane or hydrogen) and the directly injected fuel (diesel), respectively, and LHV_p and LHV_d are their low heating values.

Table 4. Engine operating conditions for the cases with methane.

	M1	M2	M3	M4
Engine speed [rpm]	1500	1500	2000	2000
IMEP [bar]	1.9	4.8	2	4.3
Pilot SOI [deg]	−16	−11.6	−21.2	−18.6
Main SOI [deg]	−6	0.3	−8	−2.4
Dwell [deg]	10	11.9	13.2	16.2
Rail pressure [bar]	615	867	700	891
Pilot and main duration [deg]	2.6	2.4	3.4	3.1
Pilot and main diesel mass injected [mg/cycle]	0.711	0.822	0.717	0.833
Thermal energy from diesel [J]	61.15	70.69	61.66	71.64
Methane mass [mg/cycle]	7.58	11.19	7.55	10.15
Thermal energy from methane [J]	379	559.5	377.5	507.5
Total thermal energy from fuels [J]	440.15	630.19	439.16	579.16
Inlet pressure [bar]	1.5	1.7	1.5	1.7
Inlet temperature [°C]	44	46	50	51
Air mass [mg/cycle]	750.9	802.4	679.3	741.2
Methane/air ER	0.174	0.240	0.192	0.236
RP [%]	86.10	88.78	85.97	87.62

Table 5. Engine operating conditions for the test cases with hydrogen.

	H1	H2	H3	H4
Engine speed [rpm]	1500	1500	2000	2000
IMEP [bar]	0.9	3	0.9	2.7
Pilot SOI [deg]	−16	−11.6	−21.2	−18.6
Main SOI [deg]	−6	0.3	−8	−2.4
Dwell [deg]	10	11.9	13.2	16.2
Rail pressure [bar]	615	867	700	891
Pilot and main duration [deg]	2.6	2.4	3.4	3.1
Pilot and main diesel mass injected [mg/cycle]	0.711	0.822	0.717	0.833
Thermal energy from diesel [J]	61.15	70.69	61.66	71.64
Hydrogen mass [mg/cycle]	2.45	4.35	2.32	3.92
Thermal energy from hydrogen [J]	295	523	279	470
Total thermal energy from fuels [J]	356.15	593.69	340.66	541.64
Inlet pressure [bar]	1.5	1.7	1.5	1.7
Inlet temperature [°C]	44	51	55	59
Air mass [mg/cycle]	727.5	796	672	738.6
Hydrogen/air ER	0.115	0.187	0.118	0.181
RP [%]	82.8	88.1	81.9	86.8

It is worth noting that the tests were characterized by similarly low values of the equivalence ratio (ranging from 0.115 to 0.240) and of the premixed ratio (above 81%). In particular, the high RP values in Table 4 indicated that most of the energy supply was provided by the primary fuel (methane or hydrogen). Figures 1–6 display the experimental in-cylinder pressure, the rate of heat release and the accumulated heat release for methane and hydrogen. The curves in the diagrams corresponding to the experimental cases with the same diesel mass flow and injection timings (SOI and duration of pilot and main injections) have the same color.

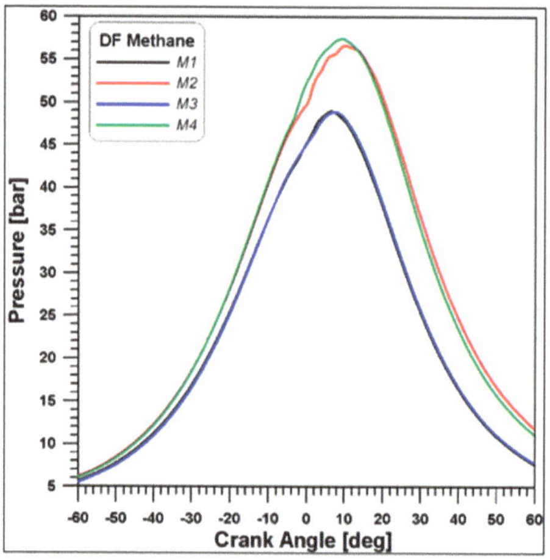

Figure 1. Experimental in-cylinder pressure. Cases with methane.

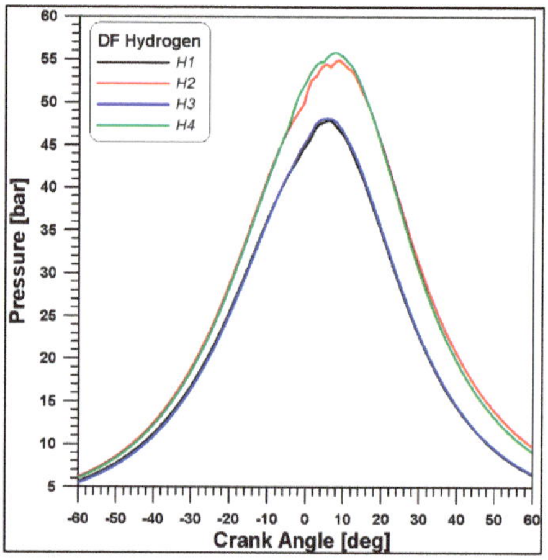

Figure 2. Experimental in-cylinder pressure. Cases with hydrogen.

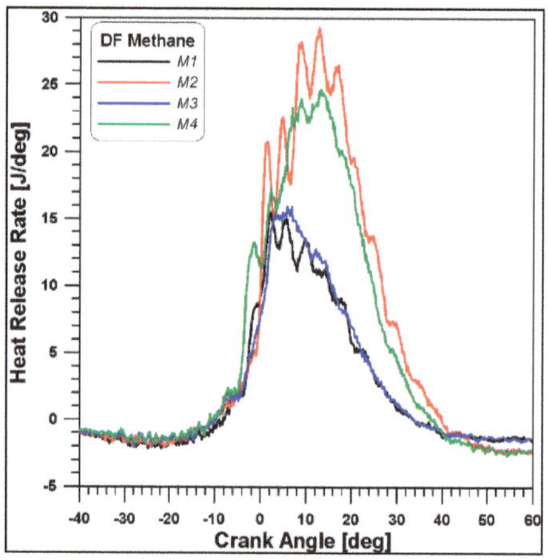

Figure 3. Experimental heat release rate. Cases with methane.

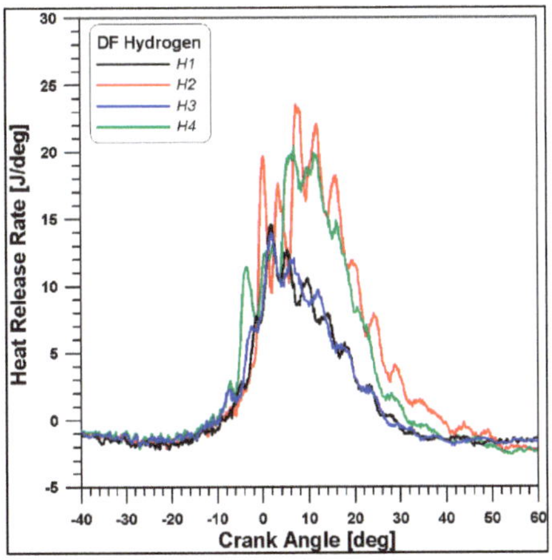

Figure 4. Experimental heat release rate. Cases with hydrogen.

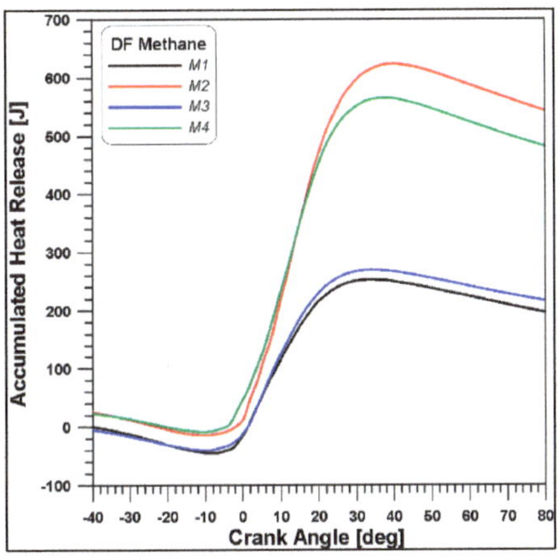

Figure 5. Accumulated heat release. Cases with methane.

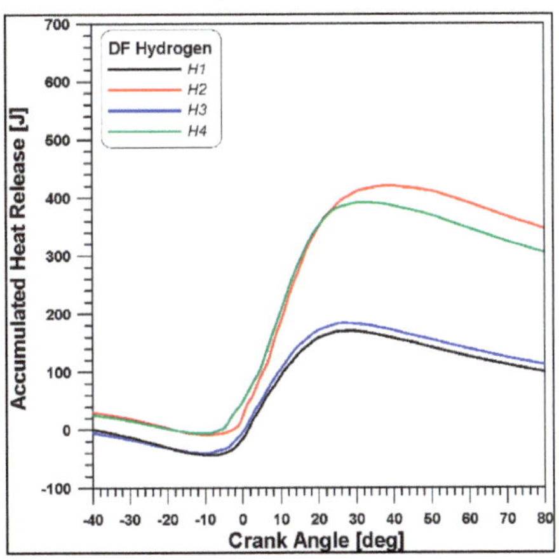

Figure 6. Accumulated heat release. Cases with hydrogen.

Similar considerations apply to both fuels, whose curves present the same trends. The pressure graphs (Figures 1 and 2) show that the difference in the pressure peak was mainly due to the different IMEP values (the red and green curves are at a higher load). However, the engine speed also affected the pressure, since higher speeds (blue and green curves are at 2000 rpm) led to higher peaks of pressure for similar IMEP values.

The rate of heat release shown in Figures 3 and 4 reflects typical dual-fuel combustion behavior: in the first part of the combustion process, only a small amount of the pilot diesel fuel burns and this enables control of the ignition timing of the air-fuel mixture that occurs when the main diesel injection is provided. However, when the pilot injection occurs long in advance, and the energy provided by the primary fuel is sufficiently high, the pilot fuel causes the ignition of a part of the methane–air or hydrogen–air mixture (green curves). Furthermore, the heat release rate indicates that the higher energy contribution provided by the primary fuel led to a more rapid and intense combustion (greater slope of the green curves at 5° BTDC and higher heat release rate peaks).

The accumulated heat release rate diagrams (Figures 5 and 6) confirm this trend, showing how cases with higher energy inputs correspond to the major final values of the total energy.

Finally, a comparison between the case M3 (methane) and H4 (hydrogen) is reported in Figure 7. These two cases are taken as the basis for a first comparison since they have the same engine speed, and the values of the equivalence ratio and RP are very similar. For a better comparison, the graph of the heat release rate is normalized with respect to the total fuel energy.

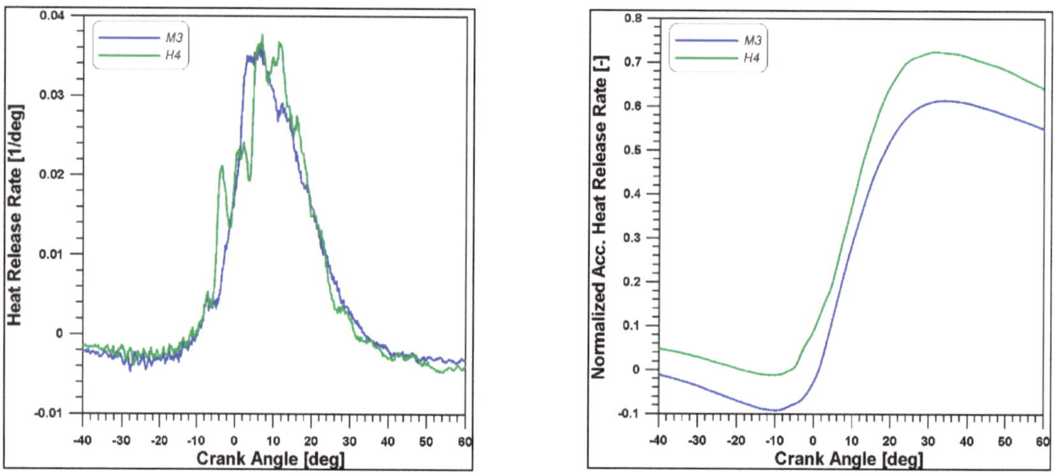

Figure 7. Comparison of normalized heat release rate and accumulated heat release (M3 vs. H4).

It was observed that the ignition of the pilot diesel fuel occurred at about the same crank angle although the injection of diesel fuel occurred later for the H4 case (SOI pilot of $-21.2°$ for M3 vs. SOI pilot of $-18.6°$ for H4). Thus, the presence of hydrogen accelerated the diesel combustion reactions. Consequently, the pilot injection was sufficient to burn part of the hydrogen–air mixture whereas the main injection was necessary for the combustion of the methane. Moreover, the hydrogen combustion appeared faster since the slope of the curve was greater.

A further comparison between the M2 and H2 cases (Figure 8) confirmed that, even when the energy inputs were similar (559.5 J of methane potential thermal energy vs. 523 J of hydrogen potential thermal energy), the ignition of hydrogen occurred in advance, confirming its higher reactivity.

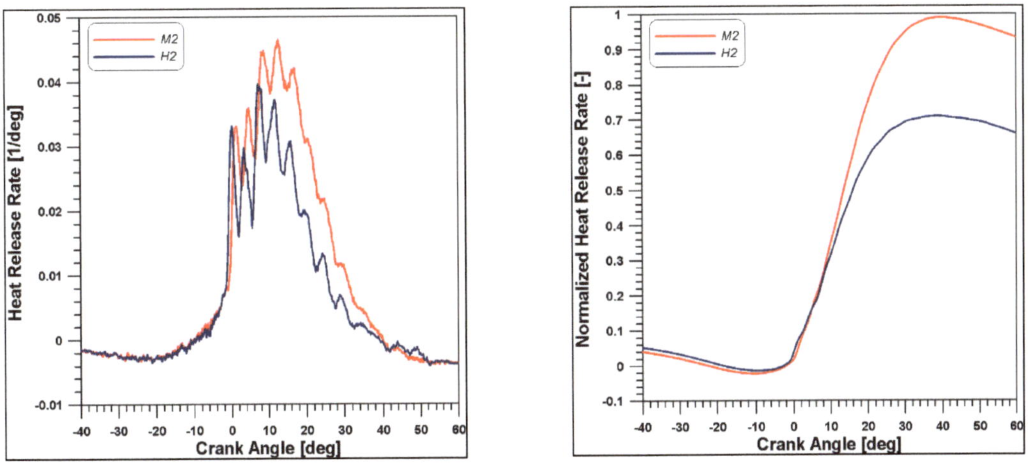

Figure 8. Comparison of normalized heat release rates and accumulated heat release (M3 vs. H4).

3. Combustion Modeling

3.1. Methodology

As explained in Section 1, the authors constructed a kinetic mechanism model based on GRIMECH 3.0 with a detailed scheme proposed by Ra and Reitz [26] for n-dodecane as a surrogate for diesel fuel. The mechanism, containing 124 species and 660 reactions, coupled with an autoignition-induced flame propagation model, was implemented in ANSYS Forte® code as described in previous papers [27,28]. The initiation of flame propagation induced by the autoignition kinetics was represented by the oxidation and the relative reaction scheme for n-dodecane, while the autoignition-induced flame propagation measurements were used by the software to track the position of the flame front. Using the GRIMECH 3.0 mechanism in ANSYS Chemkin-Pro, tables with values of laminar flame speed (LFS) as a function of conditions of pressure, temperature and equivalence ratio were obtained. The tables were used by ANSYS Forte® to evaluate the laminar flame speed values by interpolation. The accuracy of this approach was verified by comparing the trend in the laminar flame speed derived from tables with the LFS obtained with the use of LFS correlations. The RANS RNG k-ε model was used to simulate the turbulence. Finally, for the diesel spray simulation, the KHRT atomization model and a discrete multi-component (DMC) fuel vaporization model were used. For study of the combustion process, the simulations described below were performed with closed valves.

3.2. Mesh Sensitivity Analysis

The meshes and calculations were carried out on a sector angle of the engine of 51.43°. The number of cells and the average size of the tetrahedral cells of the meshes at TDC are reported in Table 6. With the aim of identifying the mesh which provided the best compromise between computational time and solution accuracy, for each mesh, a simulation in full diesel mode was performed using the settings reported in Table 7. The mesh sensitivity analysis was performed by comparing trends of the in-cylinder pressure and chemical heat release (Figures 9 and 10). To better understand the in-cylinder pressure difference, in Figure 11 the evolution of the mass inside the cylinder is reported. It is possible to notice that, in the coarse meshes, the lost mass through the crevices was greater during the compression phase, while, during the expansion phase, the mass returned more quickly. Moreover, when the mesh was not sufficiently fine, the vaporization and atomization phenomena were not well described, with consequent lower pressure during the combustion. Although the meshes 4, 6 and 7 each provided the most accurate results, mesh 4 (Figure 12) was chosen because it involved reduced computational time.

Table 6. Mesh characteristics.

#Mesh	Average Size of Cells at TDC [mm]	Number of Cells at TDC
1	0.730	12,516
2	0.543	30,336
3	0.535	31,776
4	**0.514**	**35,748**
5	0.486	42,368
6	0.424	63,693
7	0.401	75,488

Table 7. Simulation settings for the full diesel simulations.

Parameter	Data
Engine speed [rpm]	2000
SOI pilot [deg]	15°BTDC
SOI main [deg]	1.2°ATDC
Duration of pilot injection [deg]	7
Duration of main injection [deg]	10
Pilot and main diesel mass [mg/cycle]	12

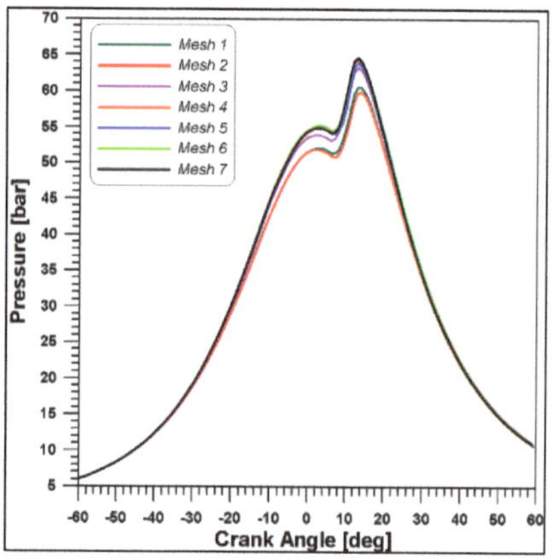

Figure 9. Pressure for the different meshes.

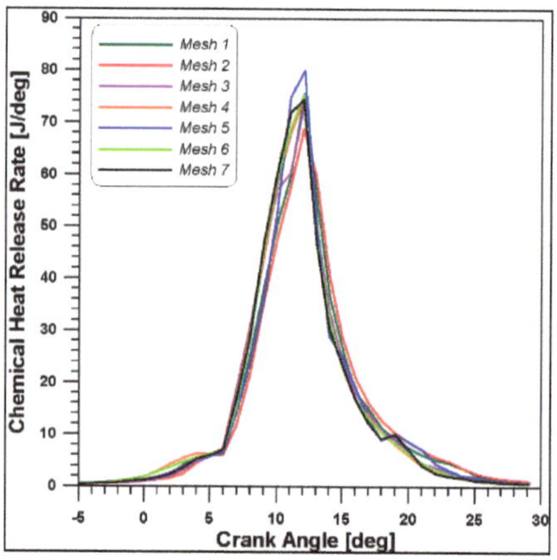

Figure 10. Rate of heat release for the different meshes.

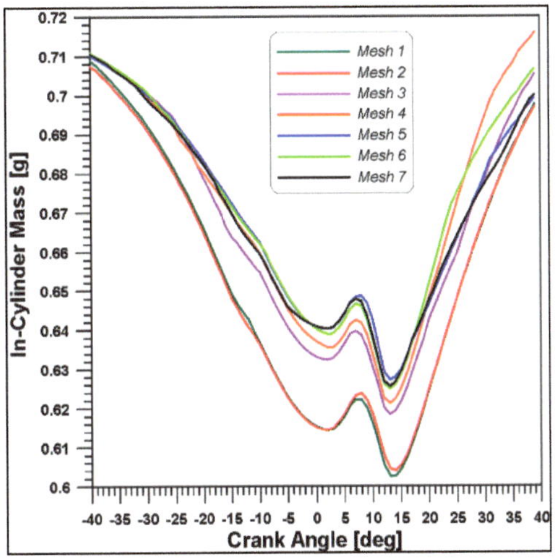

Figure 11. Mass inside the cylinder.

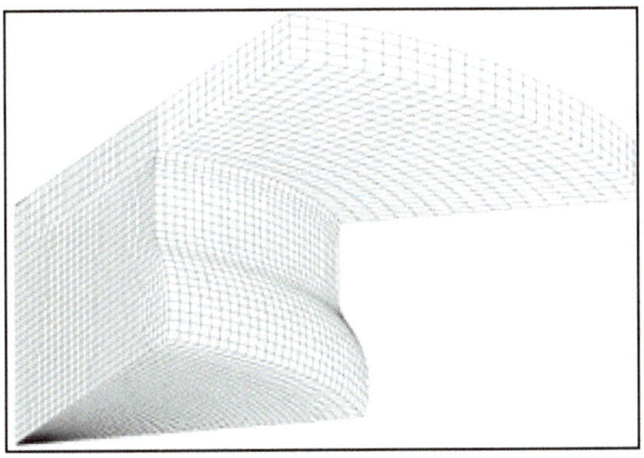

Figure 12. Mesh#4.

3.3. Model Tuning

Validation of the model was performed by simulating the test case H4 (see Table 5), reproducing the same experimental operating conditions. Several simulations were carried out by varying certain model parameters and empirical constants. All the simulations were conducted with closed valves (from 132° BTDC to 116° ATDC); the pilot and main SOI were 15° BTDC and 1.2° ATDC, respectively. The simulation results reported below refer to the best tuning performed (Table 8).

Table 8. Simulation settings.

Pilot and main injection duration [deg]	3.1°
Pilot and main mass injected [mg/cycle]	0.83
Turbulent kinetic energy [cm^2/s^2]	3.42×10^4
Turbulent length scale [cm]	0.2378
Size constant of KH breakup	1
Time constant of KH breakup	40
Size constant of RT breakup	0.15
Time constant of RT breakup	1

As can be seen from the graphs of pressure and heat release (Figures 13 and 14), the numerical results matched the experimental data quite well, although there were some differences between the curves of heat release rate, since, in the numerical simulations, the combustion started slightly later than in the experiments.

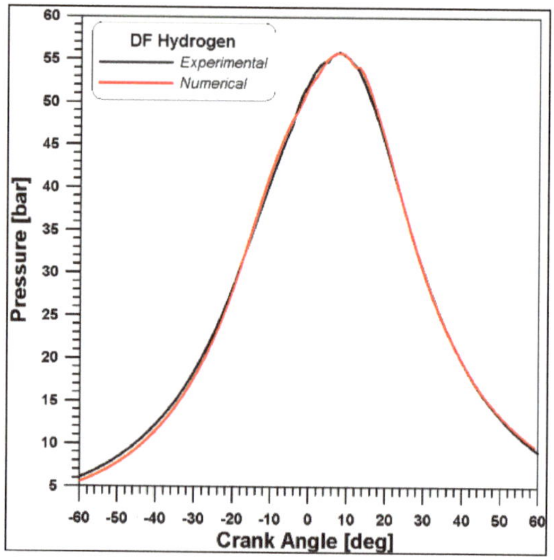

Figure 13. In-cylinder pressure. Experimental vs. Numerical.

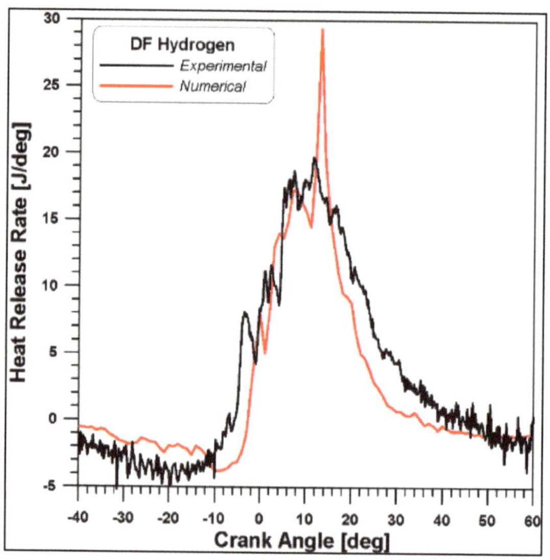

Figure 14. Rate of heat release. Experimental vs. Numerical.

Finally, Figure 15 shows that the hydrogen was not completely burned (about 35% of the hydrogen did not burn), contrary to expectations, while the diesel fuel vaporized quickly and burned entirely. The incomplete burning of hydrogen could be due to the very lean mixture used, with an equivalence ratio value of 0.2.

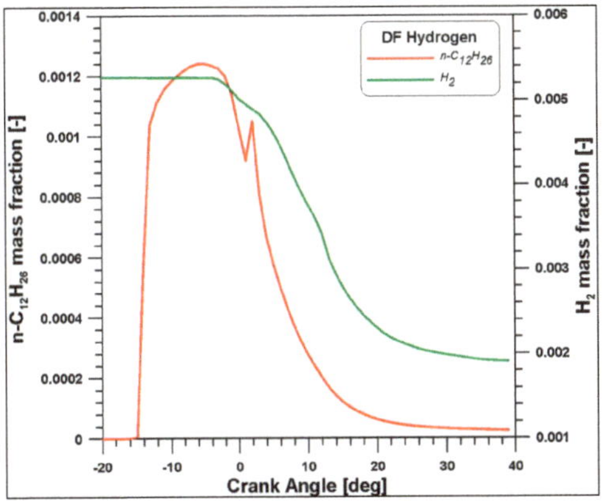

Figure 15. Numerical H_2 and n-$C_{12}H_{26}$ mass fraction.

3.4. Hydrogen-Methane Mixtures: Results and Discussion

Starting from the previous simulation with hydrogen as the gaseous fuel validated against the experimental results, several calculations were performed varying the percentage of hydrogen in the premixed mixture, while maintaining the same amount of diesel injected and the injection timing. The simulations were carried out with hydrogen–methane mixtures, always in dual-fuel mode, keeping the same fuel energy input as in the H4 case (Table 5).

In Table 9, a summary of the input mass fractions, energy and the initial conditions of pressure and temperature for each test case is reported. The test case HES100 corresponds to the previous simulation with only hydrogen as the primary fuel, while HES0 includes only methane.

Table 9. Input data for the test cases.

	HES100	HES75	HES50	HES25	HES0
E_{H_2} [J]	470	352.5	235	2.27	0
E_{CH_4} [J]	0	117.5	235	2.60	470
m_{H_2} [mg]	3.92	2.94	1.96	0.98	0
m_{CH_4} [mg]	0	2.35	4.70	7.05	9.4
x_{H_2}	0.00527	0.00395	0.00263	0.00131	0
x_{CH_4}	0	0.00316	0.00631	0.00944	0.01256
Initial pressure [bar]	1.591	1.591	1.591	1.591	1.591
Initial temperature [K]	346.4	346.4	346.4	346.4	346.4
Mass [mg]	742.6	753.2	764.2	775.4	786.9
Gaseous fuel/airER	0.184	0.203	0.212	0.221	0.184
RP [%]	86.8	86.8	86.8	86.8	86.8

The in-cylinder pressures (Figure 16) show that, for reduced hydrogen content, a lower pressure peak was observed. This was probably due to the slower reactivity of methane at high temperatures compared to that of hydrogen. Furthermore, by reducing the hydrogen share, a second peak became evident since only the pilot diesel fuel burned before TDC, while the mixture started to burn when the main injection occurred.

The curves of the rate of heat release confirmed this phenomenon (Figure 17): in the case with only hydrogen (HES100), an earlier combustion start occurred, and the mixture reacted more actively. In the HES0 case (only methane), the pilot injection did not initiate the premixed combustion. The consequence was a sharp increase following the main injection (SOI 1.2° ATDC) since a large part of the methane present burned simultaneously. The HES50 case, as expected, showed intermediate behavior between the two extreme cases of HES100 and HES0.

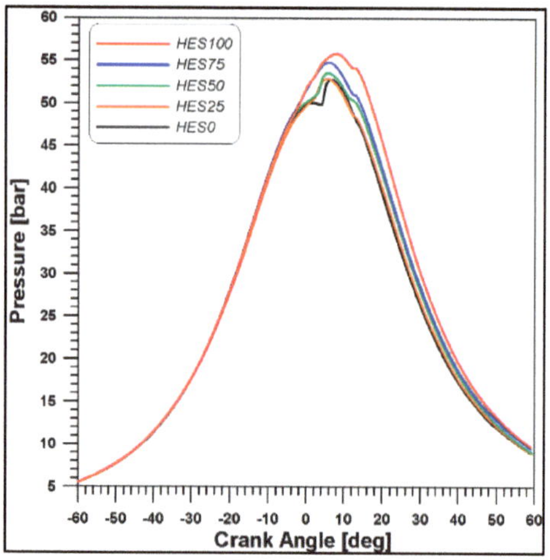

Figure 16. In-cylinder pressures.

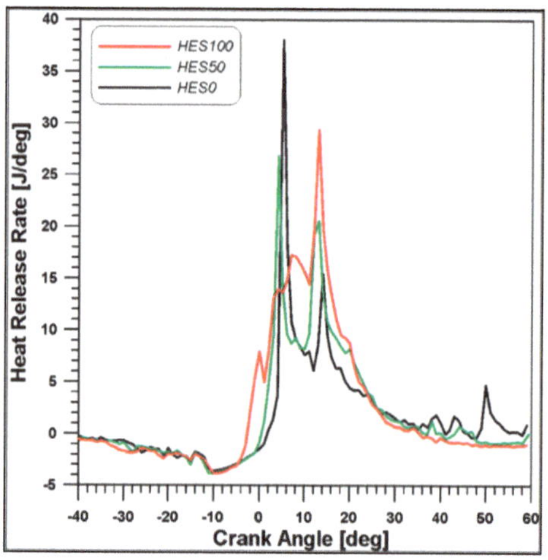

Figure 17. Rate of heat release.

The accumulated heat release in Figure 18 highlights that the combustion was not completed when the input energy value was not reached (see Table 9). Furthermore, for the HES25 and HES0 cases, the heat release increased constantly up to 80° ATDC.

The evolution of the gaseous fuel mass fraction (hydrogen and methane) shown in Figure 19 indicates that the fastest fuel consumption occurred in the HES100 case. The same figure provides additional evidence that the initiation of combustion of pure hydrogen occurred close to the pilot injection event. In the HES0 case (pure methane in the premixed mixture), the combustion of gaseous fuel was clearly activated by the main injection.

On the other hand, Figure 20 shows that a large amount of diesel fuel vapor was still present with increasing methane content. Following the main injection, more fuel vapor remained because it could not oxidize. The temperature values provide an explanation: the reduced amount of hydrogen, and, at the same time, the increase in methane, led to lower temperatures peaks (Figure 21) and the maximum temperature zones were less extensive, as shown in Figure 22, slowing the fuel oxidation. For all cases, it was observed that the high temperature zones were concentrated in the center of the combustion chamber.

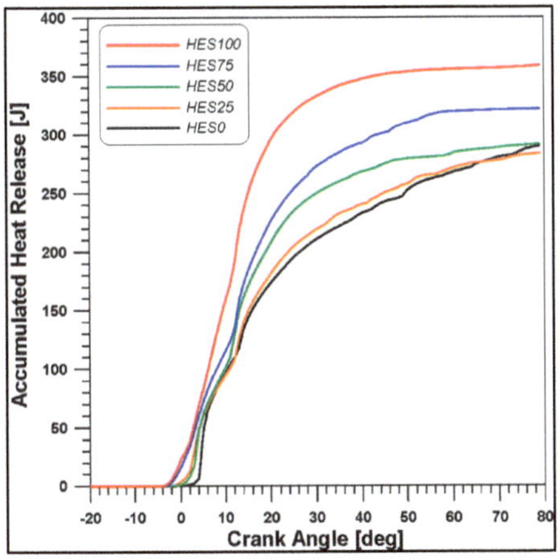

Figure 18. Accumulated chemical heat release.

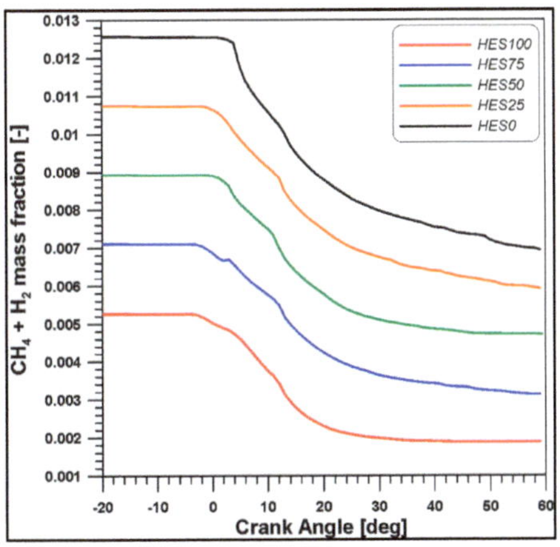

Figure 19. Fuel ($H_2 + CH_4$) mass fraction.

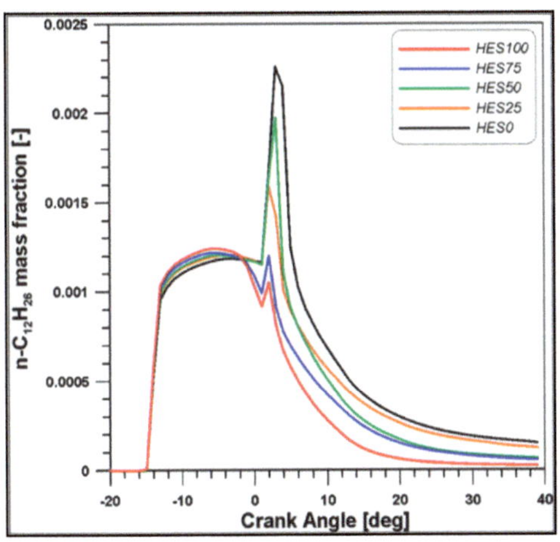

Figure 20. Diesel mass fraction.

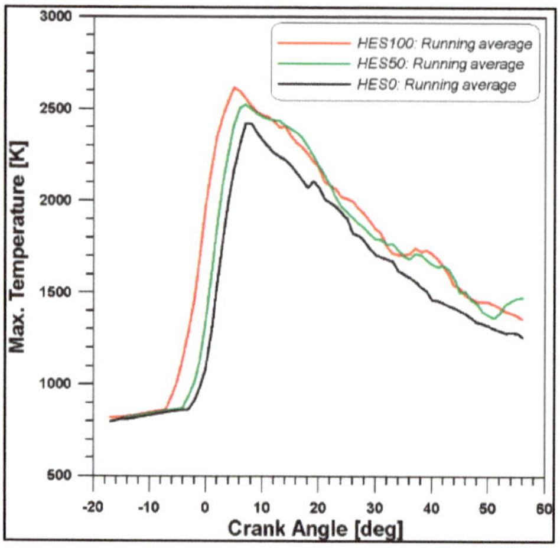

Figure 21. Maximum temperatures.

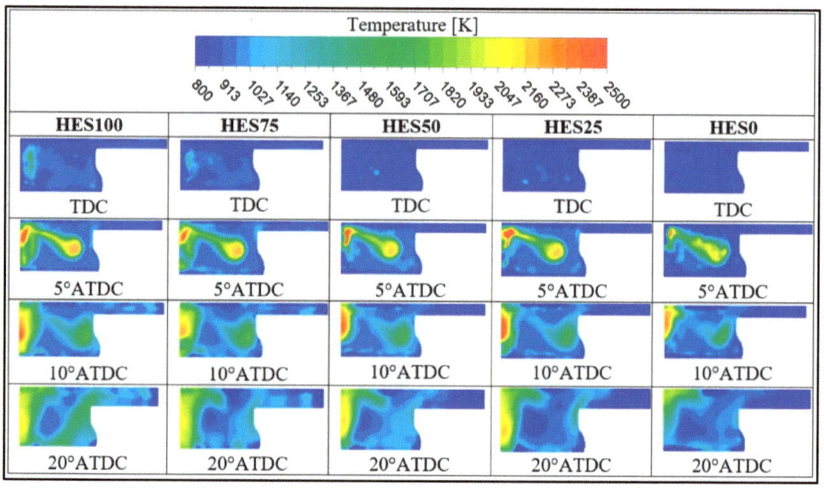

Figure 22. Temperature distributions for the different HES cases.

By examining the hydrogen contours in Figure 23, the locations of the mixture consumption can be identified; comparing the HES100 and HES0 cases, the hydrogen tended to burn even at a distance from the areas with the major concentration of diesel vapor (Figure 24). In this regard, it is useful to examine the development of the flame and, therefore, the 'G' function, which describes the flame propagation, as represented in Figure 25. The regions where $G > 0$ correspond to the zones where the premixed charge was being consumed, the zones where $G < 0$ identify the unburned mixture, while the flame front is localized where the function assumes a value $G = 0$. From this analysis, it is evident that the hydrogen combustion began early and that the flame spread more in the combustion chamber at TDC for the HES100 case, continuing its development at the following crank angles. In the case of HES0, the flame had not even started to propagate at TDC, in confirmation of the previous results.

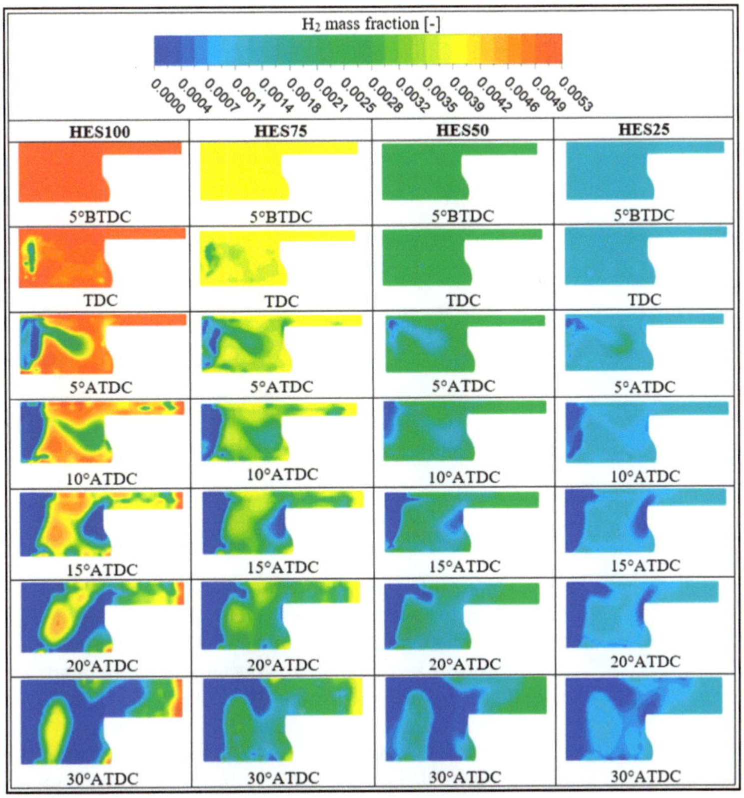

Figure 23. Hydrogen mass fraction distributions for the different HES cases.

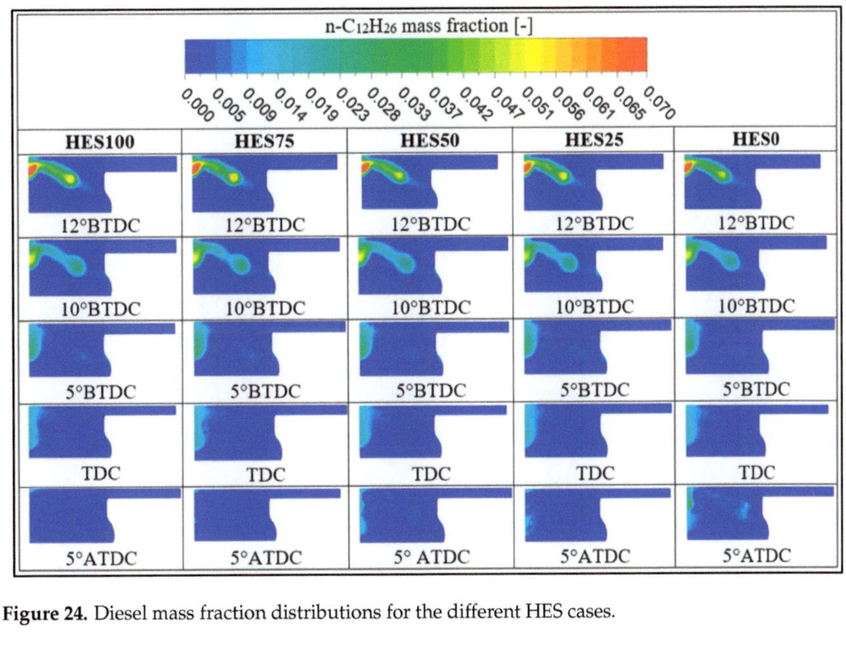

Figure 24. Diesel mass fraction distributions for the different HES cases.

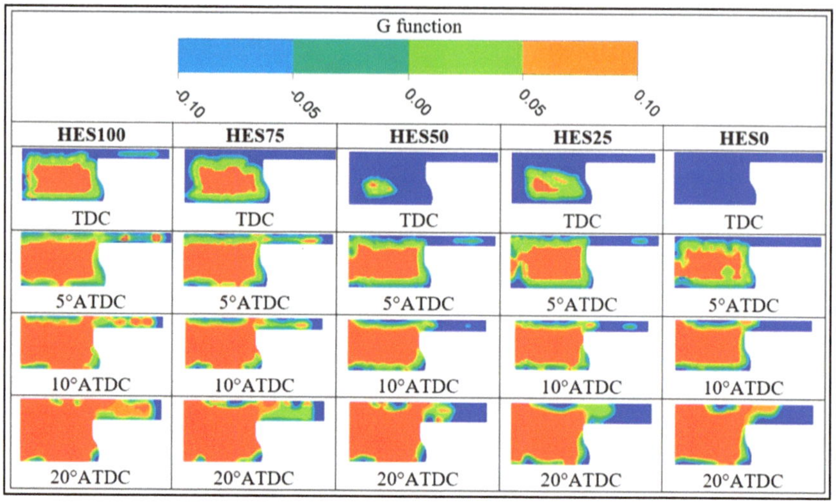

Figure 25. G function distributions for the different HES cases.

For completeness, the laminar and turbulent flame speed trends are shown in Figure 26. Only three cases (HES0, HES50, HES100) are plotted for a more readable representation. For each case, the trends of the laminar and turbulent speeds were almost the same, with higher values for the second speed, especially in the first part, confirmed by the high turbulent kinetic energy values (Figure 27). Both speeds reached higher values by increasing the hydrogen presence in the mixture, in addition to the fact that the flame propagation in the mixture started earlier, as already verified. It should be remembered that the curves in Figure 26 display the trend in the in-cylinder averaged flame speed. In fact, multiple flame fronts were generated by the diesel fuel jets. Nevertheless, the G function contours in Figure 25 clearly demonstrate that the peripheral zones inside the cylinder were not yet reached by the flame front at 20° ATDC.

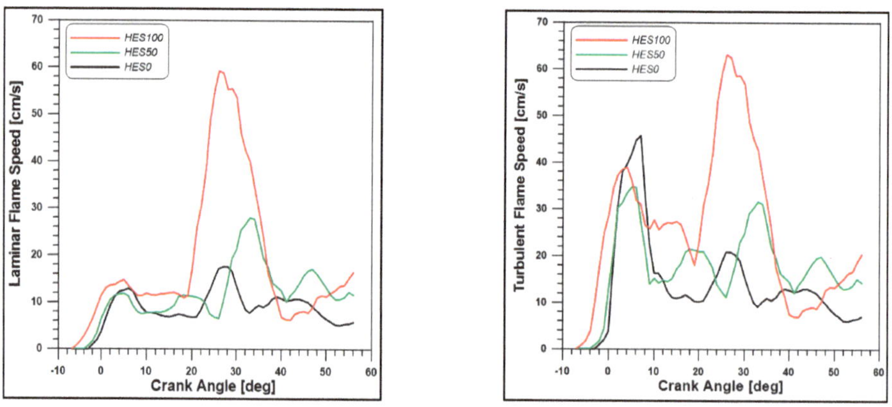

Figure 26. Laminar and turbulent flame speeds.

In Tables 10 and 11, a summary of the results is reported. The HES100 case indicates higher work (in terms of IMEP, calculated at closed valves) and power because of the better thermal and combustion efficiency. Moreover, the HES100 case involved faster combustion of lower duration (HRR10-HRR90). The results were very similar in the first half of the combustion process (from HRR10 to HRR50), while, in the second part (from HRR50 to

HRR90), the progressive replacement of hydrogen with methane led to slowing of this phase. The final values (at EVO) of the concentration of some of the most important chemical species are displayed in Table 11.

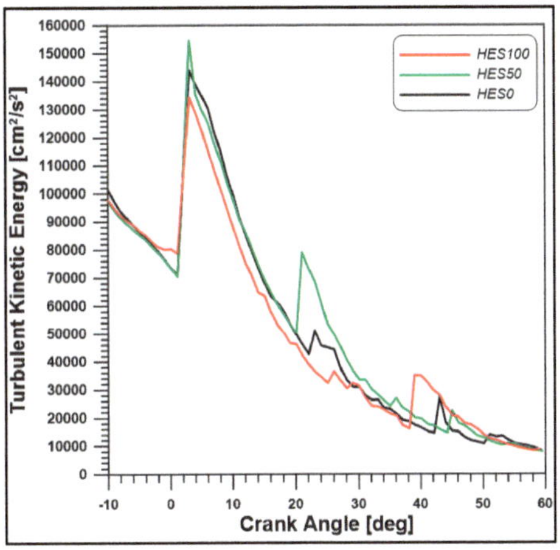

Figure 27. Turbulent kinetic energy.

Table 10. Engine summary for the different HES cases.

	HES100	HES75	HES50	HES25	HES0
Gross power [kW]	2.61	2.27	2.07	2.07	1.88
IMEP [bar]	3.0	2.60	2.38	2.21	2.16
Combustion efficiency	0.66	0.60	0.56	0.56	0.58
Thermal efficiency	0.29	0.25	0.24	0.24	0.22
Total chemical heat release [J]	361	323	294	294	296
Total wall heat transfer loss [J]	52	47	45	45	42
Total net heat [J]	309	276	249	249	254
Total net heat (from PV with variable Gamma) [J]	215	186	165	165	168
Max. pressure [bar]	55.8	54.8	53.4	53.4	52.8
Max. temperature [K]	1179	1066	1035	1035	983
Max. pressure rise rate [bar/deg]	1.44	1.44	1.74	1.74	2.36
HRR10 [deg ATDC]	2	2	4	4	5
HRR50 [deg ATDC]	12	14	13	13	16
HRR90 [deg ATDC]	27	39	38	38	60
HRR10-HRR90 Duration [deg]	25	37	34	34	55

Table 11. Final concentrations of some chemical species.

	HES100	HES75	HES50	HES25	HES0
H_2	1877	1731	1342	690	0
CH_4	0	1403	3244	4984	6404
n-$C_{12}H_{26}$	18.30	33.98	45.65	66.42	78.58
CO_2	32,648	28,082	24,590	21,502	17,649
O_2	193,871	196,419	198,408	197,990	195,845
CO	226	569	818	888	860
NO_x	2.66	2.16	1.50	1.96	0.45

The HC emissions (Figure 28) showed higher values by reducing H_2 in the mixture, also considering that, in the HES100 case, the only carbon atoms present were due to the diesel fuel. In the other test cases, the higher unburned hydrocarbon value was also due to unburned methane. The higher temperatures caused high NO_x emissions as a result (Figure 29). As expected, the introduction of hydrogen significantly reduced the production of carbon dioxide (Figure 30), as reported in many studies, e.g., [13,20,23].

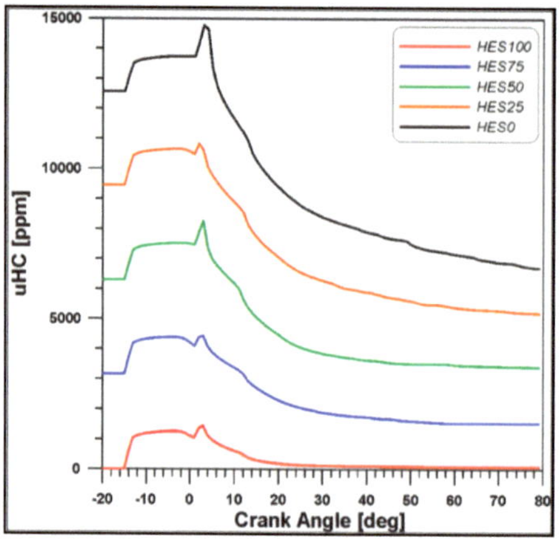

Figure 28. HC pollutant.

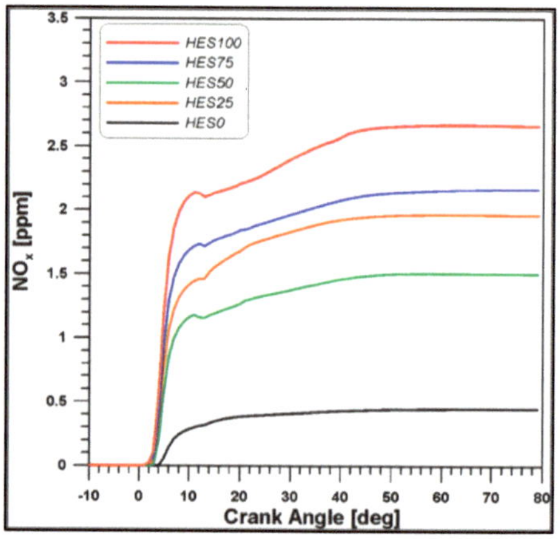

Figure 29. NO_x emissions.

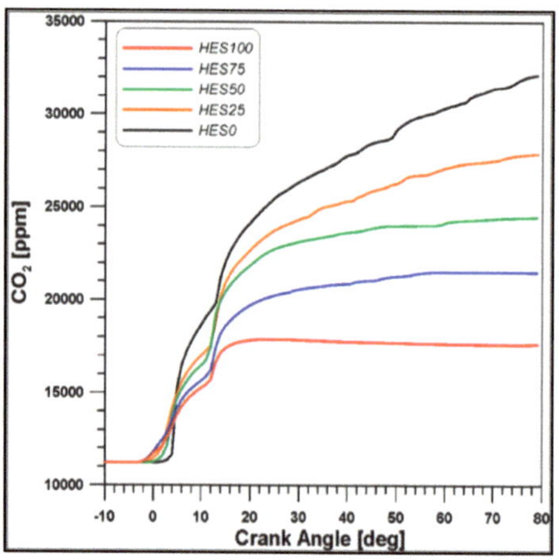

Figure 30. CO_2 emissions.

4. Conclusions

The main objective of this investigation was to numerically examine the phenomenon of dual-fuel combustion with hydrogen/methane blends used as primary fuels in a research compression ignition engine. A detailed study on combustion development was carried out to evaluate the impact of hydrogen addition on the process and emissions. After validation of the model by comparison with experimental data, several calculations were performed to estimate the effects of hydrogen in the blend on engine performance and low emissions.

The conclusions of the investigation can be summarized as follows:

- Under the operating conditions considered, the gaseous fuel did not burn completely and most of the unburned hydrogen was concentrated near the crevice region.
- The comparison between the different dual-fuel cases with the hydrogen–methane mixtures showed that the addition of methane entailed a slowing down of combustion with lower efficiency. By increasing the hydrogen amount up to 100% of the premixed fuel, the thermal and combustion efficiency increased.
- With respect to pollutant emissions, the model applied confirmed the results of previous studies, e.g., [13,20,23]: the presence of hydrogen led to less carbon-based emissions, but higher NO_x emissions due to the high temperatures.
- Future numerical studies should focus on NO_x reduction by proven EGR systems and reduce unburned hydrogen through hydrogen direct injection to avoid fuel accumulation in crevices.

Author Contributions: Conceptualization, M.C.C. and E.M.; methodology, M.C.C. and R.D.R.; software, M.P.; validation, R.D.R. and M.P.; investigation, R.D.R. and E.M.; data curation, M.P.; original draft preparation, M.C.C. and M.P.; review, R.D.R.; supervision, M.C.C. and E.M. All authors have read and agreed to the published version of the manuscript.

Funding: This research received no external funding.

Acknowledgments: Thanks to Carlo Rossi of STEMS Institute of Naples for the support provided for the experimentation. The calculations are licensed by ANSYS®.

Conflicts of Interest: The authors declare no conflict of interest.

References

1. Heywood, J.B. *Internal Combustion Engines Fundamentals*; McGraw-Hill: Graveport, OH, USA, 2018.
2. Krishnamoorthi, M.; Malayalamurthi, R.; He, Z.; Kandasamy, S. A review on low temperature combustion engines: Performance, combustion and emission characteristics. *Renew. Sustain. Energy Rev.* **2019**, *116*, 109404. [CrossRef]
3. Pham, Q.; Park, S.; Agarwal, A.K.; Park, S. Review of dual-fuel combustion in the compression-ignition engine: Spray, combustion, and emission. *Energy* **2022**, *250*, 123778. [CrossRef]
4. Zhang, W.; Feng, T.; Li, Z.; Chen, Z.; Zhao, J. EGR thermal and chemical effects on combustion and emission of diesel/natural gas dual-fuel engine. *Fuel* **2021**, *302*, 121161. [CrossRef]
5. Xu, M.; Cheng, W.; Zhang, H.; An, T.; Zhang, S. Effect of diesel pre-injection timing on combustion and emission characteristics of compression ignited natural gas engine. *Energy Convers. Manag.* **2016**, *117*, 86–94. [CrossRef]
6. Yang, B.; Wei, X.; Xi, C.; Liu, Y.; Zeng, K.; Lai, M.-C. Experimental study of the effects of natural gas injection timing on the combustion performance and emissions of a turbocharged common rail dual-fuel engine. *Energy Convers. Manag.* **2014**, *87*, 297–304. [CrossRef]
7. Zheng, J.; Wang, J.; Zhao, Z.; Wang, D.; Huang, Z. Effect of equivalence ratio on combustion and emissions of a dual-fuel natural gas engine ignited with diesel. *Appl. Therm. Eng.* **2018**, *146*, 738–751. [CrossRef]
8. Muralidharan, M.; Srivastava, A.; Subramanian, M. A Technical Review on Performance and Emissions of Compressed Natural Gas-Diesel Dual Fuel Engine. *SAE Tech. Pap.* **2019**. [CrossRef]
9. Sankaranarayanan, R.; Petersen, E.L. Laminar flame speed correlations for pure-hydrogen and high-hydrogen content syngas blends with various diluents. *Int. J. Hydrogen Energy* **2012**, *37*, 19177–19189. [CrossRef]
10. Akal, D.; Öztuna, S.; Büyükakın, M.K. A review of hydrogen usage in internal combustion engines (gasoline-Lpg-diesel) from combustion performance aspect. *Int. J. Hydrogen Energy* **2020**, *45*, 35257–35268. [CrossRef]
11. Nag, S.; Sharma, P.; Gupta, A.; Dhar, A. Experimental study of engine performance and emissions for hydrogen diesel dual fuel engine with exhaust gas recirculation. *Int. J. Hydrogen Energy* **2019**, *44*, 12163–12175. [CrossRef]
12. Castro, N.; Toledo, M.; Amador, G. An experimental investigation of the performance and emissions of a hydrogen-diesel dual fuel compression ignition internal combustion engine. *Appl. Therm. Eng.* **2019**, *156*, 660–667. [CrossRef]

13. Zhou, J.H.; Cheung, C.H.; Leung, C.W. Combustion, performance and emissions of a diesel engine with H2, CH4 and H2–CH4 addition. *Int. J. Hydrogen Energy* **2014**, *39*, 4611–4621. [CrossRef]
14. Ouchikh, S.; Lounici, M.S.; Tarabet, L.; Loubar, K.; Tazerout, M. Effect of natural gas enrichment with hydrogen on combustion characteristics of a dual fuel diesel engine. *Int. J. Hydrogen Energy* **2019**, *44*, 13974–13987. [CrossRef]
15. Liu, F.; Kang, Y.; Wu, H.; Lee, C.-F.; Li, Y. Effect of Hydrogen Volume Ratio on the Combustion Characteristics of CNG-Diesel Dual-Fuel Engine. *SAE Tech. Pap.* **2017**. [CrossRef]
16. Tripathi, G.; Sharma, P.; Dhar, A. Computational study of diesel injection strategies for methane-diesel dual fuel engine. *Clean. Eng. Technol.* **2022**, *6*, 100393. [CrossRef]
17. Li, Y.; Li, H.; Guo, H.; Li, Y.; Yao, M. A numerical investigation on methane combustion and emissions from a natural gas-diesel dual fuel engine using CFD model. *Appl. Energy* **2017**, *205*, 153–162. [CrossRef]
18. Papagiannakis, R.G.; Hountalas, D.T.; Krishnan, S.R.; Srinivasan, K.K.; Rakopoulos, D.C.; Rakopoulos, C. Numerical evaluation of the effects of compression ratio and diesel fuel injection timing on the performance and emissions of a fumigated natural gas–diesel dual-fuel engine. *J. Energy Eng.* **2016**, *142*, E4015015. [CrossRef]
19. Mousavi, S.M.; Saray, R.K.; Poorghasemi, K.; Maghbouli, A. A numerical investigation on combustion and emission characteristics of a dual fuel engine at part load condition. *Fuel* **2016**, *166*, 309–319. [CrossRef]
20. Jabbr, A.I.; Gaja, H.; Koylu, O. Multi-objective optimization of operating parameters for a H2/Diesel dual-fuel compression-ignition engine. *Int. J. Hydrogen Energy* **2020**, *45*, 19965–19975. [CrossRef]
21. Ramsay, C.J.; Dinesh, K.K.J.R.; Fairney, W.; Vaughan, N. A numerical study on the effects of constant volume combustion phase on performance and emissions characteristics of a diesel-hydrogen dual-fuel engine. *Int. J. Hydrogen Energy* **2020**, *45*, 32598–32618. [CrossRef]
22. Ekin, F.; Ozsoysal, O.A.; Arslan, H. The effect of using hydrogen at partial load in a diesel-natural gas dual fuel engine. *Int. J. Hydrogen Energy* **2022**, *47*, 18523–18550. [CrossRef]
23. Mansor, M.R.A.; Mohamad, T.I.; Sabah, O. Numerical investigation on combustion and emissions in a direct injection compression ignition engine fuelled with various hydrogen–methane–diesel blends at different intake air temperatures. *Energy Rep.* **2021**, *7*, 403–421. [CrossRef]
24. Sharma, P.; Dhar, A. Compression ratio influence on combustion and emissions characteristic of hydrogen diesel dual fuel CI engine: Numerical Study. *Fuel* **2018**, *222*, 852–858. [CrossRef]
25. Tripathi, G.; Sharma, P.; Dhar, A.; Sadiki, A. Computational investigation of diesel injection strategies in hydrogen-diesel dual fuel engine. *Sustain. Energy Technol. Assess.* **2019**, *36*, 100543. [CrossRef]
26. Wang, H.; Ra, Y.; Jia, M.; Reitz, R.D. Development of a reduced n-dodecane-PAH mechanism and its application for n-dodecane soot predictions. *Fuel* **2014**, *136*, 25–36. [CrossRef]
27. De Robbio, R.; Cameretti, M.C.; Mancaruso, E.; Tuccillo, R.; Vaglieco, B.M. Combined CFD-Experimental Analysis of the In-Cylinder Combustion Phenomena in a Dual Fuel Optical Compression Ignition Engine. *SAE Tech. Pap.* **2021**. [CrossRef]
28. De Robbio, R.; Cameretti, M.C.; Mancaruso, E.; Tuccillo, R.; Vaglieco, B.M. CFD Study and Experimental Validation of a Dual Fuel Engine: Effect of Engine Speed. *Energies* **2021**, *14*, 4307. [CrossRef]
29. Mancaruso, E.; Todino, M.; Vaglieco, B.M. Study on dual fuel combustion in an optical research engine by infrared diagnostics varying methane quantity and engine speed. *Appl. Therm. Eng.* **2020**, *178*, 115623. [CrossRef]
30. Mancaruso, E.; Todino, M.; Vaglieco, B.M. Analysis of Dual Fuel Combustion in Single Cylinder Research Engine Fueled with Methane and Diesel by IR Diagnostics. *SAE Tech. Pap.* **2019**. [CrossRef]
31. Magno, A.; Mancaruso, E.; Vaglieco, B.M. Combustion Analysis of Dual Fuel Operation in Single Cylinder Research Engine Fuelled with Methane and Diesel. *SAE Tech. Pap.* **2015**. [CrossRef]

energies

MDPI

Article

EGR and Emulsified Fuel Combination Effects on the Combustion, Performance, and NOx Emissions in Marine Diesel Engines

Elsayed Abdelhameed [1,2,*] and Hiroshi Tashima [1,*]

1 Interdisciplinary Graduate School of Engineering Sciences, Kyushu University, Fukuoka 816-8580, Japan
2 Faculty of Engineering, Kafrelsheikh University, Kafrelsheikh 33511, Egypt
* Correspondence: abdelhameed.elsayed.203@s.kyushu-u.ac.jp (E.A.); tajima.hiroshi.539@m.kyushu-u.ac.jp (H.T.)

Abstract: Techniques such as exhaust gas recirculation (EGR) and water-in-fuel emulsions (WFEs) can significantly decrease NOx emissions in diesel engines. As a disadvantage of adopting EGR, the afterburning period lengthens owing to a shortage of oxygen, lowering thermal efficiency. Meanwhile, WFEs can slightly reduce NOx emissions and reduce the afterburning phase without severely compromising thermal efficiency. Therefore, the EGR–WFE combination was modeled utilizing the KIVA-3V code along with GT power and experimental results. The findings indicated that combining EGR with WFEs is an efficient technique to reduce afterburning and enhance thermal efficiency. Under the EGR state, the NO product was evenly lowered. In the WFE, a considerable NO amount was created near the front edge of the combustion flame. Additionally, squish flow from the piston's up–down movement improved fuel–air mixing, and NO production was increased as a result, particularly at high injection pressure. Using WFEs with EGR at a low oxygen concentration significantly reduced NO emissions while increasing thermal efficiency. For instance, using 16% of the oxygen concentration and a 40% water emulsion, a 94% drop in NO and a 4% improvement in the Indicated Mean Effective Pressure were obtained concurrently. This research proposes using the EGR–WFE combination to minimize NO emissions while maintaining thermal efficiency.

Keywords: combustion modeling; diesel engine; emission; emulsion; exhaust gas recirculation; IMEP; marine diesel engine; NOx; optical engine; two-color method

Citation: Abdelhameed, E.; Tashima, H. EGR and Emulsified Fuel Combination Effects on the Combustion, Performance, and NOx Emissions in Marine Diesel Engines. *Energies* **2023**, *16*, 336. https://doi.org/10.3390/en16010336

Academic Editors: Roberta De Robbio and Maria Cristina Cameretti

Received: 11 December 2022
Revised: 22 December 2022
Accepted: 23 December 2022
Published: 28 December 2022

1. Introduction

Marine transportation, primarily driven by the diesel engine, represents 90% of the global trade sector [1]. Marine transportation consumes 300 million tons of oil annually [2] and contributes 2.89% of global emissions [3]. Those exhaust gas emissions negatively impact the global environment. The main components in exhaust gas emissions are nitrogen, water vapor and carbon dioxide, which are not toxic, apart from their impact on global warming [4,5]. In addition to those emissions, the combustion process results in relatively tiny amounts of other undesirable emissions: carbon monoxide, particulate matter, and NOx. The atmosphere contains various nitrogen oxides, including N_2O, NO, NO_2, NO_3, N_2O_4, and N_2O_5 [6]. When discussing nitrogen oxides, the term "NOx" is commonly used to refer to nitrous oxide (N_2O), nitrogen monoxide (NO), and nitrogen dioxide (NO_2), all of which are considered "fresh" nitrogen oxides from a photochemical perspective as they enter the atmosphere in their specific gaseous states [7]. Combustion processes, which include the high temperatures required to create nitrogen oxides, produce these gases. Mainly, NOx emitted from diesel engines is "thermal NOx", produced from direct reactions of nitrogen and oxygen in the high-temperature atmosphere [8]. Nearly all of the NOx produced by combustion processes is NO by 95%, with only a tiny portion being NO_2 by 5% [6].

After being released to the ambient, NOx can react with organic matter releasing Ozone, leading to lung damage, and reducing lung function, especially for children and the elderly [6]. Meanwhile, respiratory exposure to nitrogen oxides for a long time may lead to asthma [9]. For those effects, the International Maritime Organization (IMO) strictly regulated the emissions as recommended in MARPOL Annex VI to reduce NOx emissions by 75% in Tier III [10,11]. Such an intensive emission reduction is almost impossible for ship engines without any technical measures. Therefore, some technologies have been implemented to decrease NOx formation in the last decade [12], including fuel additives [13], fuel blending [14], Double-Stage Turbocharging, Fuel Injection Strategy [15], Direct Water Injection (DWI), dual-fuel combustion strategy [16,17], Scavenging Air Moisturizing (SAM), Selective Catalytic Reduction (SCR) [18], nonthermal plasma (NTP) [19], Seawater Flue Gas Desulphurization (SWFGD) [20,21], exhaust gas recirculation (EGR) [22–24], and water-in-fuel emulsions (WFE) [25–27].

Exhaust gas recirculation (EGR) is a technique used to reduce nitrogen oxide (NOx) concentrations in internal combustion engines [28]. It recirculates a portion of the engine's exhaust gas back to the combustion chamber, mixed with the incoming air and fuel mixture. This process lowers the amount of oxygen available for combustion, reducing the peak combustion temperatures that can lead to NOx formation. EGR can reduce NOx emissions by up to 90% [29] through three main mechanisms [30]: the thermal effect, which refers to the reduction in peak combustion temperature due to the presence of EGR [31]; the dilution effect, which refers to the dilution of the air–fuel mixture with exhaust gas, lowering oxygen concentration and combustion temperatures [32]; and the chemical effect, which refers to the reaction of EGR with NOx in the exhaust gas to form less harmful compounds such as nitrogen and water [32]. However, EGR can also increase the risk of knocking, prolong after-burning, and lead to higher levels of carbon monoxide, soot, unburned hydrocarbons, and particulate matter. Using water-in-fuel emulsions can increase brake thermal efficiency and reduce NOx emissions, but also increase CO emissions. Both EGR and water approaches have drawbacks, including low combustion rates and flame propagation. The specific conditions of the engine, including temperature history, boundary conditions, and operating characteristics, can also affect the combustion process and emissions production. Zhang et al. [33] studied a simulation of the properties of the in-cylinder soot on the oxidation performance in the EGR condition. The simulation concentrated on the oxidation-dominant phase reporting an increase in the in-cylinder soot oxidation rate. EGR slows soot oxidation by decreasing flame temperature and oxygen and OH concentrations throughout the combustion. Using a variety of biodiesels, Jaliliantabar et al. [34] examined the effects of pilot diesel timing and the EGR system on decreasing exhaust emissions. The study's findings showed that the engine's reaction to varying injection strategies and EGR applications was affected by the fuel's different chemical and physical properties. Using the coffee methyl-ester reduced the combustion duration by 17.7% compared to diesel. EGR and pilot injection (up to 86% and 29.3%, respectively) reduced NOx emission, but the combustion quality was altered due to changes in the combustion efficiency.

Mukhtar et al. [35] used an emulsified fuel with different biodiesel and ethanol percentages as a fuel for a single-cylinder diesel engine. The results revealed that the in-cylinder pressure and the heat released rate for the test fuel exceeded the pure diesel's values. Because of evaporation, the HRR profile showed a negative value in the ignition delay phase. The study's most important finding is that the load shortening of the test fuel ignition delay period exhibited proof of the micro-explosion phenomena happening without load, peak HRR, or combustion pressure. Park and Oh [36] studied flame visualization, fundamental combustion characteristics, and emissions of the emulsion fuel. Emulsion fuel reduced nitrogen oxide emissions by 19.6% and smoke by 66.3%, respectively. The emulsion fuel showed higher in-cylinder pressure and heat release rate than the diesel fuel. Additionally, the combustion duration was shorter by 12.3% than the pure diesel. Abdollahi et al. [37] studied the effect of the nano-emulsion biodiesel fuel on the combustion and emission characteristics of the single-cylinder diesel engine. The results indicate that nano-emulsion fuel boosted diesel engine power and torque by 4.84% and 4.65% compared to diesel fuel. Emissions of CO, UHC, NOx and soot opacity were reduced by approximately 10% using nano-emulsion fuel. However, there was a 7% increase in CO_2 emissions. Gad et al. [38] performed a comprehensive study on the influence of different blends of mandarin peel waste oil with diesel on the diesel engine. Then, the study was extended to predict the impact of enriching the blends with propanol considering EGR implementation. The results showed a decreased HHR, CO, HC, and smoke opacity, and the NOx emissions increased significantly.

Although some researchers studied WFEs and EGR, the detailed combustion process and NOx reduction mechanism in the WFE–EGR combination are still not well understood. As known, it is a competitive and challenging matter to calculate the NOx distribution or in-cylinder gas compositions versus crank angle by only analyzing measured data from experiments. Hence, A numerical calculation code was used to reveal the combustion mechanisms with the WFE–EGR combination. This paper is organized as follows: Early model calibration involved testing various EGR and WFE modes in an RCEM (Rapid Compression Expansion Machine) and obtaining shots, pressure measurements, and emissions data from in-cylinder combustion. Additionally, study two EGR system methods—Constant Air Suction (CAS) and Constant Oxygen Charge (COC)—utilizing a 1D GT power simulation model and next, initiating and verifying the KIVA-3V code against the experimental data and using the calibrated model to simulate the combustion process and NOx emissions. The novelty of this study was primarily demonstrated in two following aspects. Firstly, a simulation platform based on the KIVA-3V code was established and utilized to calibrate the EGR and WFEs with a newly developed code to simulate the emulsified fuel. Secondly, the produced code was utilized to investigate a variety of EGR–WFE combinations and their impacts on in-cylinder combustion and NOx emissions. The objectives of this study are to (a) reveal combustion processes and NOx reduction mechanisms under the WFE/EGR conditions; (b) calculate the emulsified fuel concentration to shorten the afterburning duration in the EGR method; (C) and determine the effect of injection timing, pressure, and nozzle diameter on the WFE examined. From these techniques, the WFE is expected to solve the problem of afterburning and improve thermal efficiency with a high NOx reduction ratio.

2. Materials and Methods

In this study, the combination of the EGR and water-in-fuel emulsion (WFE) techniques has been studied with a constant total heat release. This study starts with an experimental investigation of the EGR and WFEs and their combinations then this study was extended to cover a wide range of percentage combinations using the KIVA-3V code. The EGR cases were performed experimentally using three oxygen concentrations (21%, 18.3%, and 16.3%) by diluting the intake air with pure nitrogen at a constant injection pressure rating of 150 MPa, as shown in Table 1.

This research work considered an engine outline to calculate the amount of inert gas and oxygen in the intake, exhaust, and recirculated exhaust gases. Figure 1 illustrates the mass balance of the intake and exhaust of a non-EGR supercharged diesel engine operating with an air intake mass flow rate of 100 units. To make this methodology feasible, some assumptions were considered. (1) The mass oxygen concentration in the air is assumed to be 20%, and the other is considered inert gas with 80%. (2) The mass increase due to fuel is ignored. (3) The combustion is assumed to be complete, and the generated H_2O and CO_2 were added to the inert gas, ignoring the mass of C and H. (4) The mass oxygen concentration after complete combustion is assumed to be 10%, neglecting the effect of load. It is worth highlighting that the numerator represents the amount of inert gas, and the denominator represents the amount of oxygen. However, it does not mean that it is a fraction.

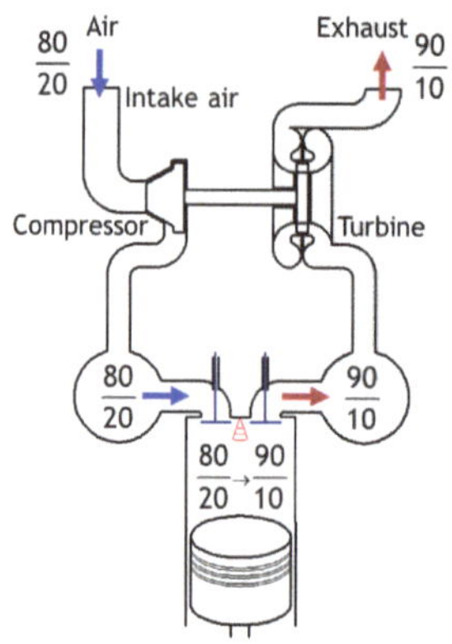

Figure 1. The mass balance for a non-EGR supercharged diesel engine.

- Constant Air Suction: CAS

The engine intake in CAS is kept the same as in the non-EGR case. Additionally, the mass ratio between the inert gases and the oxygen in the recirculated exhaust gas is as same as in the non-EGR case. The increased coefficient of flow passing through the engine is set to α. The CAS mass balance and two examples of α (1.25 and 1.5) are shown in Figure 2. When the increase coefficient (α) equals 1.25, the mass balance shows that 25% of EGR is required to reach 18% oxygen concentration in the intake air. While the calculations show that to reach an oxygen concentration of 16.7% with α equals 1.5, a 50% EGR should be ensured.

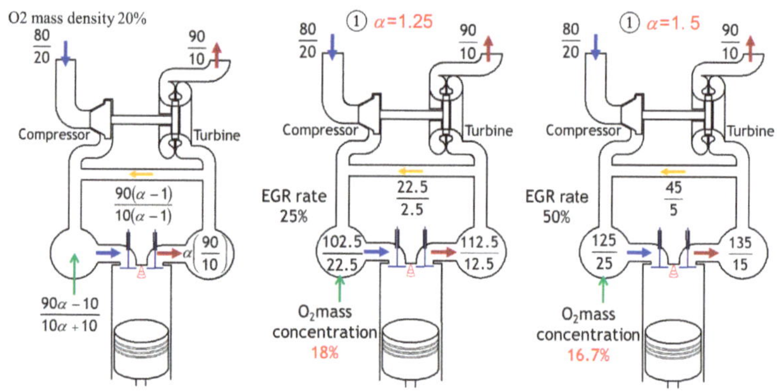

Figure 2. Mass balance of a diesel engine with EGR in the CAS case.

- Constant Oxygen Charge (COC)

In this case, the intake air volume can be reduced until the oxygen mass in the air supply is equal to the oxygen mass in the air intake in the case of non-EGR. In the COC, the air intake can be reduced until it becomes equal to that of non-EGR, and this reduction factor is considered β. At the same time, the x and y could be calculated from Equations (1) and (2). Figure 3 shows the mass balance in the case of COC with two examples of reduction factor (β) 0.9 and 0.86. from the figure, when applying a reduction factor β = 0.9 and recirculating 20% of the exhaust gases, 18% of the oxygen concentration could be reached. The most valuable results were obtained when applying a reduction factor of 0.86, which allows using a low rate of EGR, reaching 28% to 16.7% of the oxygen concentration in the in-cylinder content.

$$x = \frac{80\beta + 10}{2\beta - 1} \quad (1)$$

$$y = \frac{200(8\beta + 1)(1 - \beta)}{(2\beta - 1)} \quad (2)$$

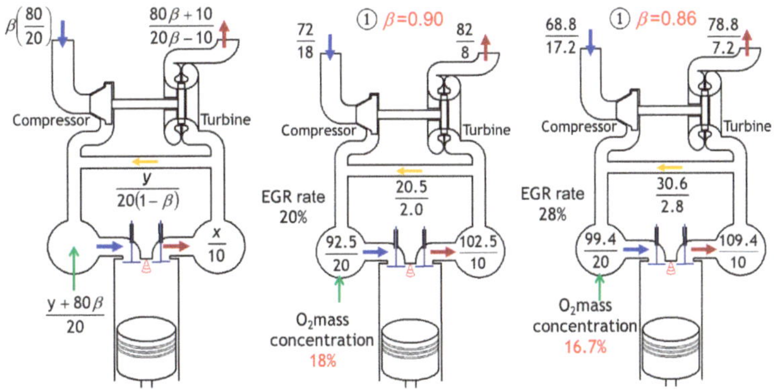

Figure 3. Mass balance of a diesel engine with EGR in the COC case.

In this section, the GT power environment was used to construct a 1D simulation model of a marine diesel engine [31]. Indicated thermal efficiency (ITE), brake-specific fuel consumption (BSFC), and a residual gas fraction (RGF) can all be reliably calculated using GT power [30]. Based on the configuration of the RCEM in the lab, a full 1D simulation model was constructed. The test RCEM's 1D GT power simulation model is shown in Figure 4. The engine is modeled using GT power code, and the simulation results are compared to those from an actual engine test bench. Generally, the 1D GT power model depends heavily on geometry, regulation, and performance. The 1D simulation model, therefore, includes boundary conditions based on data, including the bore, stroke, compression ratio, valve diameter, valve lift, valve timing, and turbocharger maps [32] of the test engine. The results of a 1D GT power simulation are susceptible to boundary conditions, such as shape, regulation, and performance. Therefore, in this study, the software was fed geometry and design data from the construction charts and the RCEM sensors, including the test engine's bore, stroke, compression ratio, and injection timing. This study applied the Shell Global Combustion Model (SGCOM), which is a detailed, multi-zone combustion model that can accurately predict combustion analytics, as well as performance and emissions, in both spark-ignition (SI) and compression-ignition (CI) engines.

Two cases of WFEs with 0, 25, and 40 $v/v\%$ of water presence were applied. The fuel is premixed with 0.1% glycerin, while the water is premixed with 0.9% pharmaceutically active ingredient (Sorbitan monolaurate) to stabilize the water dispersion [39]. As shown in Figure 5, the water-in-fuel emulsion was performed mechanically by adding the precious percentages of additives in each tank and mixing both mixtures mechanically. Moreover, a circular homogenizer was applied to instantaneously mix the fuel with the water before being supplied to the RCEM. In the last cases of the experimental study, four cases of EGR–WFE combinations were tested. When applying WFEs, as the total volume of emulsified fuel is larger than the gas oil, the injection period, injection pressure, or diameter of the nozzle hole have to be increased. All three cases were applied in this study, as shown in Table 2. The injection duration was extended from 25.4 to 33.3 or 37.4 °CA in Case 1. The injection pressure of 80 to 125 and 150 MPa in Case 2. The nozzle hole diameter is from 0.5 to 0.6 mm in Case 3. In Case 3, only injectors with 0.5 or 0.6 mm nozzle holes were available for the experiment.

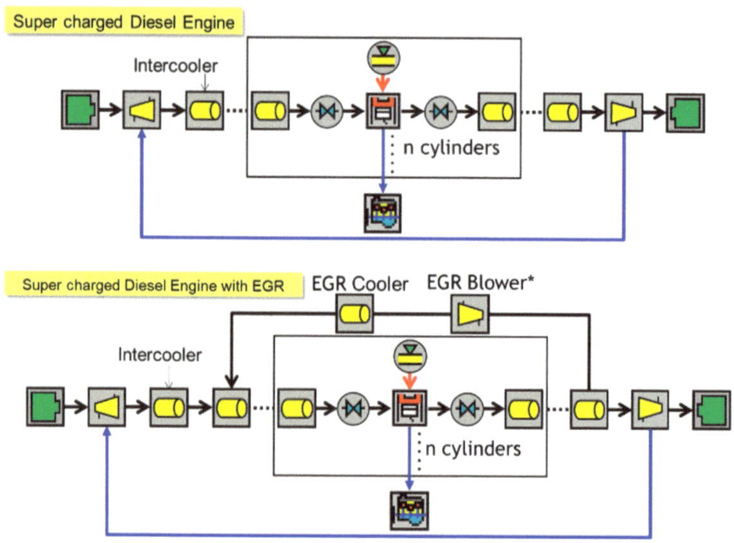

Figure 4. Diesel engine model simulated using G.T. Power (with and without EGR); The 'EGR Blower*' has not been applied in this model.

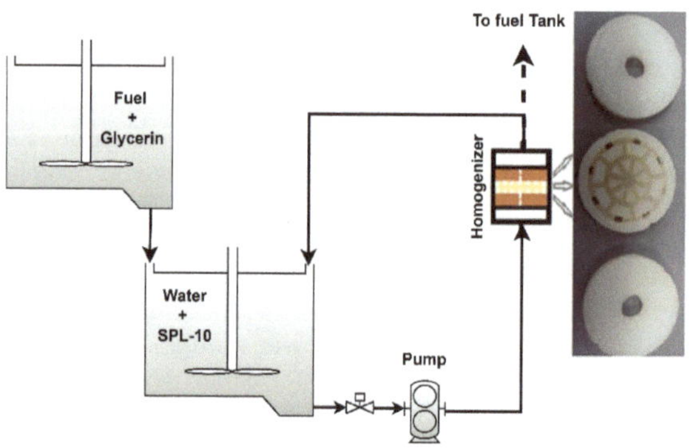

Figure 5. The instantaneous emulsion system before injection.

Table 1. Experimental conditions in EGR.

	Exhaust Gas Recirculation			
Initial oxygen concentration [vol%]	21.0	18.3		16.3
Timing of injection start [deg. ATDC]	−5.2	−5.1		−5.1
Injection duration [deg.]	20.3	20.7		20.6
	EGR–WFE Combinations			
Initial Oxygen concentration [vol%]	18.3		16.3	
Fuel	WFE25	WFE40	WFE25	WFE40
The ratio of water to fuel [vol%]	25	40	25	40
Timing of injection start [deg. ATDC]	−4.8	−5.2	−4.8	−5.2
Injection duration [deg.]	24.6	29.3	25.1	29.4

Table 2. Experimental conditions of the WFE.

	Base	Case 1		Case 2		Case 3
Initial Oxygen Concentration [vol%]	**21.0**					
Fuel	**Gas Oil**	**WFE25**	**WFE40**	**WFE25**	**WFE40**	**WFE40**
The ratio of water to fuel [vol%]	0	25	40	25	40	40
Timing of injection start [deg. ATDC]	−3.1	−4.4	−4.4	−6.3	−7.0	−3.1
Injection duration [deg.]	25.4	33.3	37.4	28.7	29.6	25.9
Injecting pressure [Mpa]	80	80		125	150	80
Diameter of nozzle hole [mm]			0.5			0.6

Figure 6a shows the applied RCEM as an experimental apparatus in this study, while Figure 6b illustrates its schematic layout. The RCEM with specifications listed in Table 3 was designed to visualize the combustion process in medium or large-sized ship engines. The RCEM is speeded up to 350 rpm using an auxiliary (75 kW) electric motor. An inverter (Model FRENIC 5000P11 from Fuji Electric Co., Ltd., Tokyo, Japan) is mounted on the electric motor to regulate its speed. A preheating and pressurizing heat tank is connected to the RCEM to maintain constant initial conditions (temperature and pressure) inside the cylinder (within ± 0.1 bar ± 5 °C). A Kistler piezoelectric pressure sensor, Model 6061B, is used to measure the pressure inside the cylinder. The RCEM has two injectors mounted on the cylinder head. In this study, one of them is actuated. Both injectors incorporate electronic fuel injection, which allows for precise regulation of injection parameters such as pressure, timing, period, and duration via adjustments to the injectors' medium working pressure and pulse width. The RCEM is attached to an exhaust gas analyzer model HORIBA MEXA-7100FX to measure the emission concentrations in the exhaust gases.

Table 3. RCEM specifications.

Engine Type	**4-Stroke**
Number of cylinders	1
Bore × stroke	240 × 260 mm
Clearance volume	200 × 66 × 80 mm^3
(Upper, lower)	240 mm × 5.4 mm
Size of the quartz window	200 × 50 × 100 mm^3 (×2)
Engine speed	350 rpm
Compression ratio	9.54
Initial charge conditions	0.59 MPa, 423 K

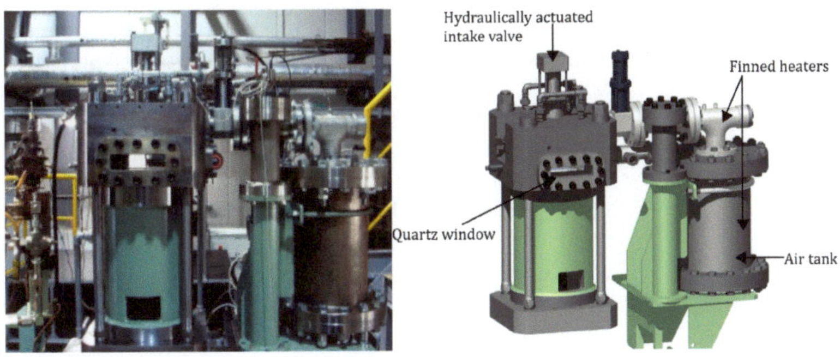

(**a**) (**b**)

Figure 6. Experimental layout: (**a**) front picture of RCEM (**b**) RCEM schematic.

The high-speed camera with specifications listed in Table 4 was used for shooting the direct flame in the combustion chamber. Figure 7 shows the outline of the optics configuration. Parallel light beams are generated by passing the laser beam through a concave mirror before entering the combustion chamber. For this reason, the initial concave mirror was located about 3 m from the primary laser. A half mirror was placed between the first concave mirror and the combustion chamber so the high-speed camera could take direct photos of the reflected beams from the ongoing combustion inside the cylinder.

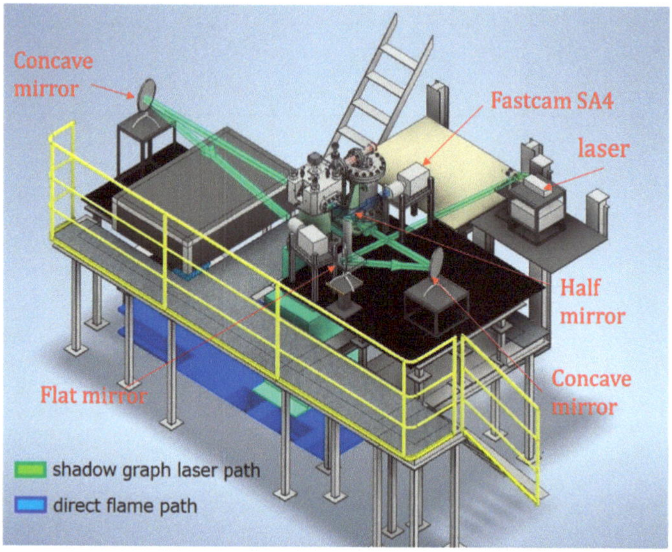

Figure 7. The optical system configuration.

Table 4. Specifications of the high-speed camera and optical settings.

Camera Type	FASTCAM SA4
Lens	AF-S VR Micro-Nikkor F2.8 f105 mm
Image sensor	CMOS image sensor
Resolution	832 × 225 pixels
Frame rate/Exposure	20,000 fps/5.8 μs
Recording color depth	RBG, each 12 bit
Shutter method	Electronic shutter

Despite the widespread use of optical experiments meant to improve experiment analysis, the data received directly from combustion pictures remain exceedingly scarce. Many scientists have used the two-color method (TCM) [40,41] on combustion pictures to get around this restriction. Combustion pictures may be used to determine flame temperature using the TCM. Processing the combustion images using the TCM can provide a local concentration value for the flame temperature, soot, and NOx emissions [40]. To measure flame temperature, this study uses two luminance temperatures at specific wavelengths emitted by soot particles in diesel flame. Because the analysis is performed at two distinct wavelength spots, it is known as the 'TCM'. Figure 8a illustrates applying the TCM by analyzing the emitted light to the blue and red radiation. Figure 8b shows an example of producing the two-color method photos from the direct photos captured from the in-cylinder combustion using a high-speed camera Model FASTCAM SA4. The captured photos were analyzed based on the color of each pixel. The exposing condition in filming the flame was adjusted appropriately, considering the digital RGB values in the actual flame pictures. RGB values are 0–255, and they could occasionally reach their peaks if the selected condition is inappropriate. So, the exposing condition was decided so that the value obtained from the brightest point in the flame picture would be kept lower than 255. The temperature differences between soot particles and the surrounding combustion gas were considered to be minor. The temperature of soot particles is calculated from radiations from two different wavelengths as in Equation (3) which is a derivative form from Hottel-Broughton equation where C_2 is second radiation constant in Plank's law, α: constant (1.38 at visible light range), T is the absolute temperature of the soot particles, λ_1 is the blue wavelength which was set to 450 nm. λ_2 is the red color wavelength which was set to 620 nm.

$$\lambda_1{}^{\alpha} \ln\left[1 - exp\left\{-\frac{C_2}{\lambda_1}\left(\frac{1}{T_{a1}} - \frac{1}{T}\right)\right\}\right] = \lambda_2{}^{\alpha} \ln\left[1 - exp\left\{-\frac{C_2}{\lambda_2}\left(\frac{1}{T_{a2}} - \frac{1}{T}\right)\right\}\right] \tag{3}$$

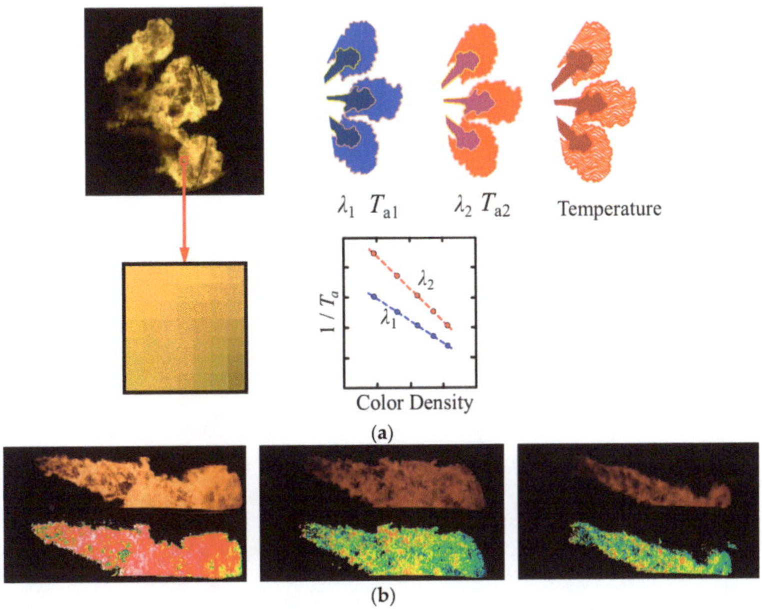

Figure 8. Two-color method definition and the application process. (**a**) the explanation of the picture analysis (**b**) an example of the results with the source photo.

3. The Numerical Study

In the simulation study, the simulation study was performed with four chosen cases in which the ratio of water to fuel was 0, 10, 25, and 40. In contrast, the initial oxygen concentration span was 16–21% with a unity step. This chosen span provided four intersection cases between the experimental and simulated study. KIVA is a 3D Computational Fluid Dynamics (3D-CFD) program created by Los Alamos National Laboratory. KIVA-3V is the third Fortran77-based version of KIVA which can calculate combustion processes in various engines and engine running parameters. Although KIVA is a helpful program for combustion calculations, combustion under WFE conditions cannot be estimated using the standard KIVA code because the emulsified fuel with WFEs is not considered. As a result, a novel model explaining emulsified fuel was established in this work. This is known as the fuel–water concentric spherical model. Figure 9 depicts emulsified fuel droplets as concentric spheres with two layers; the outer layer is the fuel that evaporates first. The water droplet in the inner layer evaporates when the fuel is entirely vaporized. The fuel–water concentric sphere model describes a heterogeneous combination of fuel and water and emulsified fuel evaporation. This model assumed two types of droplets (fuel and water/fuel) droplets as shown in Figure 9, and the ratio between the fuel and water/fuel droplets describes the water emulsion percentage.

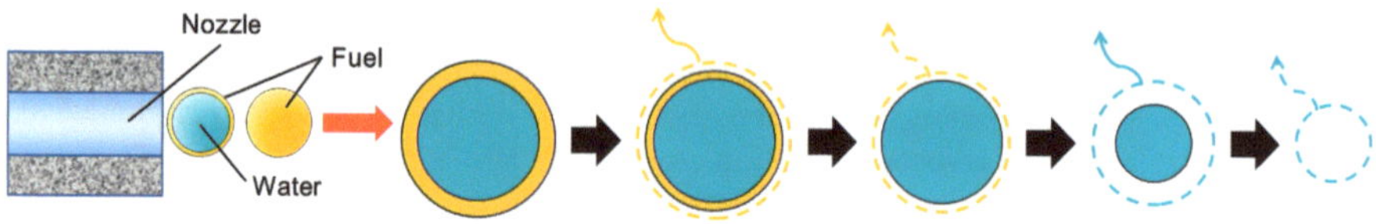

Figure 9. Fuel–water evaporation model process.

KIVA uses the Blob Method that assumes fuel is injected as spheres whose diameter is the same as the nozzle hole of the fuel injector. The radius of the water sphere to the fuel–water concentric sphere is set to a constant, determined by users' input data, and which fuel sphere or fuel–water concentric sphere is injected. The total water-to-fuel ratio in WFEs is adjusted by changing the injection frequency of fuel–water concentric spheres. The fuel–water concentric sphere model modified the break-up model used in KIVA-3V. The break-up model consists of two analysis models: the Kelvin-Helmholtz (KH model) [42] and the Rayleigh–Taylor break-up model (RT model) [43]. KH model is the primary break-up model in which fuel droplets break up due to unstable surface waves made by the aerodynamic forces, and new tiny droplets are formed. The wavelength of the unstable surface wave in the KH model is so small that the new droplets which split off the parent droplet are treated as the spheres of only fuel. RT model is the secondary break-up model in which the droplets break up due to the drag force made by the relative velocity between the droplet and air. In the modified evaporation model, if the amount of the vaporized fuel is larger than the amount of existing fuel in the droplet, the time step is divided into smaller ones not to break the mass balance of fuel. The subroutine program calculating evaporation was modified to implement this study's evaporation process.

Validity Assessments of Numerical Analysis Calculation

This study used an RCEM to conduct the experimental tests for EGR, WFEs, and their combinations. The simulation was performed using KIVA-3V with the fuel–water concentric sphere model. The values of the constants in the analysis model were considered the same among all the calculation conditions. The initial conditions, such as the water–fuel ratio and initial oxygen concentration, were set to the same values as those in the experiments. The computational domain of RCEM in a KIVA-3V calculation is shown in Figure 10. The total number of the calculated cells is 76,000, and the minimum length of the cells in the combustion chamber is 5 mm. This study set the maximum time step in a calculation to 4 μs. The collision of droplets was not considered for the calculations.

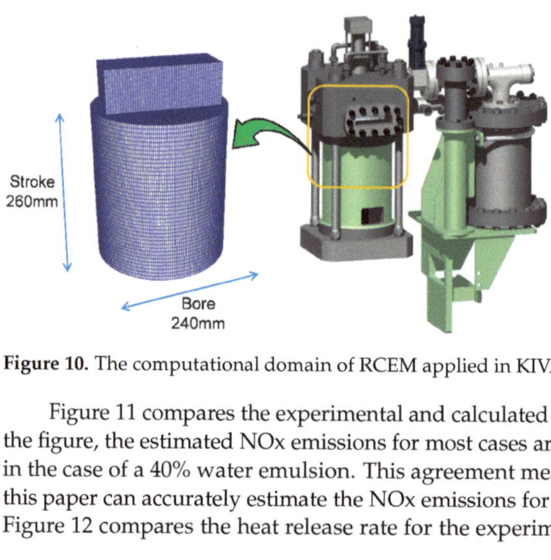

Figure 10. The computational domain of RCEM applied in KIVA-3V.

Figure 11 compares the experimental and calculated values for NOx emissions. From the figure, the estimated NOx emissions for most cases are the same, with a marginal shift in the case of a 40% water emulsion. This agreement means that the applied KIVA-3V in this paper can accurately estimate the NOx emissions for the water emulsion. Meanwhile, Figure 12 compares the heat release rate for the experimental and calculated cases. The figure shows that the Heat Release rate profile for the calculations is compatible with the experiments' cases with some slight error peaks at approximately 15 °CA which can not affect the significant profile. The figure indicates that the heat release rate decreased in all WFE cases after the injection. Using WFEs improved the afterburning period in both the experiments and calculations.

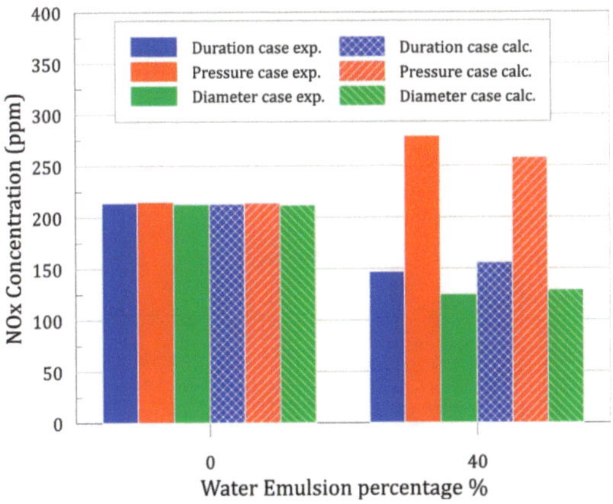

Figure 11. NOx emissions in the WFE calculations versus the experiments.

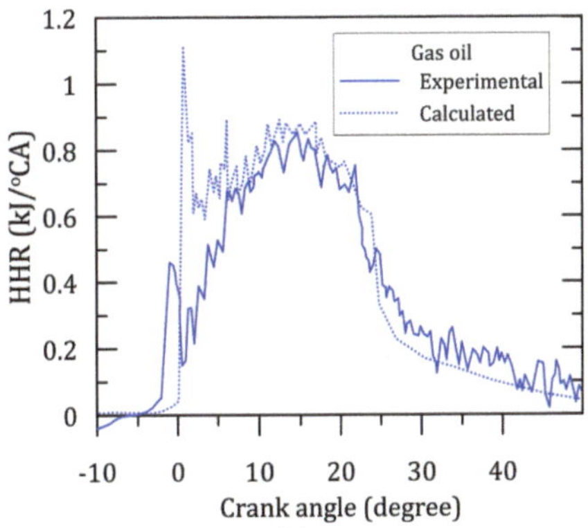

(**a**)

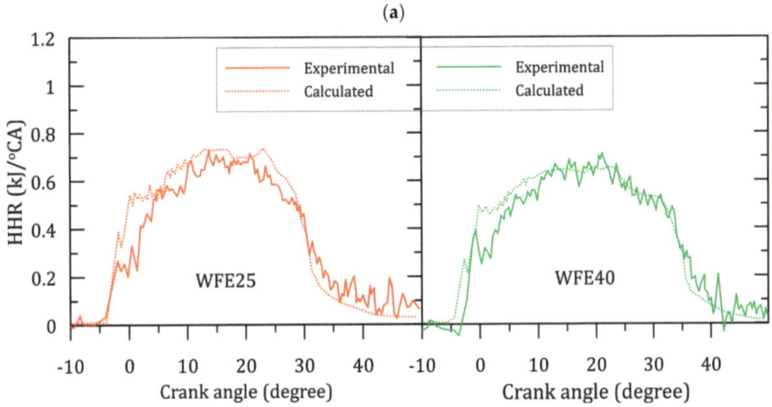

(**b**)

Figure 12. *Cont.*

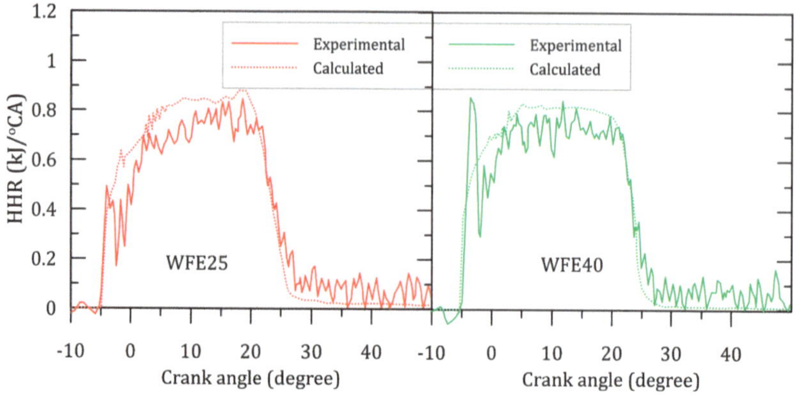

(**c**)

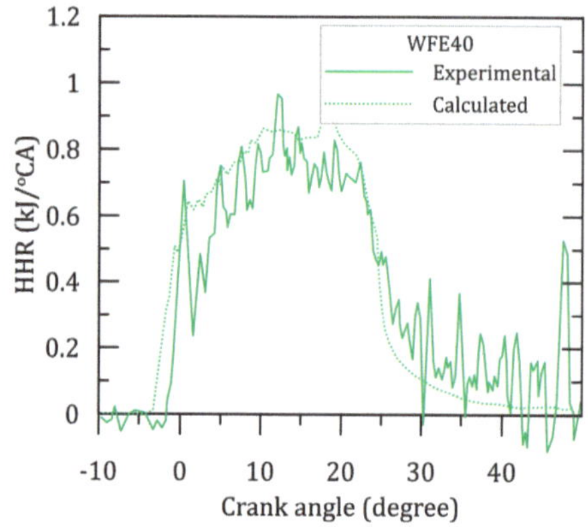

(**d**)

Figure 12. Heat release rate versus crank angle in both experiment and calculated cases: (**a**) the heat release rate for: (**a**) the gas oil as a base case, (**b**) the duration case, (**c**) the pressure case, and (**d**) the diameter case.

4. Uncertainty Analyses

Experimental test results are subject to some error and uncertainty. The precision of an experiment can be influenced by factors such as the observations, the design of the experiment, the testing equipment used, the experimental conditions, the calibration of the equipment, the environment, and the results obtained. An uncertainty analysis was conducted to verify the experimental results' reliability and accuracy. A summary of the experimental errors and uncertainties is provided in Table 5. Analytical methods were used to foretell the random fluctuations in the measured values using Equation (4).

$$w_R = \left[\left(\frac{\partial R}{\partial X_1} \cdot w_1 \right)^2 + \left(\frac{\partial R}{\partial X_2} \cdot w_2 \right)^2 + \ldots\ldots + \left(\frac{\partial R}{\partial X_n} \cdot w_n \right)^2 \right]^{1/2} \quad (4)$$

where w_R is the uncertainty of the measured parameters; w_1, w_{21}, ……, w_n is the error in the independent variables.

Table 5. The uncertainty in the measurements and the calculations.

Instrument	Range	Accuracy	Uncertainty (%)
Injection pressure sensor	0–5000 bar	±5 bar	±0.4
Intake pressure sensor	0–2 bar	±0.006 bar	±0.25
In-cylinder pressure sensor	0–250 bar	±0.125 bar	±0.5
CO	0–100 ppm	5 ppm	±0.2
HC	10–20,000 ppm	10 ppm	±0.2
NOx	10–10,000	10 ppm	±0.25
Calculated quantities			**Uncertainty (%)**
Flame temperature (TCM)			0.3%
Flame temperature (calculated)			2.5%
IMEP			1.3%
NO			1.5%

5. Results and Discussion

Typically, diesel fuel, once injected into the engine cylinder, is self-ignited due to the high temperature and pressure of the ambient. Before the start of the combustion, the air entrained the spray body providing the oxidizing oxygen. On the surface of the diesel spray, the entrained air is quite intensive, leading to a lean mixture, while the core of the combustion is rich due to the minor air entrained in the spray body [44]. The air–fuel ratio variety over the spray body is the main reason for producing the soot and NOx. Several researchers have extensively studied and mapped this variation for local temperatures versus equivalence ratio [45]. Based on this map, the NOx and soot emissions highly depend on the local temperatures and the local equivalence ratio, roughly the start to be emitted at 2200 K at rich spots. Figure 13 illustrates the flame temperature distribution for the direct experimental photos analyzed by TCM. The temperature distributions in the combustion flame have some voids area because the TCM calculated the distribution based on the summation from back to front.

The figure shows that the gas oil's high-temperature spots became less dense as the initial oxygen concentration decreased. For instance, 95% of pixels on the spray body in the case of no EGR (21% O_2) recorded more than 2200 K, while 65% of the spray body reached this extent at 18.3% O_2, and only 8% of the spray combustion temperature recorded more than 2200 k at 16.3% O_2 case. Additionally, the same effect can be noticed in the water emulsion cases, which is evident when comparing the (WFE25 with 21% O_2) case with the other cases. The number of pixels recording 2200 K decreased from 95% to 74% and 6% in the case of the gas oil, WFE25 with 21% O_2, and (WFE25 with 21% O_2), respectively. The superpositioning (combining the EGR and WFE) leads to a significant reduction in the flame temperature, recording 3% of the pixels exceeding 2200 K in the case of WFE25 with 18.3% O_2, while in the other three cases, the local temperatures were decreased below 2100 K. When applying EGR, the amount of carbon dioxide and water molecules increased [46]. As carbon dioxide and water vapor heat are higher than the air, they absorb more heat from the flame, decreasing the combustion temperature. Additionally, introducing the inert gases (carbon dioxide and water vapor) decreases the interaction between the spray surface and the oxidizer (oxygen into the ambient), which leads to a slower oxidation reaction slowing flame speed [47]. On the other hand, the flame temperature is reduced with WFEs because the latent heat of additional water evaporation decreases the flame temperature and diminishes the high-temperature spots [36]. When combining both EGR and WFEs, the local spot temperatures of the flame show a reduction due to the intensive effect of the EGR on the flame speed and the high latent heat of the water.

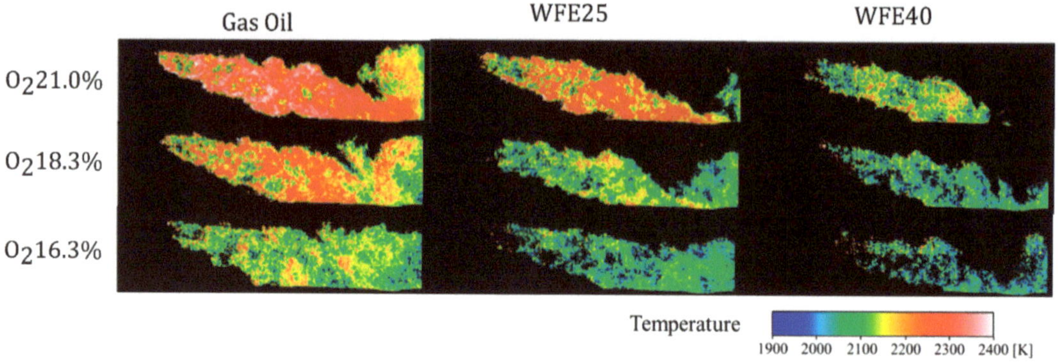

Figure 13. Flame temperature distributions calculated by the two-color method at 10 °CA ATDC.

Figure 14 shows the temperature and NO fraction distributions if the injection duration time changed in the WFE. The upper graph shows the flame temperature calculated by the TCM from the direct photographs of the combustion flame. The mid and lower graphs indicate the temperature and NO mass fraction distributions, which were imaged from the calculation results by the visualization software, Ensight 8.2. From the figure, it can be noticed that the temperatures in the calculations did not show a decrease as the water ratio increased, even though in the two-Color method, the flame temperature has been noticed to decrease drastically. The calculations showed that the temperature was high near the right-side wall of the combustion chamber as the combustion area extended, growing the burned gas at high temperatures. Although the flame temperature in the upper pictures seems lower in the EGR and WFE cases than the pure gas oil, the NO emissions were utterly different in both cases. The distributions of NO mass fraction that KIVA-3V calculates indicate that most NO was produced at the front edge of the combustion flame. At the same time, the amount of produced NO was reduced uniformly in the case of EGR. It was found from the KIVA-3V calculation results that there are significant differences in NOx reduction mechanisms between EGR and WFEs. The amount of NO product was reduced dramatically and uniformly with the EGR. Lower oxygen concentration in EGR conditions leads to lower flame temperature. It results in the synergetic effect of reducing the amount of NO formation because it depends on the temperature and concentration of nitrogen and oxygen, according to the Zeldovich mechanism [28]. Although the flame temperature in the WFE cases looks lower, it is found that the area where the temperature was high and the amount of NO product was large existed at the front edge of the combustion flame. The NO formation area was shifted to the front edge of the combustion flame as the water ratio increased in the WFE case.

Figure 15 shows the in-cylinder pressure profiles in the CAS and COC cases versus the crank angle. From the figure, the in-cylinder profile is highly affected by the EGR rate in the CAS case compared to the COC case. The charging pressure affects the maximum in-cylinder peak pressure. Additionally, introducing the EGR in the intake manifold can affect the in-cylinder combustion as the high content of carbon dioxide increases the heat capacity of the in-cylinder ambient, decreasing the flame speed and leading to lower combustion temperature. In the CAS case, the charging pressure increase leads to an impact on the combustion peak pressure. In contrast with the COC case, the increase in the charging pressure leads to a slight increase in the combustion peak pressure.

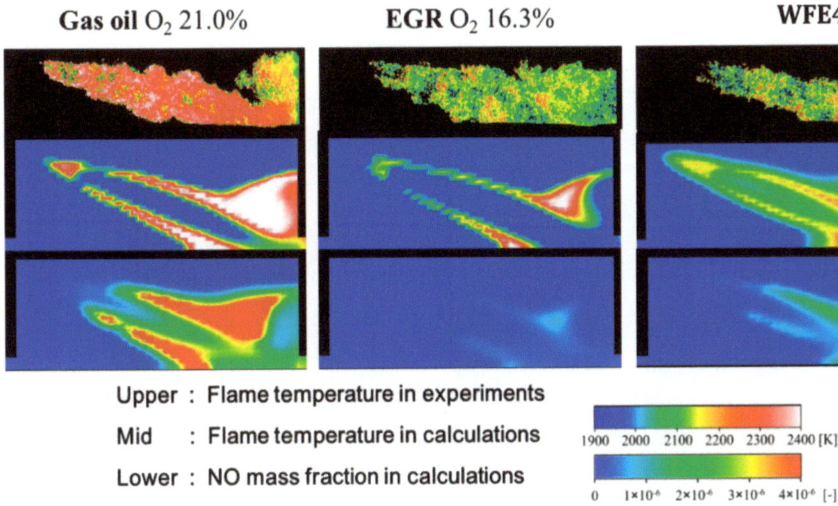

Figure 14. The calculated flame temperature and NO compared to the actual photos at 10 °CA ATDC.

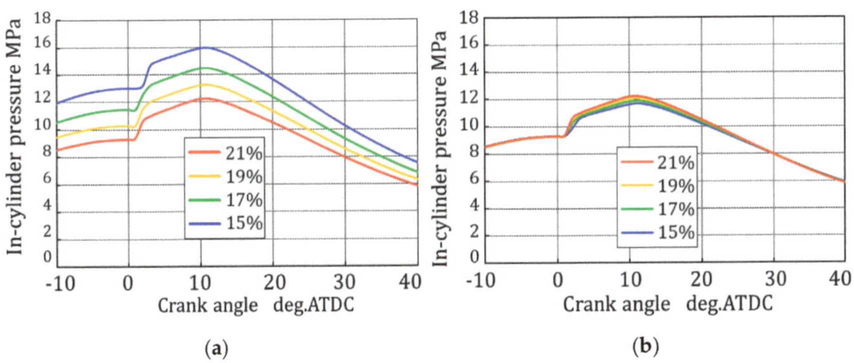

Figure 15. In-cylinder pressure versus crank angle for different EGR cases (**a**) CAS (**b**) COC.

Figure 16 illustrates the heat release rate versus crank angle for EGR and WFE cases. Moreover, the heat release rate is calculated from the first low in thermodynamics for the single-zone combustion model [1]. The heat release at the late crank angle indicates afterburning, affecting thermal efficiency. The afterburning becomes longer as the oxygen concentration decreases when applying the EGR method. The low oxygen concentration slows the fuel-oxygen chemical reaction rate, leading to increased residual fuel till the late crank angle, at which the cylinder temperature is significantly cooled down. As a result, the heat release continued until the late crank angle. In contrast, the afterburning phenomenon is diminished when applying WFEs. WFEs improve fuel–air mixing due to the increased momentum of emulsified fuel by additional water mass. At the same time, WFEs have a secondary effect that can slightly reduce the contact area between the fuel and ambient, leading to decreased fuel–air mixing. Meanwhile, the dominant factor is momentum; thus, the most fuel was consumed effectively, and a few amounts of fuel remained after the injection. Zhao et al. [48] reported that EGR increased the combustion duration and the afterburning phase. Pan et al. [49] noticed an increase in the ignition delay and the HHR peaks when applying the EGR.

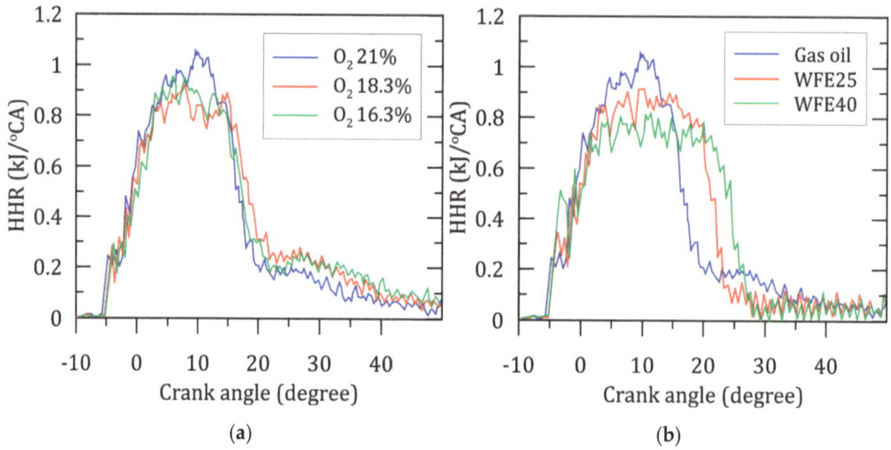

Figure 16. Experimental heat release rate versus crank angle for the two applied cases: (**a**) EGR and (**b**) WFEs.

Figure 17 shows the NOx emissions concentrations in the exhaust gas for the different EGR–WFE combination cases based on the experimental results. Generally, applying the EGR decreased the NOx emissions from 642 to 182 and 62 ppm in the case of (18.3% and 16.3) oxygen concentration, respectively, recording reduction percentages of 71.7% and 90%. At the same time, applying WFEs with 25% and 40% leads to reducing NOx emissions from 642 to 630 ppm and 443 ppm with a 2% and 31% reduction rating. The decrease in the NOx emissions when applying emulsion could be referred to as the enhancement in the fuel spray momentum, which increases the fuel and air mixing, leading to better combustion. Additionally, the micro explosion of the water can enhance the fuel atomization as the water droplets in the diesel spray could reach the superheated condition before the diesel fuel. After the combustion starts, the heat releases, leading to the evaporation of any residual water. This evaporation heat decreases combustion temperature and NOx formation. In the EGR case, the reduction in the NOx formation results from the decreasing nitrogen molecules, which could interact with the oxygen in the ambient on the circumference of the diesel spray. When applying the water addition as emulsified fuel with the EgR, the NOx emissions decreased in the same trend. For instance, in the case of 16.3% oxygen content, the NOx emissions decreased by 91.7% and 93% when applying 25% and 40% of water content based on the pure diesel case. This reduction could be attributed to the decreased oxygen content in the cylinder, which decreases the possibility of nitrogen-oxygen interaction and the enhancement in the spry diesel momentum, in addition to the decrease in the flame temperature as the water absorbs the heat in the form of latent heat to evaporate. Pan et al. [49] reported that applying the EGR can decrease NOx emissions, especially in low and medium loads, and almost no effect at high loads.

Figure 18 illustrates the CO and HC emissions from the RCEM under different oxygen percentages in the RCEM. Generally, Carbon monoxide (CO) emissions are often the result of incomplete combustion, which can be caused by low cylinder temperatures and low oxygen concentrations that slow down local oxidation reactions. Unburned hydrocarbons (HC) can be emitted for several reasons [32]. One reason is that an unburned mixture in the cylinder can escape through the exhaust valve during the gas exchange process due to valve overlap. Another reason is that the unburned mixture can be compressed into crevices in the combustion chamber and released when the exhaust valves open [31]. A third reason is that the unburned mixture can be absorbed into the lubricating oil used to lubricate the piston and piston rings and released during the expansion stroke, especially during cold start conditions. Crevices in the combustion chamber can play a significant role in forming HC emissions. The CO and HC emissions increased with the increase in the

EGR (decrease in the oxygen percentage) due to the decrease in the combustion velocity and the flame quenching in the crevices [32].

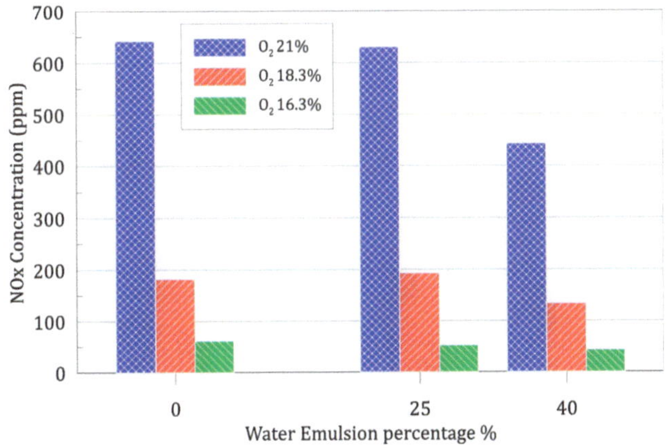

Figure 17. NOx emissions concentration in the case of EGR–WFE combination.

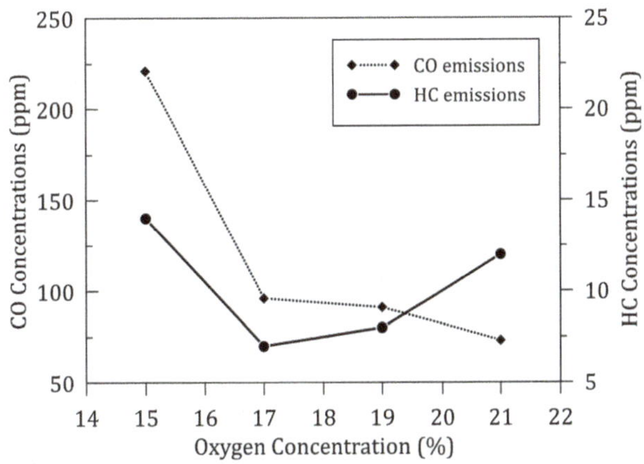

Figure 18. CO and HC emissions concentration in the case of EGR.

Figure 19 illustrates an estimated NO emission reduction percentage compared to the EGR–WFE gas oil combination. The preponderance portion of NOx released by diesel engines is "thermal NO", which is generated by direct interactions of nitrogen and oxygen in a high-temperature environment [8]. As a result, this section calculates the NO emissions in the cylinder. When the oxygen content was less than 18%, the amount of NO product was lowered by more than 75% in all scenarios. The decrease in NO became more visible at each EGR level as the amount of the water emulsion increased, implying that the impact of NO reduction when using EGR was improved when coupled with WFEs. In the pressure scenario, when combined with WFE10, WFE25, and WFE40, the reduction in NO at 16% O_2 was (91%, 91.6%, and 94%). Meanwhile, the Indicated Mean Effective Pressure (IMEP) was calculated in this study. In almost all cases, the IMEP decreased by more than 5%; this could be attributed to the longer injection period, which can result in a more dispersed fuel–air mixture in the combustion chamber, leading to a less efficient combustion process. A more dispersed mixture may not burn as effectively, resulting in a lower IMEP. In addition, a longer injection period can also result in a higher heat loss to the cylinder walls, which can also lower the IMEP [14]. The 5% IMEP reduction requires much-increased engine propulsion and fuel costs. As a result, using WFEs under 18.0% oxygen concentration is advantageous, especially in the pressure case. When the oxygen concentration was 16%, the additional water ratio was 40%, and the injection pressure increased, the amount of NO product lessened by 94%, with a 4% rise in the IMEP. This was the only scenario that fulfilled the NO reduction and the increased IMEP criteria. Wang et al. [50] reported that in some cases, when the EGR cooling temperature increases, the NOx and brake-specific fuel consumption increases.

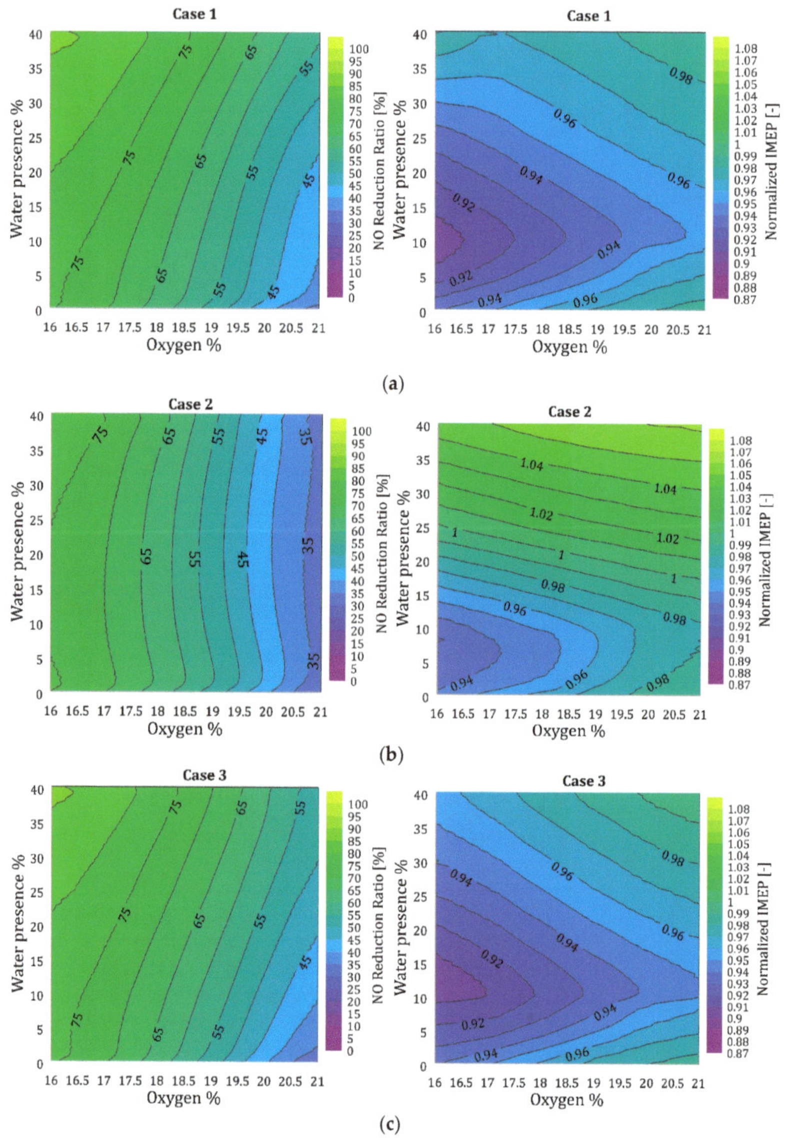

Figure 19. NO reduction and IMEP calculations for EGR–WFE combinations: (**a**) injection duration, (**b**) the pressure case, and (**c**) the diameter case.

6. Conclusions

EGR and WFE have long been regarded as effective methods for decreasing NOx emissions from diesel engines. Because of the low oxygen concentration in EGR, the duration of afterburning increases, lowering thermal efficiency. In the meantime, WFEs improves fuel–air mixing due to increased momentum from the additional water mass, resulting in better combustion and shorter afterburning. This research, experimental tests were performed to study the effect of EGR, WFEs, and their combinations on diesel combustion using an RCEM. Then, this study was extended to cover more comprehensive percentages in the EGR case implementing two EGR systems: Constant Air Suction (CAS) and Constant Oxygen Charge (COC) utilizing a 1D GT power simulation model. Meanwhile, a numerical KIVA-3V code was used to reveal the WFE–EGR combination combustion mechanisms. Lastly, the calibrated model simulated the combustion process and NOx emissions. The results are as follows:

- In the CAS, the EGR rate rapidly increases as the required oxygen concentration decreases, resulting in higher charging pressure.
- In COC, the decrease in intake volume at a high EGR rate alleviates the problem, and in EGR, the problem is the decrease in thermal efficiency at a high EGR.
- The EGR rate in the COC case is highly affected by the in-cylinder profile compared to the CAS case.
- The flame temperature in the no EGR case (21%O_2) exceeded 2400 K while a few spots reached this extent at 17% O_2 case, and almost no spots at that high limit were noticed at 15% O_2 case.
- The KIVA-3V code calculation results reveal that the amount of NO product was reduced consistently over the whole combustion flame under the EGR condition.
- WFEs produce low NOx due to the drop in combustion temperature caused by the latent heat of water evaporation.
- In the case of an initial oxygen concentration of 16% and an extra water ratio of 40%, the calculation results revealed that the NO reduction reached 94%.
- The IMEP increased by 4%. If the WFE is combined with the EGR in the improved injection pressure configuration, it is possible to achieve a significant reduction in NO while retaining thermal efficiency.

Author Contributions: E.A., conceptualization, methodology, software, validation, formal analysis, investigation, resources, data curation, and writing—original draft preparation; H.T., conceptualization, methodology, software, validation, formal analysis, investigation, resources, data curation, writing—review and editing, supervision, and project administration. All authors have read and agreed to the published version of the manuscript.

Funding: This research received no external funding.

Data Availability Statement: Data are available upon request.

Conflicts of Interest: The authors declare no conflict of interest.

Abbreviations

ATDC	After Top Dead Center
CAS	Constant Air Suction
COC	Constant Oxygen Charge
CCD	Charge-Coupled Device
EGR	Exhaust Gas Recirculation
IMEP	Indicate Mean Effective Pressure
IMO	International Maritime Organization
RCEM	Rapid Compression Expansion Machine
TCM	Two-Color Method
TDC	Top Dead Center
WFE10	Water-in-Fuel Emulsion, 10% water, and 90% fuel, v/v%.
WFE25	Water-in-Fuel Emulsion, 25% water, and 75% fuel, v/v%.
WFE40	Water-in-Fuel Emulsion, 40% water, and 60% fuel, v/v%.

Nomenclature

$\frac{dQ}{d\theta}$	Heat release rate based on crank angle, J/deg.	κ	Specific heat ratio
$\frac{dP}{d\theta}$	Pressure-change per crank angle, MPa/deg.	θ	Crank angle, degree
$\frac{dV}{d\theta}$	Volume-change rate per crank angle, m^3/deg	V	Instantaneous in-cylinder volume, m^3
P	Instantaneous in-cylinder pressure, MPa	T_b	The black body temperature, K
N_o	Monochromatic radiance, W/(m^2·μm)		

References

1. Abdelhameed, E.; Tashima, H. Experimental investigation on methane inert gas dilution effect on marine gas diesel engine performance and emissions. *Energy Sources, Part A Recover. Util. Environ. Eff.* **2022**, *44*, 3584–3596. [CrossRef]
2. Zincir, B.; Deniz, C.; Tunér, M. Investigation of environmental, operational and economic performance of methanol partially premixed combustion at slow speed operation of a marine engine. *J. Clean. Prod.* **2019**, *235*, 1006–1019. [CrossRef]
3. Bilgili, L.; Celebi, U.B. Developing a new green ship approach for flue gas emission estimation of bulk carriers. *Measurement* **2018**, *120*, 121–127. [CrossRef]
4. De Simio, L.; Iannaccone, S. Gaseous and particle emissions in low-temperature combustion diesel–HCNG dual-fuel operation with double pilot injection. *Appl. Energy* **2019**, *253*, 113602. [CrossRef]
5. Cameretti, M.C.; De Robbio, R.; Mancaruso, E.; Palomba, M. CFD Study of Dual Fuel Combustion in a Research Diesel Engine Fueled by Hydrogen. *Energies* **2022**, *15*, 5521. [CrossRef]
6. Talebizadeh, P.; Babaie, M.; Brown, R.; Rahimzadeh, H.; Ristovski, Z.; Arai, M. The role of non-thermal plasma technique in NOx treatment: A review. *Renew. Sustain. Energy Rev.* **2014**, *40*, 886–901. [CrossRef]
7. Førby, N.; Thomsen, T.B.; Cordtz, R.F.; Bræstrup, F.; Schramm, J. Ignition and combustion study of premixed ammonia using GDI pilot injection in CI engine. *Fuel* **2023**, *331*, 125768. [CrossRef]
8. Wang, Y.; Zhou, X.; Liu, L. Theoretical investigation of the combustion performance of ammonia/hydrogen mixtures on a marine diesel engine. *Int. J. Hydrogen Energy* **2021**, *46*, 14805–14812. [CrossRef]
9. Hossain, A.K.; Refahtalab, P.; Omran, A.; Smith, D.I.; Davies, P.A. An experimental study on performance and emission characteristics of an IDI diesel engine operating with neat oil-diesel blend emulsion. *Renew. Energy* **2020**, *146*, 1041–1050. [CrossRef]
10. Puškár, M.; Kopas, M.; Puškár, D.; Lumnitzer, J.; Faltinová, E. Method for reduction of the NOX emissions in marine auxiliary diesel engine using the fuel mixtures containing biodiesel using HCCI combustion. *Mar. Pollut. Bull.* **2018**, *127*, 752–760. [CrossRef]
11. Theotokatos, G.; Stoumpos, S.; Bolbot, V.; Boulougouris, E. Simulation-based investigation of a marine dual-fuel engine. *J. Mar. Eng. Technol.* **2020**, *19*, 5–16. [CrossRef]
12. Deng, J.; Wang, X.; Wei, Z.; Wang, L.; Wang, C.; Chen, Z. A review of NOx and SOx emission reduction technologies for marine diesel engines and the potential evaluation of liquefied natural gas fuelled vessels. *Sci. Total. Environ.* **2021**, *766*, 144319. [CrossRef] [PubMed]
13. Jiaqiang, E.; Zhang, Z.; Chen, J.; Pham, M.; Zhao, X.; Peng, Q.; Zuo, W.; Yin, Z. Performance and emission evaluation of a marine diesel engine fueled by water biodiesel-diesel emulsion blends with a fuel additive of a cerium oxide nanoparticle. *Energy Convers. Manag.* **2018**, *169*, 194–205. [CrossRef]
14. Kim, H.J.; Park, S.H.; Lee, C.S. Impact of fuel spray angles and injection timing on the combustion and emission characteristics of a high-speed diesel engine. *Energy* **2016**, *107*, 572–579. [CrossRef]
15. Yoon, W.H.; Kim, J.Y.; Ha, J.S. Optimization of Fuel Injection Nozzles for the Reduction of NOx Emissions in Medium-speed Marine Diesel Engines (Diesel Engines, Performance and Emissions, NOx Strategies). *Proc. Int. Symp. Diagn. Model. Combust. Intern. Combust. Engines* **2004**, *6*, 81–86. [CrossRef]
16. Hariharan, D.; Krishnan, S.R.; Srinivasan, K.K.; Sohail, A. Multiple injection strategies for reducing HC and CO emissions in diesel-methane dual-fuel low temperature combustion. *Fuel* **2021**, *305*, 121372. [CrossRef]
17. Cameretti, M.C.; De Robbio, R. Raffaele Tuccillo Performance Improvement and Emission Control of a Dual Fuel Operated Diesel Engine. In Proceedings of the 13th International Conference on Engines & Vehicles, Capri Napoli, Italy, 10–14 September 2017; pp. 24–66.
18. D'Aniello, F.; Arsie, I.; Pianese, C.; Stola, F. Development of an Integrated Control Strategy for engine and SCR system basedon effective EGR rate. *IFAC-PapersOnLine* **2020**, *53*, 14034–14039. [CrossRef]
19. Kuwahara, T.; Yoshida, K.; Kuroki, T.; Hanamoto, K.; Sato, K.; Okubo, M. Pilot-Scale Aftertreatment Using Nonthermal Plasma Reduction of Adsorbed NOx in Marine Diesel-Engine Exhaust Gas. *Plasma Chem. Plasma Process.* **2014**, *34*, 65–81. [CrossRef]
20. Yang, S.; Pan, X.; Han, Z.; Zhao, D.; Liu, B.; Zheng, D.; Yan, Z. Removal of NOx and SO2 from simulated ship emissions using wet scrubbing based on seawater electrolysis technology. *Chem. Eng. J.* **2018**, *331*, 8–15. [CrossRef]
21. Zhou, J.; Ma, J.; Wang, Z. Experimental study on removal performance of SO2 and NOx in marine exhaust gas using seawater/urea peroxide solution and analysis of ions concentration change. *Fuel Process. Technol.* **2022**, *227*, 107133. [CrossRef]

22. Wang, P.; Tang, X.; Shi, L.; Ni, X.; Hu, Z.; Deng, K. Experimental investigation of the influences of Miller cycle combined with EGR on performance, energy and exergy characteristics of a four-stroke marine regulated two-stage turbocharged diesel engine. *Fuel* **2021**, *300*, 120940. [CrossRef]
23. Zhang, W.; Feng, T.; Li, Z.; Chen, Z.; Zhao, J. EGR thermal and chemical effects on combustion and emission of diesel/natural gas dual-fuel engine. *Fuel* **2021**, *302*, 121161. [CrossRef]
24. Ni, P.; Wang, X.; Li, H. A review on regulations, current status, effects and reduction strategies of emissions for marine diesel engines. *Fuel* **2020**, *279*, 118477. [CrossRef]
25. Yuan, X.; Ding, X.; Leng, L.; Li, H.; Shao, J.; Qian, Y.; Huang, H.; Chen, X.; Zeng, G. Applications of bio-oil-based emulsions in a DI diesel engine: The effects of bio-oil compositions on engine performance and emissions. *Energy* **2018**, *154*, 110–118. [CrossRef]
26. Gopidesi, R.K.; Premkartikkumar, S.R. Evaluating the hythane/water diesel emulsion dual fuel diesel engine characteristics at various pilot diesel injection timings. *Mater. Today Proc.* **2021**. [CrossRef]
27. Aghbashlo, M.; Tabatabaei, M.; Khalife, E.; Najafi, B.; Mirsalim, S.M.; Gharehghani, A.; Mohammadi, P.; Dadak, A.; Shojaei, T.R.; Khounani, Z. A novel emulsion fuel containing aqueous nano cerium oxide additive in diesel–biodiesel blends to improve diesel engines performance and reduce exhaust emissions: Part II—Exergetic analysis. *Fuel* **2017**, *205*, 262–271. [CrossRef]
28. Wang, Z.; Zhang, F.; Xia, Y.; Wang, D.; Xu, Y.; Du, G. Combustion phase of a diesel/natural gas dual fuel engine under various pilot diesel injection timings. *Fuel* **2021**, *289*, 119869. [CrossRef]
29. Liu, X.; Wang, H.; Zheng, Z.; Yao, M. Numerical investigation on the combustion and emission characteristics of a heavy-duty natural gas-diesel dual-fuel engine. *Fuel* **2021**, *300*, 120998. [CrossRef]
30. Wang, J.; Duan, X.; Wang, W.; Guan, J.; Li, Y.; Liu, J. Effects of the continuous variable valve lift system and Miller cycle strategy on the performance behavior of the lean-burn natural gas spark ignition engine. *Fuel* **2021**, *297*, 120762. [CrossRef]
31. Wang, J.; Duan, X.; Liu, Y.; Wang, W.; Liu, J.; Lai, M.-C.; Li, Y.; Guo, G. Numerical investigation of water injection quantity and water injection timing on the thermodynamics, combustion and emissions in a hydrogen enriched lean-burn natural gas SI engine. *Int. J. Hydrogen Energy* **2020**, *45*, 17935–17952. [CrossRef]
32. Duan, X.; Liu, Y.; Liu, J.; Lai, M.-C.; Jansons, M.; Guo, G.; Zhang, S.; Tang, Q. Experimental and numerical investigation of the effects of low-pressure, high-pressure and internal EGR configurations on the performance, combustion and emission characteristics in a hydrogen-enriched heavy-duty lean-burn natural gas SI engine. *Energy Convers. Manag.* **2019**, *195*, 1319–1333. [CrossRef]
33. Zhang, W.; Song, C.; Lv, G.; Bi, F.; Qiao, Y.; Wang, L.; Zhang, X. Properties and oxidation of in-cylinder soot associated with exhaust gas recirculation (EGR) in diesel engines. *Proc. Combust. Inst.* **2021**, *38*, 1319–1326. [CrossRef]
34. Jaliliantabar, F.; Ghobadian, B.; Carlucci, A.P.; Najafi, G.; Mamat, R.; Ficarella, A.; Strafella, L.; Santino, A.; De Domenico, S. A comprehensive study on the effect of pilot injection, EGR rate, IMEP and biodiesel characteristics on a CRDI diesel engine. *Energy* **2020**, *194*, 116860. [CrossRef]
35. Mukhtar, M.; Hagos, F.Y.; Aziz, A.R.A.; Abdulah, A.A.; Karim, Z.A.A. Combustion characteristics of tri-fuel (diesel-ethanol-biodiesel) emulsion fuels in CI engine with micro-explosion phenomenon attributes. *Fuel* **2022**, *312*, 122933. [CrossRef]
36. Park, J.; Oh, J. Study on the characteristics of performance, combustion, and emissions for a diesel water emulsion fuel on a combustion visualization engine and a commercial diesel engine. *Fuel* **2022**, *311*, 122520. [CrossRef]
37. Abdollahi, M.; Ghobadian, B.; Najafi, G.; Hoseini, S.; Mofijur, M.; Mazlan, M. Impact of water–biodiesel–diesel nano-emulsion fuel on performance parameters and diesel engine emission. *Fuel* **2020**, *280*, 118576. [CrossRef]
38. Gad, M.S.; He, Z.; El-Shafay, A.S.; El-Seesy, A.I. Combustion characteristics of a diesel engine running with Mandarin essential oil -diesel mixtures and propanol additive under different exhaust gas recirculation: Experimental investigation and numerical simulation. *Case Stud. Therm. Eng.* **2021**, *26*, 101100. [CrossRef]
39. Mukhtar, M.N.A.; Hagos, F.Y.; Noor, M.M.; Mamat, R.; Abdullah, A.A.; Aziz, A.R.A. Tri-fuel emulsion with secondary atomization attributes for greener diesel engine—A critical review. *Renew. Sustain. Energy Rev.* **2019**, *111*, 490–506. [CrossRef]
40. Choi, M.; Mohiuddin, K.; Kim, N.; Park, S. Investigation of the effects of EGR rate, injection strategy and nozzle specification on engine performances and emissions of a single cylinder heavy duty diesel engine using the two color method. *Appl. Therm. Eng.* **2021**, *193*, 117036. [CrossRef]
41. Jeon, J.; Park, S. Effect of injection pressure on soot formation/oxidation characteristics using a two-color photometric method in a compression-ignition engine fueled with biodiesel blend (B20). *Appl. Therm. Eng.* **2018**, *131*, 284–294. [CrossRef]
42. Kakoee, A.; Gharehghani, A. Comparative study of hydrogen addition effects on the natural-gas/diesel and natural-gas/dimethyl-ether reactivity controlled compression ignition mode of operation. *Energy Convers. Manag.* **2019**, *196*, 92–104. [CrossRef]
43. Lu, Y.; Pan, J.; Fan, B.; Otchere, P.; Chen, W.; Cheng, B. Research on the application of aviation kerosene in a direct injection rotary engine-Part 1: Fundamental spray characteristics and optimized injection strategies. *Energy Convers. Manag.* **2019**, *195*, 519–532. [CrossRef]
44. Aoyagi, T.; Abdelhameed, E.; Tsuru, D.; Tashima, H. Experimental and Numerical Considerations of Air Entrainment Process of Diesel Spray. *Int. Conf. Liq. At. Spray Syst.* **2021**, *1*. [CrossRef]
45. Glassman, I.; Yetter, R.A. *Diffusion Flames*; Academic Press: Burlington, ON, Canada, 2008; pp. 311–377.
46. Li, X.; Zhen, X.; Wang, Y.; Liu, D.; Tian, Z. The knock study of high compression ratio SI engine fueled with methanol in combination with different EGR rates. *Fuel* **2019**, *257*, 116098. [CrossRef]

47. Verma, S.; Das, L.M.; Kaushik, S.C.; Bhatti, S.S. The effects of compression ratio and EGR on the performance and emission characteristics of diesel-biogas dual fuel engine. *Appl. Therm. Eng.* **2019**, *150*, 1090–1103. [CrossRef]
48. Zhao, J.; Fu, R.; Wang, S.; Xu, H.; Yuan, Z. Fuel economy improvement of a turbocharged gasoline SI engine through combining cooled EGR and high compression ratio. *Energy* **2022**, *239*, 122353. [CrossRef]
49. Pan, M.; Qian, W.; Zheng, Z.; Huang, R.; Zhou, X.; Huang, H.; Li, M. The potential of dimethyl carbonate (DMC) as an alternative fuel for compression ignition engines with different EGR rates. *Fuel* **2019**, *257*, 115920. [CrossRef]
50. Wang, Z.; Zhou, S.; Feng, Y.; Zhu, Y. EGR modeling and fuzzy evaluation of Low-Speed Two-Stroke marine diesel engines. *Sci. Total. Environ.* **2020**, *706*, 135444. [CrossRef]

energies

Article

Data-Driven Model for Real-Time Estimation of NOx in a Heavy-Duty Diesel Engine

Alessandro Falai [1,2,*,†] and Daniela Anna Misul [1,2,*,†]

1 Department of Energy, Politecnico di Torino, Corso Duca Degli Abruzzi 24, 10129 Turin, Italy
2 Interdepartmental Center for Automotive Research and Sustainable Mobility (CARS@PoliTO), Politecnico di Torino, Corso Duca Degli Abruzzi 24, 10129 Turin, Italy
* Correspondence: alessandro.falai@polito.it (A.F.); daniela.misul@polito.it (D.A.M.)
† These authors contributed equally to this work.

Abstract: The automotive sector is greatly contributing to pollutant emissions and recent regulations introduced the need for a major control of, and reduction of, internal combustion engine emissions. Artificial intelligence (AI) algorithms have proven to hold the potential to be the thrust in the state-of-the-art for engine-out emission prediction, thus enabling tailored calibration modes and control solutions. More specifically, the scientific literature has recently witnessed strong efforts in AI applications for the development of nitrogen oxides (NOx) virtual sensors. These latter replace physical sensors and exploit AI algorithms to estimate NOx concentrations in real-time. Still, the calibration of the algorithms, together with the appropriate choice of the specific metric, strongly affects the prediction capability. In the present paper, a machine learning-based virtual sensor for NOx monitoring in diesel engines was developed, based on the Extreme Gradient Boosting (XGBoost) machine learning algorithm. The latter is commonly used in the literature to deploy virtual sensors due to its high performance, flexibility and robustness. An experimental campaign was carried out to collect data from the engine test bench, as well as from the engine electronic control unit (ECU), for the development and calibration of the virtual sensor at steady-state conditions. The virtual sensor has, since then, been tested throughout on an on-road driving mission to assess its prediction performance in dynamic conditions. In stationary conditions, its prediction accuracy was around 98%, whereas it was 85% in transient conditions. The present study shows that AI-based virtual sensors have the potential to significantly improve the accuracy and reliability of NOx monitoring in diesel engines, and can, therefore, play a key role in reducing NOx emissions and improving air quality.

Keywords: compression ignition engine; experimental data processing and analysis; model validation; machine learning; combustion modeling

Citation: Falai, A.; Misul, D.A. Data-Driven Model for Real-Time Estimation of NOx in a Heavy-Duty Diesel Engine. *Energies* **2023**, *16*, 2125. https://doi.org/10.3390/en16052125

Academic Editors: Roberta De Robbio and Maria Cristina Cameretti

Received: 30 January 2023
Revised: 18 February 2023
Accepted: 20 February 2023
Published: 22 February 2023

1. Introduction

Air pollution caused by vehicle emissions, including nitrogen oxide (NOx), is a major environmental and public health concern. Strict regulations have been implemented globally to reduce NOx emissions from vehicles, and car manufacturers are actively seeking ways to optimize the performance of their internal combustion engines (ICEs) to meet these regulations [1].

The most well-known techniques for reducing and controlling engine-out NOx basically involve the following: the use of Exhaust Gas Recirculation (EGR) [2–6]; the deployment of advanced combustion technology, such as homogeneous charge compression ignition (HCCI) [7,8] and premixed charge compression ignition (PCCI) [9,10], which can achieve low NOx emissions by carefully controlling the air–fuel mixture and the combustion process; different injection strategies targeting pressure and timing [11,12]. As far as the vehicle tailpipe emissions are concerned, the main NOx reduction techniques involve the following: filters and catalysts, such as Selective Catalytic Reduction (SCR)

[13], which uses a catalytic converter to reduce NOx emissions by injecting a reducing agent, such as urea, into the exhaust stream; Lean NOx Trap (LNT) [14], which exploits a catalytic converter to temporarily store NOx emissions for reduction at specific engine operating conditions; Diesel Particulate Filter (DPF) [15], which uses a filter to capture particulate matter from the exhaust stream and can also help in reducing NOx emissions. Still, the target NOx reduction performance can only be accomplished by means of reliable NOx sensing.

Several models can be used for NOx prediction in engines, and engine maps exploitation is a well-established approach to calibrate the model, which can be tested on-road during transient conditions [16–18]. Engine maps are the graphical representations of the relationship between various engine operating parameters (e.g., load, speed, fuel flow rate) and engine performance characteristics (e.g., power output, emissions). These maps can be used to predict the NOx emissions of an engine under a given set of operating conditions [19,20]. A method exploiting the engine maps for NOx prediction was developed in [21]. An engine map was implemented in the electronic control unit (ECU) and represented by mathematical models to compute NOx emissions as a function of the observed operating conditions of the engine. The model was calibrated using engine test bench data, and tested on-road to determine its prediction capability under real-world conditions. Applications related to the virtual NOx sensing, exploiting the engine map operating points at steady-state conditions [22], can be found in the literature, together with applications derived from experimental test bench acquisition [23]. However, little research has been conducted on the virtual sensing and prediction of NOx during on-road real-world driving conditions, accounting for the effects of transient phenomena.

The optimization of NOx emissions remains a challenge in the present and virtual sensing techniques are widely used [24,25] to avoid the disadvantages of physical solid-state sensors to retrieve engine-out NOx information in diesel engine applications. Recently, several works have aimed at designing virtual sensors exploiting machine learning algorithms [26,27] in order to predict the engine-out NOx levels, based on the engine operating conditions, as from the ECU. Machine learning algorithms appear to be very promising as virtual measurement tools due to their ability to detect strongly non-linear behavior in the analyzed physical system [28,29]. Several machine learning algorithms have been applied to the prediction of engine NOx for virtual sensor applications, including linear regression [30], support vector machines (SVM) [31], and artificial neural networks (ANNs) [32]. Among these machine learning models, the XGBoost has proved to be particularly effective, outperforming other machine learning models, such as gradient boosting (GBT) and random forest (RF) [33]. This is due to the XGBoost's ability to handle large datasets and to cope with 'missing values', thus making it well-suited for virtual sensor applications where data may be incomplete or noisy. Moreover, the XGBoost was found to be a powerful and efficient machine learning tool for a variety of applications, including the prediction of engine-out NOx levels [34]. Recent studies also explored the use of ensemble models, such as random forest and adaptive boosting, in addition to the XGBoost [35,36]. These models combine the predictions of multiple individual models to achieve improved accuracy, whilst attaining a robustness equivalent to the that of the single models. Moreover, the ensemble methods mainly rely on randomization techniques, thus creating many different solutions to the considered problem. In this framework, the XGBoost algorithm has exhibited great prediction performance when compared to other ensemble models, such as GBT and RF, [37] causing to be selected for real-time applications. Moreover, some Extreme Gradient Boosting Regression Tree (XGBoost) models have been created to investigate the accuracy in estimating physical parameters, such as tailpipe NOx emissions [38,39]. However, the literature has limited research on the control and monitoring of engine-out NOx emissions under transient conditions.

The present paper, hence, focused on the use of the gradient boosting algorithm known as extreme gradient boosting (XGBoost) for real-time engine-out NOx sensing in real-world driving conditions. The XGBoost-based virtual sensor was calibrated in steady-state conditions and then validated in transient on-road traces following an experimental campaign conducted on the test bench and on the road.

2. Materials and Methods

Virtual sensing techniques were exploited as valid alternatives to physical measurement instruments, due to economic or application limits of the latter. The present study focused on the machine learning-based virtual sensor development for nitrogen oxides prediction in a diesel engine application. At the very beginning, a robustness analysis was conducted on the acquired steady-state data for the performance evaluation of the machine learning algorithms. Then, XGBoost prediction capabilities were verified and validated over a real-world driving mission through an experimental campaign conducted on-road. The case study and the involved datasets, as well as the proposed method for model training and validation, were investigated.

In this section, the proposed method, composed of sequential steps, is discussed and shown in Figure 1.

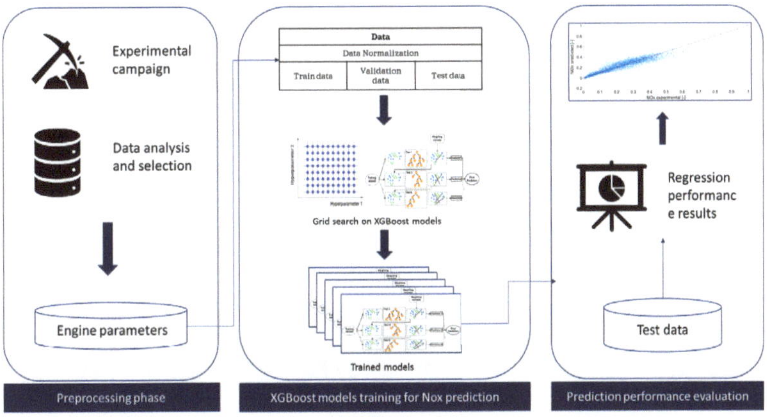

Figure 1. The proposed method for the model's performance evaluation.

In the preprocessing phase, the engine data values acquired from the experimental campaigns were analyzed, handled and 'cleaned'. In the following steps, related to the XGBoost models training phase, the data were normalized and split into training, validation and test sets when necessary, and then employed to perform the learning process of several XGBoost architectures. The grid search algorithm was used as a powerful hyperparameters tuning technique to find the most accurate architecture for each specific case study. Given that the performance of an AI model is heavily dependent on the hyperparameter values, a combination of grid search and cross validation approaches [40] was used to identify the optimal values for a specific model. As with any machine learning model, the XGBoost was defined by a set of hyperparameters that had to be specified in order to tailor the model for the specific application. In the final step, the performance of the best XGBoost architecture was evaluated by considering a test dataset according to different metrics, i.e., $RMSE$ and coefficient of determination R^2.

2.1. Preprocessing Phase

The case study for the deployed AI-based virtual sensor was an 11 L diesel engine for heavy-duty application, tested both at steady state and under transient conditions. The details of the experimental campaign, together with a deep insight into the experimental set-up, as well as the engine behavior, can be found in [41]. The main engine parameters at steady-state conditions were acquired by means of an engine test bench equipped with physical sensors mounted on the acquisition system and through the variable acquired by the electronic control unit (ECU) sensors and maps, whereas in the transient case a prototype heavy-duty (HD) vehicle was tested on-road and the main engine parameters were acquired by the on-board system. The main characteristics of the case study are shown in Table 1.

Table 1. Main engine parameters and test characteristics.

Segment of Application	Displacement	Turbocharger
Heavy-duty vehicles & trucks	11 L	VGT type
Fuel Injection System	**Engine Params**	**n° Engine Points**
High pressure common rail	14	4711 [1]

[1] The operating engine points are related to the steady state tests over the engine map.

The main engine parameters acquired for the different operating conditions follow:

- ω_{eng}: engine speed in rpm;
- Q_{tot}: total amount of injected fuel inside the cylinders in $mm^3/(cycle \times cylinder)$;
- P_{rail}: pressure in the rail system in bar;
- Q_{main}: amount of injected fuel during the main injection in $mm^3/(cycle \times cylinder)$;
- Q_{pil}: amount of injected fuel during the pilot injection in $mm^3/(cycle \times cylinder)$;
- SOI_{main}: start of the injection of the main injection in degree;
- SOI_{pil}: start of the injection of the pilot injection in degree;
- $IMAP$: pressure value in the intake manifold in bar;
- $IMAT$: temperature value in the intake manifold in K;
- O_2: oxygen concentration in the chamber in %
- Q_{air}: amount of air introduced in the chamber in $Kg/(cycle \times cylinder)$;
- λ: ratio between air and fuel quantities.
- EGR: amount of exhaust gas recirculated in $Kg/(cycle \times cylinder)$;
- NOx: amount of nitrogen oxides emitted in ppm

The experimental campaign was carried out at Politecnico di Torino on an engine test bench equipped with an ELIN APA 100 AC dynamometer. The raw engine-out gaseous emissions were measured by means of an AVL AMAi60- endowed with two complete trains for the simultaneous measurement of the gaseous concentrations of the main species, both at the intake and exhaust manifolds. The test engine was equipped with thermocouples and piezoresistive pressure transducers to gather temperature and pressure at many points, including upstream and downstream from the turbine, compressor, and intercooler, as well as in the intake manifold and EGR circuit. [17] The experimental acquisitions were conducted according to different engine strategies that are not presented, due to confidentiality issues. The data are, hence, consistently reported in a normalized form. The data from the engine experiment were exploited for the machine learning model and the real-time virtual NOx sensor development. The engine steady state operating points are presented in terms of total injected fuel quantity versus engine speed in Figure 2.

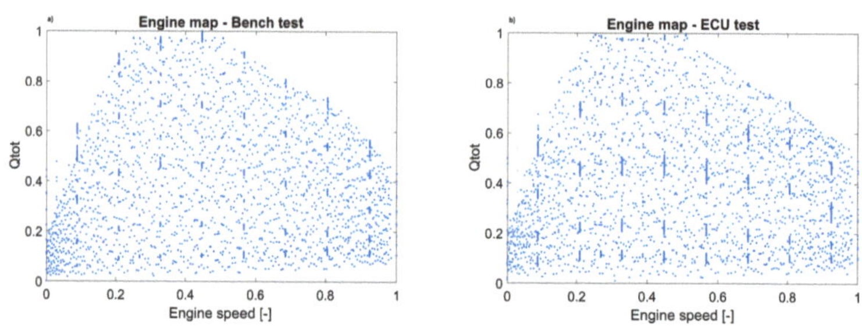

Figure 2. Engine operating points in stationary conditions through (**a**) test bench and (**b**) ECU acquisition systems.

The common engine map shape and the full load (FL) curve can be recognized in the graphs, as well as the presence of different strategies adopted during the experimental campaign bringing an unusual grouping phenomena. Given that the experimental acquisition occurred at steady-state testing conditions, the characteristic time associated with transient events was basically extinguished for each individual engine point depicted on the map. The road tests, on the other hand, were dominated by transient events with a large variation rate. Therefore, the sampling frequency which the engine variables were acquired at was set to 100 Hz.

2.2. XGBoost Models' Training

The development and assessment of the NOx predictor were based on the XGBoost algorithm and moved the steady state analysis, where the ML models were defined and trained on the engine operating points detected with the two different acquisition systems, i.e., test bench and ECU. A sensitivity analysis was carried out over the size of datasets exploited for the models' learning processes. Specifically, among the whole available data, the size of training and test datasets were changed in order to assess for the robustness of the selected models through a physical data fitting procedure. Considering that machine learning algorithms are data-driven models, the more data supplied to the training phase, the better the learning performance. Still, increasing the number of points during the training might lead to overfitting. As a result, reducing the training dataset dimension in favor of an increased number of test datasets allows for assessing the models' robustness, despite a loss in accuracy. Furthermore, in the creation of the predictors, a feature extraction approach was employed to identify the most influential and weighted variables. This approach, also known as feature importance, is regarded as a critical preprocessing step in the creation of the ML models [42].

In the processing phase, all datasets employed over the different analysis were split into train, validation and test sets in order to ensure a reliable learning process for the models and to evaluate the prediction performance. This particular mathematical problem belongs to the class of supervised learning, due to the presence of desired output exploited in the learning process [43]. The logic for the developed ML model, together with the XGBoost architectures and the hyperparameters involved, are shown in Figure 3.

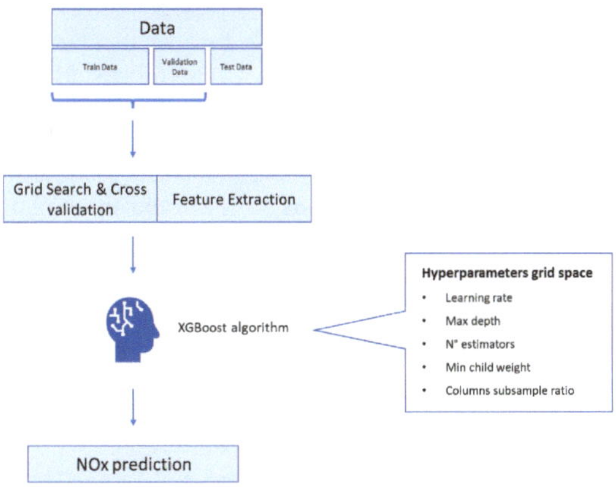

Figure 3. The workflow of XGBoost learning process. The involved Hperparameters in the XGBoost model are the following: the *n_estimators*, which specifies the number of decision trees to be boosted; the *max_depth*, which limits the depth in each tree's growth; the *Learning_rate*, which is a regularization parameter that shrinks feature weights in each boosting step; the *min_child_weight*, which corresponds to the minimum number of instances needed to be in each node of the tree; the *columns_subsample_ratio*, which is the subsample ratio of columns when constructing each tree [44,45].

Grid search and k-fold cross-validation techniques were combined to perform hyperparameter tuning in order to determine optimal values for the given model [40]. All different parameters were fed into a parameter grid, and based on a scoring metric (accuracy), the best combination was identified. The k-fold cross-validation process was allowed to perform the models' learning, evaluating each parameter combination over different datasets, i.e., validation set. The procedure involves partitioning the original training data into k subsets, where k is a positive integer, typically 5 or 10. The model is then trained on k-1 folds before being evaluated on the remaining fold. This procedure is repeated k times, with each test set fold being utilized once. The model's performance is measured by its average performance throughout all k iterations. In this investigation, the k parameter was assigned a value of 10. This strategy prevents overfitting, a typical issue that arises in machine learning when a model performs well on training data but fails to generalize to new, unknown data. Using k-fold cross-validation, the model is trained and assessed on multiple different subsets of data, thus yielding a more accurate performance estimation [46].

The XGBoost (eXtreme Gradient Boosting) is a popular and powerful machine learning algorithm that has been widely used for classification and regression tasks. The XGBoost is an optimized gradient-boosting decision tree algorithm, specifically designed to handle large datasets and to perform well with a high number of features. The algorithm is an ensemble method that combines decision trees to build a more powerful model. At its core, the XGBoost predicts the target variable by generating decision trees and optimizing the predictions by modifying the weights applied to each decision tree. Gradient descent is used in the optimization process to minimize the loss function, which assesses the difference between the anticipated and actual target values. During each iteration of the gradient descent optimization procedure, the XGBoost updates the weights of each decision tree, based on the gradient of the loss function. The technique employs a special algorithm for creating trees that divides the data into subsets, and the tree is built in a recursive fashion until a stopping criterion is reached [47]. There are a variety of loss functions that can be used, depending on the mathematical problem task. In the present study, the XGBoost-based predictor was laid on a regression task. Therefore, common loss functions included the mean squared error (MSE) and mean absolute error (MAE). These loss functions measure the difference between the predicted and the actual values and penalize large errors more heavily than small ones. In the present study, the MSE loss function was employed for the optimization process of the XGBoost parameters.

Feature selection was exploited as a machine learning approach that helps to determine which features in the experimental data are most significant in delivering the predictions. This is particularly useful if the number of variables is large and finding the most relevant ones is an objective of the analysis, or if is important to know which features are driving the model predictions. In the case of the XGBoost algorithm, feature importance is relevant for several reasons:

- Model interpretability: by identifying the most important features, a better understanding of how the model is producing the predictions, and how different features are interacting with each other, can be gained. This can be helpful to interpret the results of the models and to make more informed decisions based on its predictions.
- Feature selection: identifying the most important features can also be useful for feature selection, which is the process of selecting a subset of features to use in the models. By selecting the most important features, the performance of the models can be potentially improved by eliminating less important features that may be adding noise or reducing the model's ability to generalize.
- Model debugging: if poor model performance is experienced, identifying the most important features can be helpful to debug the model and identify potential issues. For example, if a particular feature is not statistically very important, it can be removed from the model.

Basically, the feature importance algorithm identifies, for each feature, a corresponding weight that represents the total gain of splitting data along the feature. It is calculated as the sum of the improvement in the loss function for all splits that use the selected feature. Features with higher weights are generally considered to be more important. The feature extraction technique was applied to the engine parameters listed in Section 2.1. The output of the algorithm was the computation of the relative importance of each parameter for NOx formation, and the results are presented in the Result section.

2.3. *Prediction Performance Evaluation*

Each selected model was exploited as a virtual predictor of NOx pollutant and the prediction performance investigated through different metrics. Together with the test data, the performance of all trained XGBoost architectures for each case study was analyzed, based on the following:

- the coefficent of determination R^2;

$$R^2 = 1 - \frac{\sum_{i=1}^{n}(x_i - \hat{x}_i)^2}{\sum_{i=1}^{n}(x_i - \bar{x}_i)^2} \tag{1}$$

- the normalized RMSE considering the test dataset;

$$RMSE_{norm} = \sqrt{\frac{\sum_{i=1}^{n}(x_{n,i} - \hat{x_{n,i}})^2}{n}} \tag{2}$$

- the real RMSE in ppm considering the test dataset;

$$RMSE = \sqrt{\frac{\sum_{i=1}^{n}(x_i - \hat{x}_i)^2}{n}} \tag{3}$$

where x_i is the experimental data, $\hat{x}_i$ is the model estimated value, $\bar{x}_i$ is the mean of experimental measured data, $x_{n,i}$ are the normalized experimental data, $\hat{x_{n,i}}$ are the normalized estimated value and n is dataset sample size. The performance evaluation of the machine learning models and the selection of the best hyperparamenter values are widely discussed and analysed in the Results and Discussion section.

Using the high-level Python programming language and a PC with an Intel(R) Core(TM) i7-8700 processor at 3.20 GHz and 64GB RAM architecture, the virtual sensor was designed and the complete data processing phase was executed.

2.4. *Virtual NOx Sensing in Steady-State Conditions*

Virtual sensing is a frequently adopted approach to ensure sensorless solutions in the automotive field capable of properly calibrating the main engine parameter, thus allowing for consumption and emissions reduction. Still, considering the transient dynamic conditions that an engine encounters during regular on-road driving missions, the virtual NOx prediction sensor should be calibrated in stationary conditions throughout the entire engine operating range. Such operations require a considerable amount of experimental readings but would, in turn, enable the electronic control unit (ECU) to properly control the engine during real driving missions. The present paper, hence, moved from the development and calibration of an AI-based virtual sensor through the steady-state data set. It is worth remembering that two different sets of acquisition were available from the experimental campaign, namely, a set of data collected at the test bench (Bench) and a set of data directly acquired from the engine control unit (ECU). The ML virtual sensor was, hence, developed considering different combinations of the available data sets for the training and testing phase, to better investigate the algorithm response to the use of specific inputs. Moreover, a sensitivity analysis was carried out on the train and test sub-set sizes (test dataset size) in order to prove the robustness of the methodology. The different cases analyzed are reported in Table 2, whereas the XGBoost model performance results are widely discussed in the Results and Discussion section.

Table 2. Steady-state case analysis based on datasets and train–test split sensitivity.

Case Study	Train–Test	Test Dataset Size [%]
#1	Bench–Bench	[10; 50; 95]
#2	ECU–ECU	[10; 50; 95]
#3	Bench–ECU	NO [1]

[1] The datasets involved in the training and test phases were derived from two different acquisition systems. Hence, a train–test split technique was not adopted.

In the Bench–Bench and ECU–ECU examples, the datasets used for the model learning were distinct and derived from the same experiments, i.e., the Bench–Bench case was based on the engine bench test data, whereas the ECU–ECU case was based on engine data collected from the ECU. In the last scenario examined, Bench–ECU, the training dataset and the test dataset were derived from the two applications. Therefore, there was no train–test split and all dataset samples were utilized for the different learning phases. Given that the bench acquisition system detects real events with very accurate physical sensors, whereas the engine map implements data tables, the model performance evaluation in Case Study 3 was dependent on the extent to which the physical phenomena were detected by the virtual sensor, as a potential real-time estimator to be implemented on ECU.

2.5. On-Road NOx Prediction

After calibrating and testing the virtual sensor in steady-state conditions to assess for the model robustness and NOx prediction performance, validation under dynamic conditions was carried out. In order for the virtual sensor to accurately predict NOx emissions under a range of operating conditions, it was important to train the model using engine map data and to test it on real-driving missions. Engine map data, which comprises engine speed and load information, replicates a variety of steady-state scenarios the engine may experience. However, these conditions may not precisely represent the dynamic, real-world operating conditions that the engine would face when driving on a road. By testing the sensor during an on-road driving mission, it is, in turn, possible to identify how well the virtual sensor performs under actual driving scenarios and to assess its ability to properly predict real-time emissions. As for integrating the virtual sensor in a control system, any potential issues or limitations that need to be addressed can, hence, be identified.

Therefore, for this investigation, stationary experimental data on the ECU and on the bench were utilized to train the algorithm, whilst on-road-based measurements were used to validate the algorithm and examine the accuracy of NOx prediction during real-world driving missions. During experimental road tests, the vehicle powered by a compression ignition engine was equipped with a physical measuring system for engine-out management and monitoring. The development and assessment of a virtual NOx sensor for real-time applications would enable the replacement of real sensors, possibly resulting in cost and reliability improvements. The real driving mission performed by the vehicle during the experimental campaign occurred in Piedmont and accounted for a variety of operating conditions and maneuvers related to the constraints and requirements of realistic urban and extra-urban scenarios. The cases under investigation are summarized in Table 3.

Table 3. Dynamic on-road case analysis based on training phase over different acquisition systems.

Case Study	Train–Test
#1	Bench–On-road mission
#2	ECU–On-road mission

The *Bench–Onroad* case is related to the training conducted on the engine test bench at steady state conditions and *ECU–Onraod* studies the learning process on stationary engine ECU data and validation on transient conditions. Since the development of the virtual sensor was oriented toward a future real-time implementation and could, thus, be implemented in the ECU, it was crucial to examine the two data collection systems independently, given the heterogeneous nature of the variables. Thus, the engine map-based model's prediction of performance may be compared to the real collection of physical sensors.

3. Results and Discussion

In this work, an XGBoost-based NOx virtual sensor was developed considering an engine steady-state calibration for real-time NOx monitoring during on-road driving missions. A preliminary study was conducted under stationary conditions to assess the model's capability to capture NOx phenomena under various operating conditions. In addition, a robustness study was conducted by progressively lowering the training data size and reversely increasing the test data one. In this way, the magnitude of noise could be varied and the quality of the model performance inferred. Consequently, the XGBoost model for the real-time estimation of NOx in on-road applications was built by training the model under steady-state conditions, from which the entire engine map domain might be completely covered during experiments, and testing it on real-world driving data.

3.1. NOx Prediction in Steady-State Conditions

The case study results concerning the training and test phases on bench test conditions (*Bench–Bench*), and taking into account the sensitivity analysis conducted on the train–test split, are summarized in Table 4 and the regression results are shown in Figure 4.

Table 4. Summary of performance prediction results for test case *Bench–Bench*. The test size is here reported for the sensitivity analysis, considering the different percentage values 10%, 50% and 95%.

Test Size [%]	10	50	95
R^2	0.98	0.97	0.85
$RMSEnorm$	0.017	0.022	0.052
$RMSE[ppm]$	60.0	77.4	186.5
q	0.0034	0.0062	0.0224
m	0.985	0.976	0.912

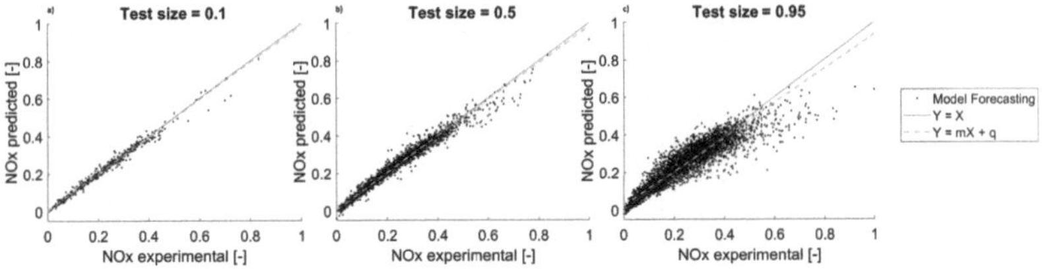

Figure 4. Regression prediction results on test dataset for bench test conditions. Sensitivity analysis on the three different sizes of test dataset: (**a**) 10%, (**b**) 50% and (**c**) 95%.

The considerable reduction in the engine points ascribed to the model learning phase, as the test dataset increased from 0.1 to 0.95, caused the prediction performance to decrease.

When 95% of the data was used for the testing, significant errors were attained, despite the moderate increase in rmse as the split switched from 10% of test size (60.0 ppm) to 50% of test size (77.4 ppm). It is, in fact, worth underlining that the model prediction performance remained high, despite the halving of the training data set, thus, indicating a remarkable capacity to properly capture the physical phenomena. Additionally, the regression line in Figure 4 demonstrates the rise in noise brought on by the test size-appropriate sensitivity analysis.

The case study results concerning the training and test phases on ECU test conditions (*ECU–ECU*) are summarized in Table 5 and the regression results are shown in Figure 5.

Table 5. Summary of performance prediction results for test case *ECU–ECU*. The test size is here reported for the sensitivity analysis, considering the different percentage values 10%, 50% and 95%.

Test Size [%]	10	50	95
R^2	0.98	0.97	0.76
$RMSEnorm$	0.021	0.024	0.067
$RMSE[ppm]$	73.3	84.9	237.7
q	0.0038	0.0047	0.0309
m	0.985	0.9802	0.862

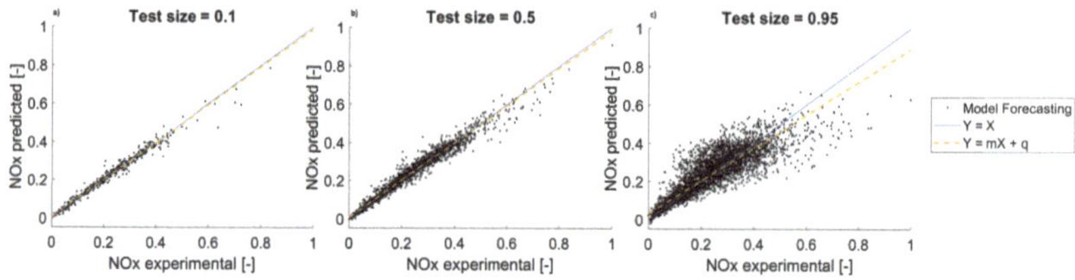

Figure 5. Regression prediction results test dataset for ECU test conditions. Sensitivity analysis on the three different size of test dataset: (**a**) 10%, (**b**) 50% and (**c**) 95%.

As for the previous case, the accuracy of NOx prediction tended to decline (and the error to grow) as the test size increased and the training size decreased. The surprisingly slight decrease in R2, 0.98 for 90% and 0.97 for 50% of the full dataset, as a result of nearly half the amount of data available for model training, is noteworthy. This was due to the high predictive power of these machine learning methods, especially considering that the data in this particular application were collected under stationary conditions. Therefore, it was more efficient to capture NOx generation phenomena beginning with steady-state engine operating conditions.

Finally, the case study results concerning the training conducted on Bench test conditions and test conducted on ECU test conditions (*Bench–ECU*) are summarized in Table 6 and the regression results are shown in Figure 6.

Table 6. Summary of performance prediction results for test case *Bench–ECU*.

R^2	0.97
$RMSEnorm$	0.024
$RMSE[ppm]$	86.7
q	0.0037
m	0.972

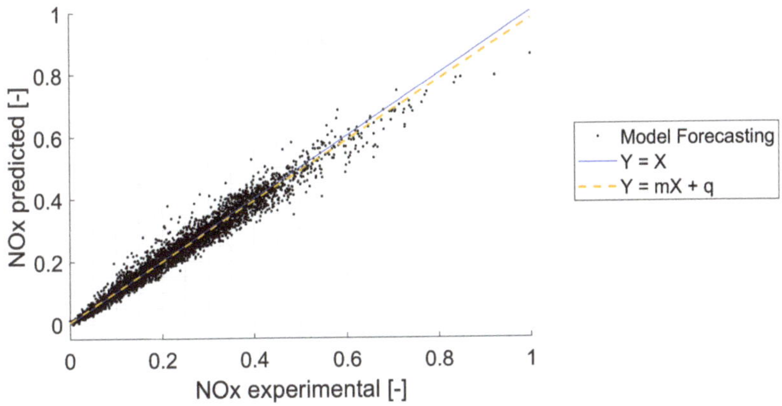

Figure 6. Regression prediction results of training conducted on bench test conditions and test carried out under ECU test conditions.

In terms of the robustness of the virtual sensor, the findings obtained in this example were consistent with those obtained in the previous cases, as can be observed. In addition, the Bench–ECU example demonstrates how the formation phenomenology of nitrogen oxides was accurately estimated (R^2 of about 98%) by the developed virtual sensor which was tested on ECU engine data.

The more features a model contains, the more complex it is (and the sparser the data), and, therefore, the more susceptible it is to variance-induced errors. For this reason, the feature importance approaches belong to the feature engineering process which entails selecting the minimum required features to generate a feasible model [48]. Given that the formation of nitrogen oxides in diesel engines is strongly influenced by the specific engine operating conditions, the feature extraction approach simplifies the models and can provide an assessment of the model performance. Indeed, the robustness of the predictions can also be verified by comparing the statistically most relevant engine variables for the models to the empirical knowledge of the physical phenomenon. Considering the full availability of data during the training phase of the models, i.e., a test size equal to 10% for both *Bench–Bench* and *ECU–ECU* case studies, the feature extraction approach outputs are shown in Table 7. For the same train–test split value, the hyperparameters of the best XGBoost architecture for each considered analysis are reported in Table 8.

Table 7. The feature importance outputs for the three steady-state cases study which highlight the relative importance of each variable. The green labeling identifies the most influencial engine variables up to reaching 90% of the threshold in the feature importance algorithm for the XGBoost application.

Variable	Bench–Bench	ECU–ECU	Bench–ECU
1	$SOI_{main}(13.9\%)$	$SOI_{main}(13.5\%)$	$SOI_{main}(14.2\%)$
2	$P_{rail}(13.4\%)$	$P_{rail}(13.4\%)$	$\omega_{eng}(13.3\%)$
3	$\omega_{eng}(13.3\%)$	$\omega_{eng}(13.0\%)$	$P_{rail}(13.3\%)$
4	$\lambda(9.5\%)$	$\lambda(10.7\%)$	$\lambda(9.7\%)$
5	$O_2(9.3\%)$	$IMAT(7.9\%)$	$O_2(9.6\%)$
6	$IMAT(7.5\%)$	$Q_{tot}(7.9\%)$	$IMAT(7.7\%)$
7	$Q_{tot}(7.4\%)$	$O_2(6.3\%)$	$EGR(7.4\%)$
8	$EGR(6.8\%)$	$Q_{main}(6.3\%)$	$Q_{tot}(6.5\%)$
9	$IMAP(5.6\%)$	$IMAP(6.1\%)$	$Q_{air}(5.9\%)$
10	$Q_{air}(5.5\%)$	$SOI_{pil}(5.4\%)$	$IMAP(5.1\%)$
11	$SOI_{pil}(4.5\%)$	$Q_{air}(4.6\%)$	$SOI_{pil}(3.8\%)$
12	$Q_{main}(2.6\%)$	$EGR(4.1\%)$	$Q_{main}(2.7\%)$
13	$Q_{pil}(0.7\%)$	$Q_{pil}(0.8\%)$	$Q_{pil}(0.8\%)$

According to the feature importance definition, the variables SOI_{pil}, Q_{main} and Q_{pil} were discarded as they proved not to be significantly affected by large variations. The algorithm, therefore, did not consider SOI_{pil} and Q_{pil} to be particularly significant in the evolution of NOx emissions. Furthermore, since the Q_{main} was strictly proportional to Q_{tot}, only one of them was taken into account.

Table 8. Best XGBoost architectures and grid values setup for the three case studies under investigation. The mean squared error (MSE) evaluation metric was used to obtain the optimal value by the GridsearchCV algorithm.

Hyperparameters	Grid Values	Bench–Bench	ECU–ECU	Bench–ECU
learning rate	[0.1, 0.05, 0.08, 0.1, 0.15]	0.08	0.08	0.08
max_depth	[3, 4, 5]	4	4	4
n° estimators	[500, 1000, 1200, 1500]	1500	1500	1500
min_child_weight	[3, 5, 7]	3	7	7
colsample_bytree	[0.5, 0.7, 1.0]	1	1	1

As can be seen, the most relevant engine variables identified by the feature extraction process were approximately the same in the three cases analyzed. First, the variables most relevant for predicting NOx emissions are likely to be consistent across different types of data sources and testing environments. For instance, variables such as engine load, fuel flow, and exhaust gas temperature are likely to be important for predicting NOx emissions, regardless of whether the data are collected from a test bench or from the vehicle's engine control unit. Moreover, the most relevant engine variables selected by XGBoost were similar across the three different cases because the data used to train and test the model were similar. For instance, the dataset used to train the model on the test bench was similar to the dataset used to test the model on the vehicle ECU. Therefore, the most relevant variables were similar in both cases. For the same reasons, the descriptive patterns of the data detected by the algorithm would be very similar. Therefore, the architectures of the XGBoost in the three cases were very similar to one another.

3.2. On-Road NOx Prediction

Once the model was assessed in terms of its high predictive performance and its robustness in capturing phenomenological events under steady-state conditions, a virtual NOx sensor was built for real-time applications taking advantage of the experimental data collected on-road. As opposed to the bench tests, where data wre collected while the engine was running under controlled conditions to allow for consistent and repeatable testing conditions, on-road data were acquired while the vehicle was being driven on actual roads under realistic scenarios. This data reflected the real working conditions of the engine, including environmental parameters, such as temperature, humidity, and altitude, as well as the dynamic load variations that occurred during the driving mission. The on-road measurements also included the vehicle's speed and acceleration, which might impact engine performance. Consequently, the design of the virtual NOx sensor was based on its calibration in stationary conditions (either bench or ECU) and validated by means of real-world experimental tests. The considered engine parameters were identified by the feature importance process during the previous steps of the analysis. Therefore, the learning phase was covered by the feature importance output from the steady-state conditions. The predictive performance results for the two case studies are listed in Table 9 outlining the same statistical metrics.

Table 9. Performance prediction results over on-road test driving mission.

Case Study	#1 Bench—On-Road Mission	#2 ECU—On-Road Mission
R^2	0.76	0.75
$RMSEnorm$	0.054	0.055
$RMSE[ppm]$	192.0	194.9
m	0.92	0.97
q	−0.007	−0.017

From the values of the metrics, a noticeable reduction in predictive performance, of up to 75% accuracy, and a major shift in the regression line could be observed (q was significantly higher than the stationary cases), as well as a large increment in the error. This was mostly due to the fact that the operating conditions encountered by the engine during a road driving mission are significantly different from those encountered during the model learning phase described by stationary conditions. The predicted NOx in both case studies are shown alongside the experimental traces in Figure 7, where three different time windows [50 s 250 s], [500 s 700 s], [850 s 1050 s] of the missions, with a length equal to 200 s, are highlighted. Three fairly large time windows that corresponded to different operating conditions are shown so that the forecasting ability of the model could be tested in almost all real-world conditions the vehicle would face.

Observably, the anticipated NOx signals accurately represented the qualitative and quantitative trend of the experimental unit, as well as its transient trend during the driving mission. In the [50,250] time window, some mission points that overestimated or underestimated the experimental results led to negative NOx values. This was likely to be ascribed to the great occurrence of operating points where the lambda values reached extremely high values beyond the definition domain of the learning phase, as well as the rapid spikes in the fuel injected quantity values. Moreover, the signal was extremely spiky due to the fact that the model was an instantaneous measurement sensor and each point of the mission had operating conditions in terms of engine variables that were very different from those of the subsequent instants. The latter, together with the fact that the NOx formation dynamics with their characteristic times were not captured in the learning phase, connected to the differences in the experimental and predicted signals.

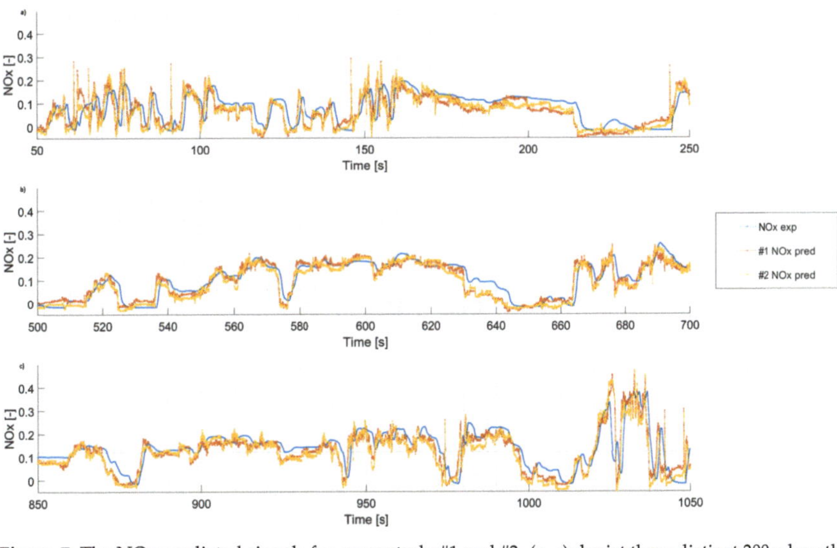

Figure 7. The NOx predicted signals for cases study #1 and #2. (**a–c**) depict three distinct 200 s-length windows of driving mission.

As seen in Table 9 and Figure 7, the lower R2 values could be attributable to the anticipation of the predicted signals relative to the observed signal. In real-time virtual sensing applications, the anticipation or delay in time of a predicted signal causes a drastic reduction in performance, leading, in this case, to incorrect emission control. Such behaviour is connected to the line and the measurement system's characteristic time delay. The latter is produced by the time lag between the observed phenomenon, i.e., NOx measured level, and the time frame such a level would refer to. As a matter of fact, the predicted NOx were referred to the implemented engine parameters at the given time-step, whereas the measured NOx would have been produced by the engine setting at a previous time frame. The latter is connected to the time needed for the exhaust gases to travel from the engine discharge to the acquisition probes, as well as to the characteristic sensor time lag. Under steady-state conditions, such delay is basically erased by the stationary characteristic of the test, whereas it stems from transient conditions. By delaying the NOx signal output by the model by 1 s [49], the anticipated signal almost perfectly matched the experimental signal (see Figure 8 referring to one time window).

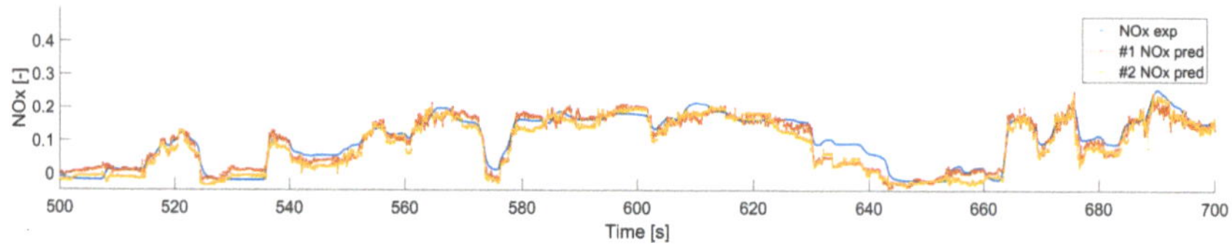

Figure 8. The NOx predicted signals for cases studies #1 and #2, taking into account the line delay of measurement system.

Taking this fact into account, the value of R2 changed to 85% for both case studies, thus emphasizing the real-time capabilities of the virtual sensor at predicting NOx emissions with a high level of accuracy [50].

4. Conclusions

The virtual sensor developed in the present study is based on the XGBoost algorithm and proved to be highly accurate in predicting NOx emissions for real-time diesel engine applications. The calibration of the virtual sensor was performed under steady-state conditions and then tested on an on-road driving mission, resulting in an accuracy of 85%. The virtual sensor was also found to be highly accurate in stationary conditions, with prediction results up to 98% in terms of accuracy when tested on the experimental data acquired at the engine test bench and through the ECU maps. The results indicate that the virtual sensor holds great potential to be a valuable tool for real-time emission monitoring and control in diesel engine applications. The virtual sensor can, in fact, be integrated into the Engine Control Unit (ECU), where it could be used, in conjunction with the ECU control system, to adjust the engine parameters in real-time to minimize NOx emissions. Future developments of the present paper may address different driving conditions (e.g., high speed, high load, and low speed) to further assess the capabilities of the virtual sensor in accurately predicting NOx emissions. The logic could also be transferred to AI deep networks, which could further improve the prediction accuracy and broaden the virtual sensor capabilities. For example, deep networks could be used to make the virtual sensor more capable of predicting NOx emissions in real-time under a broader range of conditions and to determine the characteristic time of NOx emissions phenomena by analyzing time-dependent events. In conclusion, the virtual sensor developed in the present study shows great promise as a tool for real-time emission monitoring and control in diesel engine applications.

Author Contributions: Conceptualization, A.F. and D.A.M.; methodology, A.F.; software, A.F.; validation, A.F.; formal analysis, A.F.; investigation, A.F.; resources, A.F. and D.A.M.; data curation, A.F.; writing—original draft preparation, A.F. and D.A.M.; writing—review and editing, A.F. and D.A.M.; visualization, A.F. and D.A.M.; supervision, D.A.M.; project administration, D.A.M. All authors have read and agreed to the published version of the manuscript.

Funding: This research received no external funding.

Institutional Review Board Statement: Not applicable.

Informed Consent Statement: Not applicable.

Data Availability Statement: Data is unavailable due to privacy restrictions.

Acknowledgments: We would like to thank Roberto Finesso and the research fellow Omar Marello for supporting the project.

Conflicts of Interest: The authors declare no conflict of interest.

Nomenclature

The following abbreviations were used in this manuscript:

AI	Artificial Intelligence
NOx	Nitrogen Oxides
XGBoost	Extreme Gradient Boosting
ECU	Electronic Control Unit
ICE	Internal Combustion Engine
EGR	Exhaust Gas Ricirculation
HCCI	Homogeneous Charge Compression Ignition
PCCI	Premixed Charge Compression Ignition
SCR	Selective Catalytic Reduction
LNT	Lean NOx Trap
ANN	Artificial Neural Network
SVM	Support Vector Machine
RMSE	Root Mean Squared Error
R^2	Coefficient of determination
HD	Heavy Duty
ω_{eng}	Engine speed
Q_{tot}	Total injected fuel quantity
P_{rail}	Rail pressure
Q_{main}	Fuel quantity of main injection
Q_{pil}	Fuel quantity of pilot injection
SOI_{main}	Start of injection of main injection
SOI_{pil}	Start of injection of pilot injection
$IMAP$	Intake Manifold Air Pressure
$IMAT$	Intake Manifold Air Temperature
O_2	Oxygen concentration
Q_{air}	Air quantity
λ	Air and fuel ratio
FL	Full load
ML	Machine Learning
MSE	Mean Squared Error
MAE	Mean Absolut Error

References

1. Regulation (EC) No 715/2007 of the European Parliament and of the Council of 20 June 2007 on Type Approval of Motor Vehicles with Respect to Emissions from Light Passenger and Commercial Vehicles (Euro 5 and Euro 6) and on Access to Vehicle Repair and Maintenance Information (Text with EEA Relevance). Available online: https://eur-lex.europa.eu/eli/reg/2007/715/oj (accessed on 22 December 2022)
2. Kannan, C.; Vijayakumar, T. Application of exhaust gas recirculation for NOx reduction in CI engines. In *NOx Emission Control Technologies in Stationary and Automotive Internal Combustion Engines*; Ashok, B., Ed.; Elsevier: Amsterdam, The Netherlands, 2022; pp. 189–222.
3. Baratta, M.; Finesso, R.; Misul, D.; Spessa, E. Comparison between Internal and External EGR Performance on a Heavy Duty Diesel Engine by Means of a Refined 1D Fluid-Dynamic Engine Model. *SAE Int. J. Engines* **2015**, *8*, 1977–1992. [CrossRef]
4. Jain, A.; Singh, A.P.; Agarwal, A.K. Effect of split fuel injection and EGR on NOx and PM emission reduction in a low temperature combustion (LTC) mode diesel engine. *Energy* **2017** *122*, 249–264. [CrossRef]
5. Maiboom, A.; Tauzia, X.; Hètet, J.-F. Influence of high rates of supplemental cooled EGR on NOx and PM emissions of an automotive HSDI diesel engine using an LP EGR loop. *Int. J. Energy Res.* **2008**, *32*, 1383–1398. [CrossRef]
6. Maiboom, A.; Tauzia, X. NOx and PM emissions reduction on an automotive HSDI Diesel engine with water-in-diesel emulsion and EGR: An experimental study. *Fuel* **2011**, *90*, 3179–3192. [CrossRef]
7. Stanglmaier, H.R.; Roberts, C.E. Homogeneous Charge Compression Ignition (HCCI): Benefits, Compromises, and Future Engine Applications. *SAE Int.* **1999**, *108*, 2138–2145.
8. Puškár, M.; Kopas, M. System based on thermal control of the HCCI technology developed for reduction of the vehicle NOX emissions in order to fulfil the future standard Euro 7. *Sci. Total Environ.* **2018**, *643*, 674–680. [CrossRef]
9. Torregrosa, A.J.; Broatch, A.; Garcìa, A.; Mònico, L.F. Sensitivity of combustion noise and NOx and soot emissions to pilot injection in PCCI Diesel engines. *Appl. Energy* **2013** *104*, 149–157. [CrossRef]
10. Aoyama, T.; Hattori, Y.; Mizuta, J.; Sato, Y. *An Experimental Study on Premixed-Charge Compression Ignition Gasoline Engine;* SAE Technical Paper; SAE: Warrendale, PA, USA, 1996. [CrossRef]
11. Cao, D.N.; Hoangm A.T.; Luu, H.Q.; Bui, V.G.; Tran, T.T.H. Effects of injection pressure on the NOx and PM emission control of diesel engine: A review under the aspect of PCCI combustion condition. *Energy Sources* **2020**, 1–18. [CrossRef]
12. Buyukkaya, E.; Muhammet, C. Experimental study of NOx emissions and injection timing of a low heat rejection diesel engine. *Int. J. Therm. Sci.* **2008**, *47*, 1096–1106. [CrossRef]
13. Fang, H.L.; DaCosta, H.F.M. Urea thermolysis and NOx reduction with and without SCR catalysts. *Appl. Catal. B Environ.* **2003** *46*, 17–34. [CrossRef]
14. Nova, I.; Lietti, L.; Forzatti, P. Mechanistic aspects of the reduction of stored NOx over Pt–Ba/Al_2O_3 lean NOx trap systems. *Catal. Today* **2008** *136*, 128–135. [CrossRef]
15. Jiao, P.; Li, Z.; Shen, B.; Zhang, W.; Kong, X.; Jiang, R. Research of DPF regeneration with NOx-PM coupled chemical reaction. *Appl. Therm. Eng.* **2017**, 110, 737–745. [CrossRef]
16. Hofmann, L.; Rusch, K.; Fischer, S.; Lemire, B. Onboard Emissions Monitoring on a HD Truck with an SCR System Using Nox Sensors. *J. Fuels Lubr.* **2004** *113*, 559–572.
17. Finesso, R.; Hardy, G.; Maino, C.; Marello, O.; Spessa, E. A New Control-Oriented Semi-Empirical Approach to Predict Engine-Out NOx Emissions in a Euro VI 3.0 L Diesel Engine. *Energies* **2017** *10*, 1978. [CrossRef]
18. Baratta, M.; Kheshtinejad, H.; Laurenzano, D.; Misul, D.; Brunetti, S. Modelling aspects of a CNG injection system to predict its behavior under steady state conditions and throughout driving cycle simulations. *J. Nat. Gas Sci. Eng.* **2015**, *24*, 52–63. [CrossRef]
19. Guardiola, C.; Pla, B.; Blanco-Rodriguez, D.; Calendini, P.O. ECU-oriented models for NOx prediction. Part 1: A mean value engine model for NOx prediction. *J. Automob. Eng.* **2015**, *229*, 992–1015. [CrossRef]
20. Finesso, R.; Misul, D.; Spessa, E. Estimation of the Engine-Out NO2/NOx Ratio in a EURO VI Diesel Engine. *SAE Int.* **2013**. [CrossRef]
21. Atkinson, C.; Mott, G. Dynamic model-based calibration optimization: An introduction and application to diesel engines. In Proceedings of the 2005 SAE World Congress, Detroit, MI, USA, 11–14 April 2005.
22. Donmez, N.; Ozener, O. Modeling of NOx emissions in internal combustion engine. *Int. J. Eng. Res. Adv. Technol.* **2019**, *5*, 36–43. [CrossRef]
23. D'Ambrosio, S.; Finesso, R.; Fu, L.; Mittica, A.; Spessa, E. A control-oriented real-time semi-empirical model for the prediction of NOx emissions in diesel engines. *Appl. Energy* **2014**, *130*, 265–279. [CrossRef]
24. Winkler-Ebner, B.; Hirsch, M.; del Re, L.; Klinger, H.; Mistelberger, W. Comparison of Virtual and Physical NOx-Sensors for Heavy Duty Diesel Engine Application. *SAE Int. J. Engines* **2010**, 3, 1124–1139. [CrossRef]
25. Stadlbauer, S.; Alberer, D.; Hirsch, M.; Formentin, S.; Benatzky, C.; del Re, L. Evaluation of Virtual NOx Sensor Models for Off Road Heavy Duty Diesel Engines. *SAE Int. J. Commer. Veh.* **2012**, *5*, 128–140. [CrossRef]
26. Fechert, R.; Bäker, B.; Gereke, S.; Atzler, F. Using machine learning methods to develop virtual NOx sensors for vehicle applications. In Proceedings of the Internationales Stuttgarter Symposium, Stuttgart, Germany, 18 August 2020.
27. Kempema, N.; Sharpe, C.; Wu, X.; Shahabi, M.; Kubinski, D. Machine-Learning-Based Emission Models in Gasoline Powertrains Part 2: Virtual Carbon Monoxide. *SAE Int. J. Engines* **2022**, *16*, 2023. [CrossRef]

28. Yuan, X.; Li, L.; Wang, Y. Nonlinear Dynamic Soft Sensor Modeling With Supervised Long Short-Term Memory Network. *IEEE Trans. Ind. Inform.* **2019**, 16, 3168–3176. [CrossRef]
29. Shao, W.; Tian, X. Semi-supervised selective ensemble learning based on distance to model for nonlinear soft sensor development. *Neurocomputing* **2017**, *222*, 91–104. [CrossRef]
30. Liu, C.; Guo, M.; Yang, Y.; Zhang, S. NOx prediction for diesel engine using improved GA-SVR. *Measurement* **2020**, *173*, 108187.
31. Zhang, H.; Zhang, Z.; Liu, Y. NOx emission prediction of diesel engines based on support vector machine. *Measurement* **2018**, *125*, 338–346.
32. Li, X.; Liu, Y.; Zhang, H.; Zhang, Z. NOx prediction for diesel engine using multi-input deep learning network based on LSTM. *Measurement* **2019**, *145*, 36–43
33. Wakjira, T.; Rahmzadeh, A.; Alam, M.S.; Tremblay, R. Explainable machine learning based efficient prediction tool for lateral cyclic response of post-tensioned base rocking steel bridge piers. *Structures* **2022**, *44*, 947–964. [CrossRef]
34. Chen, T.; Guestrin, C. XGBoost: A scalable tree boosting system. In Proceedings of the 22nd ACM Sigkdd International Conference on Knowledge Discovery and Data Mining, San Francisco, CA, USA, 13–17 August 2016; pp. 785–794.
35. Friedman, H.J. Greedy function approximation: A Gradient Boosting machine. *Ann. Stat.* **2001**, *29*, 1189–1232. [CrossRef]
36. Breiman, L. Random Forests. *Mach. Learn.* **2001**, *45*, 5–32. [CrossRef]
37. Bentejac, C.; Csorgo, A.; Munoz, M.G. A Comparative Analysis of XGBoost. *Artif. Intell. Rev.* **2021**, *54*, 1937–1967. [CrossRef]
38. Altug, K.B.; Kucuk, S.E. Predicting Tailpipe NOx Emission using Supervised Learning Algorithms. In Proceedings of the 3rd International Symposium on Multidisciplinary Studies and Innovative Technologies, Ankara, Turkey, 11–13 October 2019.
39. Hu, L.; Wang, C.; Ye, Z.; Wang, S. Estimating gaseous pollutants from bus emissions: A hybrid model based on GRU and XGBoost. *Sci. Total Environ.* **2021**, *783*, 146870. [CrossRef] [PubMed]
40. Great Learning. Hyperparameter Tuning with GridSearchCV. 2022. Available online: https://www.mygreatlearning.com/blog/gridsearchcv/ (accessed on 12 December 2022).
41. Finesso, R.; Marello, O. Calculation of Intake Oxygen Concentration through Intake CO_2 Measurement and Evaluation of Its Effect on Nitrogen Oxide Prediction Accuracy in a Heavy-Duty Diesel Engine. *Energies* **2022**, *15*, 342. [CrossRef]
42. Towards Data Science. Feature Selection Techniques in Machine Learning with Python. 2018. Available online: https://towardsdatascience.com/feature-selection-techniques-in-machine-learning-with-python-f24e7da3f36e (accessed on 27 December 2022).
43. Towards Data Science. A Brief Introduction to Supervised Learning. 2019. Available online: https://towardsdatascience.com/a-brief-introduction-to-supervised-learning-54a3e3932590 (accessed on 27 December 2022).
44. Towards Data Science. A Guide to XGBoost Hyperparameters. 2021. Available online: https://towardsdatascience.com/a-guide-to-xgboost-hyperparameters-87980c7f44a9 (accessed on 27 December 2022).
45. DMLC XGBoost. XGBoost Documentation. Available online: https://xgboost.readthedocs.io/en/stable/index.html (accessed on 27 December 2022).
46. Machine Learning Mastery. A Gentle Introduction to k-Fold Cross-Validation. Available online: https://machinelearningmastery.com/k-fold-cross-validation/ (accessed on 8 February 2023).
47. Medium. XGBoost: A BOOSTING Ensemble. Available online: https://medium.com/almabetter/xgboost-a-boosting-ensemble-b273a71de7a8 (accessed on 27 December 2022).
48. Yellowbrick. Feature Importances. 2022. Available online: https://www.scikit-yb.org/en/latest/api/model_selection/importances.html (accessed on 27 December 2022).
49. Peckham, M.; Parnell, J.; Hammond, M.; Mason, B. The Measurement of Fast Transient Emissions During Real World Driving. *Frontiers* **2020**, *6*, 19. [CrossRef]
50. Sellerei, T.; Ferrarese, C.; Franzetti, J.; Suarez-Bertoa, R.; Manara, D. Real-Time Measurement of NOx Emissions from Modern Diesel Vehicles Using On-Board Sensors. *Energies* **2022**, *15*, 8766. [CrossRef]

energies

MDPI

Article

Analysis of Improved In-Cylinder Combustion Characteristics with Chamber Modifications of the Diesel Engine

Arun Teja Doppalapudi [1], Abul Kalam Azad [1,*] and Mohammad Masud Kamal Khan [2]

1 School of Engineering and Technology, Central Queensland University, Melbourne Campus, 120 Spencer Street, Melbourne, VIC 3000, Australia
2 School of Engineering, Computer and Mathematical Sciences, Auckland University of Technology, Auckland 1010, New Zealand
* Correspondence: azad.cqu@gmail.com or a.k.azad@cqu.edu.au

Abstract: This study numerically analyses the effects of chamber modifications to investigate the improvement of in-cylinder combustion characteristics of the diesel engine using a computational fluid dynamics (CFD) approach. Five different modified chambers, namely, the double swirl combustion chamber (DSCC), bathtub combustion chamber (BTCC), double toroidal re-entrant combustion chamber (DTRCC), shallow depth combustion chamber (SCC), and stepped bowl combustion chamber (SBCC) were developed and compared with a reference flat combustion chamber (FCC). The effects of chamber modifications on temperature formation, velocity distribution, injection profiles, and in-cylinder turbulent motions (swirl and tumble ratio) were investigated. During the compression stroke, near top dead centre, the SCC showed a peak temperature of 970 K, followed by the FCC (968 K), SBCC (967 K), and DTRCC (748 K to 815 K). The DSCC and the SCC showed a high swirl ratio above 0.6, whereas the DTRCC and the BTCC showed a high tumble ratio of approximately 0.4. This study found that the SCC, BTCC, and DSCC have better combustion rates than the FCC in terms of temperature, heat release rate, and velocity distribution. However, the DTRCC showed poor temperature formation rates and rapid heat release rates (approx. 150 J/°CA), which can lead to rapid combustion and knocking tendencies. In conclusion, the DSCC and the SCC showed better combustion rates than the other chambers. In addition, turbulent motions inside the chambers avoided combustion in crevice regions. This study recommends avoiding chambers with wider bowls in order to prevent uneven combustion across the cylinder. Furthermore, split bowls such as the DSCC, along with adjusted injection rates, can provide better results in terms of combustion.

Keywords: combustion chamber modification; combustion simulation; heat release rate; cylinder temperature; CFD analysis

Citation: Doppalapudi, A.T.; Azad, A.K.; Khan, M.M.K. Analysis of Improved In-Cylinder Combustion Characteristics with Chamber Modifications of the Diesel Engine. *Energies* **2023**, *16*, 2586. https://doi.org/10.3390/en16062586

Academic Editors: Roberta De Robbio and Maria Cristina Cameretti

Received: 1 February 2023
Revised: 2 March 2023
Accepted: 6 March 2023
Published: 9 March 2023

1. Introduction

Air–fuel mixture formation inside the engine cylinder is mainly responsible for engine performance and emissions [1]. With the improper mixture, fuel accumulation can occur inside the chamber due to an uneven distribution of air and fuel, which can cause incomplete combustion. To address these issues, engine modifications such as nozzle modifications [2] and port flow design have been investigated during recent decades to improve the overall in-cylinder combustion [3–5]. In addition, injection modifications [6] and piston bowl modifications [7] have been experimentally investigated to improve the combustion behaviour of diesel engines. Among these, the piston bowl modification technique is well-documented in the literature as one good active technique that directly affects air–fuel mixture formation.

Piston bowl modifications have gained significant attention in the last decade due to their active interactions inside the cylinder during the air–fuel mixture formation [8]. The piston bowl influences the air–fuel mixture through turbulence, which is governed by swirl and tumble motions. Wall-wetting and the formation of fuel pockets can be avoided using chamber modifications. Biodiesel tends to form fuel pockets inside the chamber as biodiesel viscosity is higher than diesel fuel [9–12]. Moreover, biodiesel is proven as the prospective replacement for diesel fuel to reduce greenhouse gas emissions [13–17]. Several experimental results have shown increased engine performance and combustion with piston bowl geometry modifications using biodiesel as a fuel. For instance, Jaichandar and Annamalai [18,19] and Jaichandar et al. [20] conducted an experimental analysis with combustion chamber modifications on different piston bowl geometries. The study reported that the modified toroidal re-entrant combustion chamber (TRCC) showed improved combustion characteristics, reduced fuel consumption, and increased thermal efficiency compared with the conventional test engine chamber, the Hemispherical combustion chamber (HCC). In addition, Channappagoudra et al. [21] conducted an experimental analysis with the TRCC, HCC, Straight Sided Piston Bowl Geometry (SSPBG), and TCC. The study reported that the TRCC showed better results than the others because of improved fuel atomisation, increased cylinder temperature, and high turbulent kinetic energy. However, these experimental studies do not provide a detailed understanding of the effects of chamber modifications on air–fuel mixture formation and combustion rates.

Optical and computational studies have helped to understand in-cylinder flow fields with respect to modified geometries and operating parameters. Optical studies have allowed researchers to understand flame movements, flow patterns, and combustion evolution [22]. On the other hand, computational studies have assisted in developing new design geometries. They helped to investigate the in-cylinder fluid flow behaviour with respect to the chambers, along with the engine characteristics and emission estimations. Previous simulation studies have described the chamber modifications based on injection behaviour. For instance, Li et al. [23] proposed a double swirl combustion chamber (DSCC) where the injected fuel splits in the chamber grooves and spreads towards the squish and bowl region. Later, Li et al. [24] introduced a lateral swirl combustion chamber (LSCS) to accommodate the fuel inside the chamber, so the combustion occurs near the bowl region. These studies have indicated better combustion with the modified chambers with reduced soot emissions. However, the thermal profile of the combustion chamber has not been explained clearly enough to understand the intrinsic role of modified chambers. Furthermore, modified chambers greatly support low-temperature combustion (LTC) engines [25,26] and dual-fuel engines [27,28]. For example, Han et al. [29] analysed a longer air–fuel mixture period formed in the swirl direction due to the bowl chamber, which led to low-temperature combustion.

In the past few decades, several experimental studies have been conducted on piston bowl modifications, suggesting that the bowl chamber has influenced engine characteristics such as engine performance and emissions [30–33]. However, very few studies have focused on investigating flow combustion behaviour inside the chamber in relation to bowl modifications [30,34,35]. Hence, the novelty of this study is to examine and analyse combustion behaviour with respect to different combustion chamber modifications. This study used a CFD approach to investigate the combustion regime by analysing in-cylinder temperature and velocity profiles. In addition, to investigate rapid combustion and heat release rates, multiple injection pulses were used for all piston bowl geometries.

The scope of this study was to investigate flow parameters inside the chamber with respect to chamber modifications. This study considered six different bowl chambers and performed the simulation at set operating conditions to evaluate key results from the modified chambers. The following section explains our simulation modelling methodology, along with the designs of different chamber profiles. Furthermore, this study presents critical results and discusses the effect of chamber modifications on cylinder compression rates, temperature, and velocity profiles with respect to crank angles. Finally, the study's conclusions and recommendations are presented.

2. Methodology

The combustion sector simulation was carried out with Ansys IC engine fluent tool. Firstly, a flat combustion chamber was prepared similar to the Kubota V3300 engine model by maintaining the engine's bore, stroke, compression ratio, and displacement. Figure 1 presents the process flowchart for engine simulation. The simulation methodology contains three main parts—geometry preparation, simulation, and post-processing. Geometry preparation is essential to fluid volume extraction and meshing. Data related to valve positions, valve seats, port alignments, and positions were gathered and accommodated in preparing the combustion chamber. Later, the chamber was updated with respect to fluid extraction and meshing errors. A computational mesh was created to calculate the results at each time step [36]. The input variables and initialisation process were carried out during the simulation setup, and the analysis was performed during the post-processing stage. Later, bowl shapes were designed by adjusting the clearance volume to maintain the engine compression ratio and displacement. A detailed explanation of these stages is explained in the following sections.

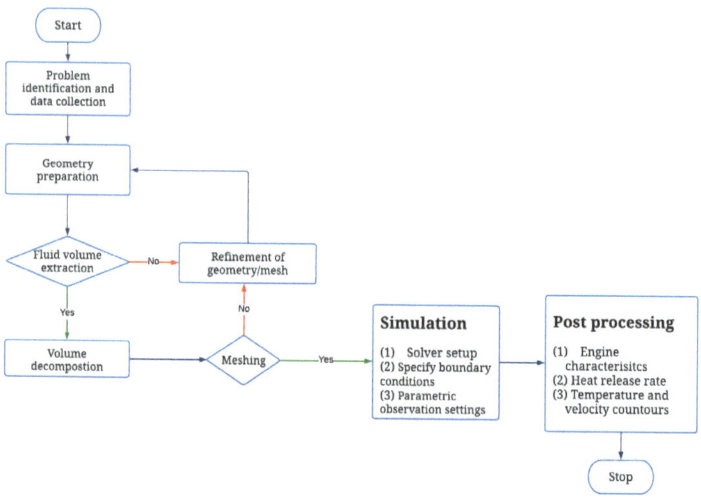

Figure 1. Flowchart for modelling and simulation of the engine.

2.1. Geometry Preparation and Meshing

In this simulation, the geometry was prepared based on the Kubota V3300 engine, as shown in Figure 2a. The detailed engine specifications are listed in Table 1. The Kubota V3300 engine is a four-cylinder heavy-duty tractor engine with two inlet valves and one exhaust valve. Moreover, it has a pre-chamber, where the combustion takes place in a separate combustion chamber and, from there, combusted gases travel to the main chamber. The modelling experiments were carried out on this Kubota V3300, a four-cylinder, four-stroke, compression ignition engine. The engine has a bore of 98 mm, a stroke of 110 mm, and a compression ratio of 22.6:1. The simulation was conducted at a speed of 1500 rpm.

A combustion chamber was created with two inlet valves and ports, and one exhaust valve and port. The cylinder displacement volume and pre-chamber rates were calculated, categorised, maintained, and modelled as direct injection engines. Arbitrary dimensions of the chamber components, such as valve heights and cylinder thickness, were assumed for the fluid extraction part as these have negligible impact on the combustion process. The fluid volume of the chamber was extracted by removing all the solid components except the valves, as shown in Figure 2b. A sector adaptive combustion simulation was carried out with an angle of 60°. The decomposed sector component of the cylinder chamber is presented in Figure 2c. During the decomposition process, the valve sets were deleted; the engine port was divided into a given sector angle of 60°; and the clearance volume and crevice were adjusted with respect to the compression ratio, as shown in Figure 2c. In the next stage, adaptive refinement technology was used to generate the mesh for the model. Figure 2d presents the mesh component of a sector of a flat combustion chamber.

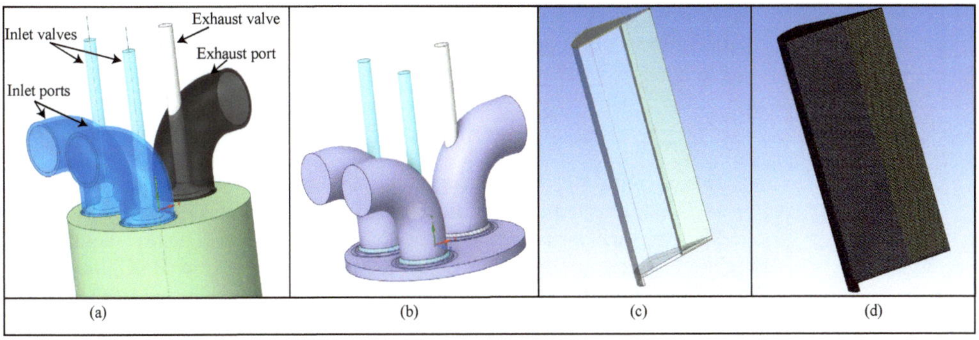

Figure 2. Geometry setup: (**a**) chamber with inlet and exhaust valves and ports; (**b**) fluid volume body; (**c**) 60° decomposed sector geometry; (**d**) meshing grid for sector geometry.

Table 1. Engine Specifications.

Make	**Kubota V3300**
Engine type	Vertical, 4-cycle liquid cooled diesel
Total displacement	3.318 (litres)
Fuel	Diesel surrogate (n-heptane)
Bore	98 mm
Stroke	110 mm
Compression ratio	22.6:1
Speed	1500 rpm
Fuel	Diesel
Cooling system	Water-cooled

2.2. Simulation Setup and Post-Processing

Ansys IC engine fluent software was used to conduct computational fluid dynamics (CFD) analysis. It provides an integrated environment with capabilities to set up powerful IC engine designs. Modelling an internal combustion engine involves dynamic interactions between moving geometries, fuel injection, and combustion processes. CFD involves jet formation, wall impingements, and combustion. The governing equations for the CFD analysis follow Navier–Stokes equations, which include conservation of energy, momentum, and continuity equations [37,38]. The governing equations are as follows.

The continuity equation:

$$\underbrace{\frac{\partial \rho}{\partial t}}_{material\ derivative} + \underbrace{\nabla(\rho V)}_{convective\ acceleration} = 0 \tag{1}$$

where ρ denotes the density of the fluid and V is the flow velocities in m/s.

The energy equation:

$$\frac{\partial(\rho e)}{\partial t} + \underbrace{\nabla(\rho e V)}_{kinetic\ energy} = \underbrace{\frac{\partial}{\partial x}\left(k\frac{\partial T}{\partial x}\right) + \frac{\partial}{\partial y}\left(k\frac{\partial T}{\partial y}\right)}_{conductive\ term} \tag{2}$$

Here, 'e' refers to the internal energy, k indicates the thermal conductivity, and T is the temperature.

The momentum equation (2-Dimensional):

$$\underbrace{\frac{\partial(\rho u)}{\partial t}+\frac{\partial(\rho u^2)}{\partial x}+\frac{\partial(\rho uv)}{\partial y}}_{inertia\ force}=\underbrace{-\frac{\partial(p)}{\partial x}}_{pressure\ gradient}+\underbrace{\frac{\partial}{\partial x}\left(\lambda\left(\frac{\partial v}{\partial x}+\frac{\partial u}{\partial y}\right)V+2\mu\frac{\partial u}{\partial x}\right)+\frac{\partial}{\partial y}\left(\mu\left(\frac{\partial v}{\partial x}+\frac{\partial u}{\partial y}\right)\right)}_{viscous\ force} \quad (3)$$

$$\frac{\partial(\rho v)}{\partial t}+\frac{\partial(\rho v^2)}{\partial y}+\frac{\partial(\rho uv)}{\partial x}=-\frac{\partial(p)}{\partial y}+\frac{\partial}{\partial x}\left(\mu\left(\frac{\partial v}{\partial x}+\frac{\partial u}{\partial y}\right)\right)+\frac{\partial}{\partial y}\left(\lambda\nabla V+2\mu\left(\frac{\partial v}{\partial x}\right)\right) \quad (4)$$

Here, u and v are velocity magnitudes in the x and y directions, λ and μ are the shear stress (viscous effect) and dynamic viscosity, and p indicates pressure in Pascals.

The empirical k-ε model was used in this simulation. The turbulent motions inside the chamber were analysed using the Renormalization Group (RNG) k-ε model based on Reynolds averaged Navier–Stokes equations. The study adopted the RNG k-ε turbulent model to couple the stream fields and converge the solution into a continuity equation. This two-equation model statistically averages the multiscale eddies that are formed during combustion and can capture the horizontal velocity profiles along the free-flow streams. The turbulent kinetic energies (k) and the dissipation rates (ε) of these eddies are calculated using k-ε model [38,39].

To observe the turbulent kinetic energy (k):

$$\frac{\partial(\rho k)}{\partial t}+\frac{\partial(\rho k\mu_i)}{\partial x_i}=\frac{\partial}{\partial x_j}\left[\frac{\mu_i}{\sigma_k}\frac{\partial k}{\partial t}\right]+2\mu_t E_{ij}E_{ij}-\rho\varepsilon \quad (5)$$

To observe the dissipation rate (ε):

$$\frac{\partial(\rho\varepsilon)}{\partial t}+\frac{\partial(\rho\varepsilon\mu_i)}{\partial x_i}=\frac{\partial}{\partial x_j}\left[\frac{\mu_t}{\sigma_e}\frac{\partial\varepsilon}{\partial x_j}\right]+C_{1\varepsilon}\frac{\varepsilon}{k}2\mu_t E_{ij}E_{ij}-C_{2\varepsilon}\rho\frac{\varepsilon^2}{k} \quad (6)$$

Apart from these, the Kelvin–Helmholtz (KH) instability model was used to predict the primary break-up model. A sector grid with periodic boundaries was adopted in this closed study from intake valve opening to exhaust valve closing. Dynamic mesh parameters, boundary conditions, materials, monitor setup, and relaxation factors were assigned during the simulation setup and initialised the solution.

The simulation setup procedure and boundary conditions were kept the same for all the piston bowl chambers. The simulation was conducted from the inlet valve closing period to the exhaust valve opening period. A constant swirl number of 1.3 was given to the inlet profiles with an injection angle of 70°. The injection was given with respect to the crank angle degree of 344 °CA to 370 °CA, as shown in Figure 3a. As represented in Figure 3b, the injection was provided with mass flow rates at three different intervals as pilot, main, and post injections. The natural atmospheric air composition was added as the intake air, and the cylinder walls were assigned to 440 K. The engine speed was kept constant at 1500 rpm.

Monitor definitions were created at the chamber bottom, chamber top and chamber fluid portion of the different cut planes. Max-vel-monitor was created to observe the maximum velocity, vol-avg-temp-monitor was created to observe the volume integral of temperature. During the post-processing stage, the results were recorded with respect to the monitor definitions according to the assigned cut planes. Before initialisation, the model setup, mixture formation, dynamic meshing, injections solution methods and solution controls were adjusted to solve the simulation. To verify the discretised equations, the convergence criteria for the velocity, k and ε were set to 1×10^{-4}, and energy was set to 1×10^{-6} with over 1400 iterations.

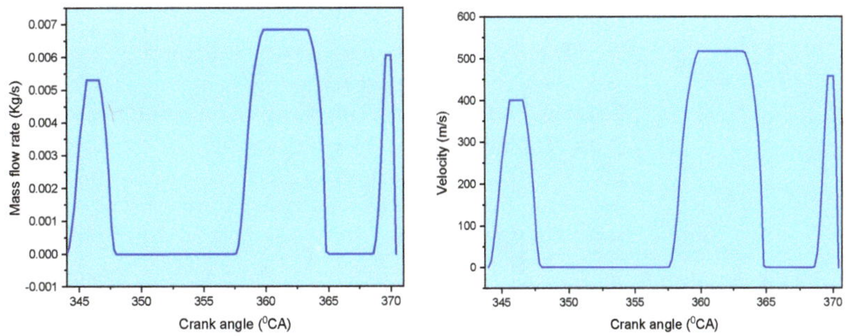

(**a**) (**b**)

Figure 3. Fuel injection profiles: (**a**) injected fuel mass vs. °CA; (**b**) injected fuel velocity vs. °CA.

2.3. Grid Sensitivity Analysis

A computational grid independence study was performed for the FCC chamber to analyse the impact of the mesh grid on the simulation process. Firstly, a finer mesh was created with a reference mesh size of 0.611 and with five inflation layers. The fine mesh created 1,948,328 nodes and 1,878,444 elements. Similarly, the coarser mesh was created with a reference mesh size of 0.917 and with three inflation layers. The coarse mesh generated 597,824 nodes and 566,470 elements. Figure 4 presents the apparent heat release rate for the fine and coarse mesh. It can be seen from Figure 4 that the type of mesh used has a slight impact on the simulation results. Hence, the study progressed with the finer mesh attributes.

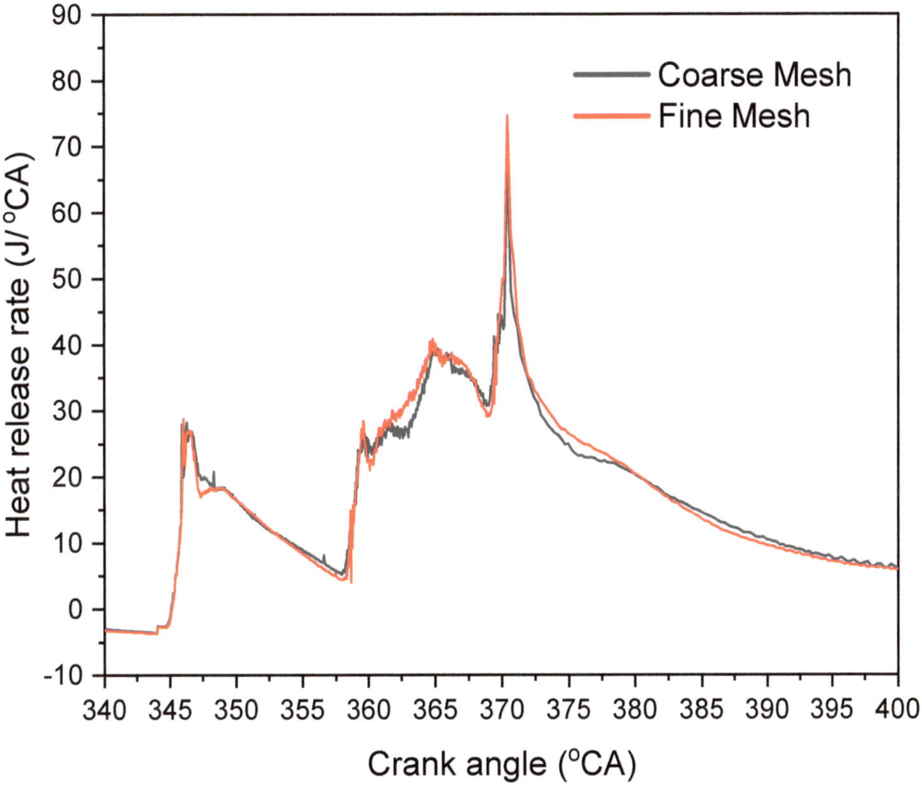

Figure 4. Comparison of HRR using coarse and fine mesh for grid sensitivity analysis.

2.4. Model Validation

The FCC simulation setup was validated against experimental data considering the heat release rate. The purpose of this validation was to examine the in-cylinder combustion behaviour of the modelling setup at constant boundary conditions with respect to the combustion condition of the test engine. The model was validated with a single injection profile (with an injection period between 356 °CA and 376 °CA), similar to the test engine injection. As presented in Figure 5, a similar trend is noted in both the experimental and simulation setup. The trend was observed during the combustion stroke between crank angles of 350 °CA and 400 °CA. The results are validated by adjusting the injection profile with respect to the crank angles. The error percentage between the actual and simulated output was calculated as 3.8%. Though there is some discrepancy in both graphs, the overall error percentage is less than 5%, and the heat release rate trend is good.

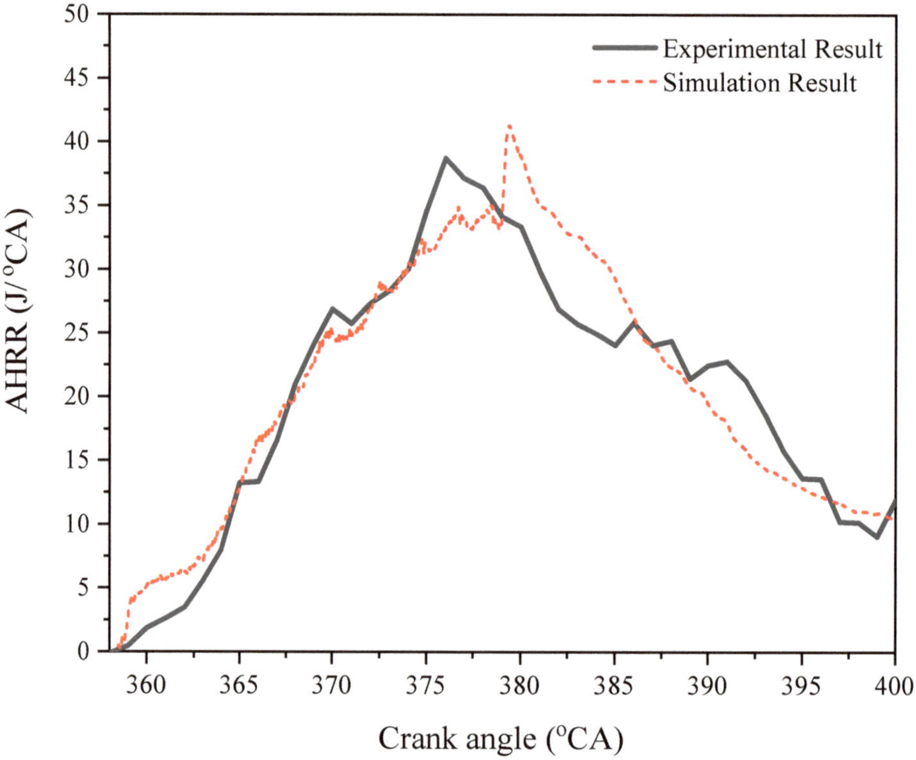

Figure 5. Model validation with experimental HRR result with simulation results.

2.5. Chamber Geometry Modification

As discussed before, the clearance volume of the flat chamber was adjusted by maintaining the same compression ratio and displacement volume to the design of the combustion chambers with respect to the test engine. Bowl geometries such as the double swirl combustion chamber (DSCC), bathtub combustion chamber (BTCC), double toroidal re-entrant combustion chamber (DTRCC), shallow depth combustion chamber (SCC), and stepped bowl combustion chamber (SBCC) were designed as shown in Figure 6. The simulation was carried out for these modified chambers with the simulation steps discussed above by maintaining the same operating parameters and boundary conditions to maintain consistency.

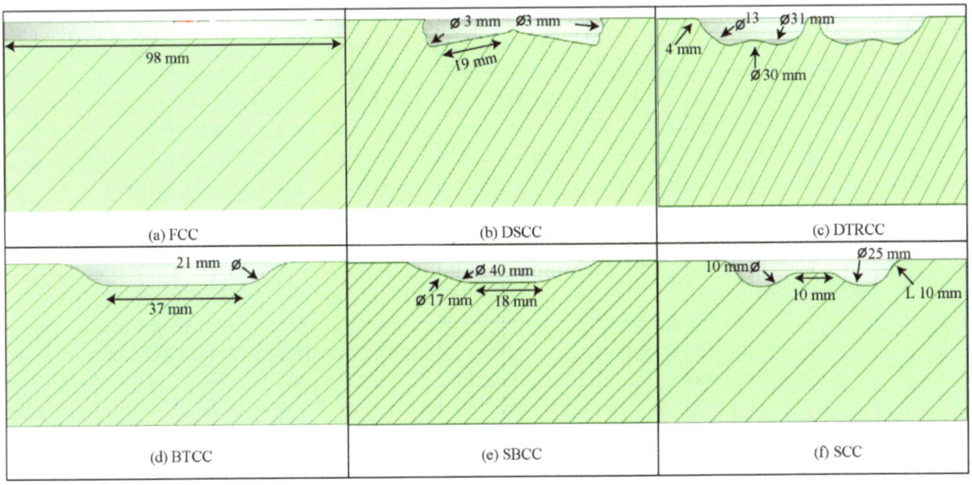

Figure 6. Representation of different piston bowl geometries.

3. Results and Discussions

3.1. Effect of Chamber Modifications on Compression Behaviour

Figure 7 illustrates the temperature distribution of the chamber at the end of the compression stroke at 344 °CA. The results show that varying the bowl shape has a visible impact on the compression rate, which is one of the primary operating parameters for a compression ignition engine. As shown in Figure 7, the SCC (970 K) showed a high-temperature formation than the reference piston FCC (968 K), followed by the SBCC (967 K) and BTCC (966 K). The DTRCC demonstrated inferior temperature formation (748 K to 815 K) during the compression stroke, which may be due to a longer bowl diameter. For the DTRCC, it can be seen from the temperature plots that the crevice region has high-temperature formation, which can cause rapid and poor combustion in crevice regions. Several studies have reported high unburnt hydrocarbons and NO_x emissions when combustion occurs in crevice regions of the chamber [40,41].

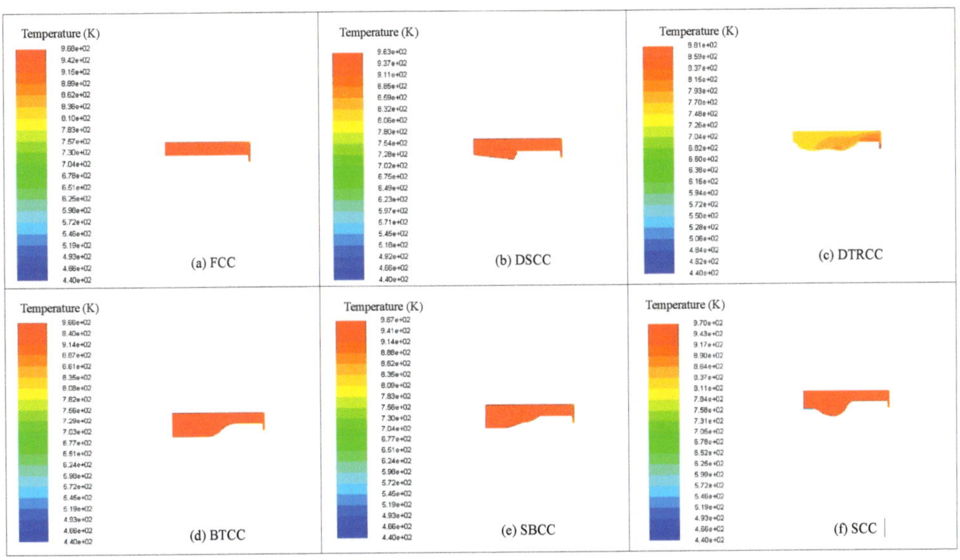

Figure 7. Temperature contours at the end of the compression stroke at 344 °CA.

During the compression stroke, particles become closer together, and the highly inhomogeneous flow of the intake stroke becomes relatively homogeneous during the compression stroke, as shown in Figures 7 and 8 [42]. The velocity distribution of different piston bowl chambers at 344 °CA is illustrated in Figure 8. It is observed from the results that high-velocity rates are formed in the squish region near the cylinder wall. Even though the FCC has no bowl shape, high-velocity motions (8.97 m/s) are observed near the wall region due to surface interactions between the compressed air and moving surfaces of the piston and wall. This phenomenon indicates that the injected air is pushed away from the cylinder axis and makes space for the injected fuel. As predicted, low-velocity rates were observed in the crevice regions as they have very confined spaces. Interestingly, the DTRCC (9.18 m/s) showed high-velocity formation at the end of the compression stroke. Usually, entropy (degree of randomness) should increase under high-temperature conditions; however, compared to the other chambers, the DTRCC formed high-velocity flow motions in the squish region. This is because the DTRCC has less squish space and longer bowl regions; meanwhile, the fluid and wall surface interactions also create high velocities near the wall region. Hence, higher turbulent kinetic energy(TKE) was noticed at the confined squish region. Similar behaviour was also seen in the BTCC (9.24 m/s), with higher TKE than the SCC (9 m/s) because of the shorter squish region.

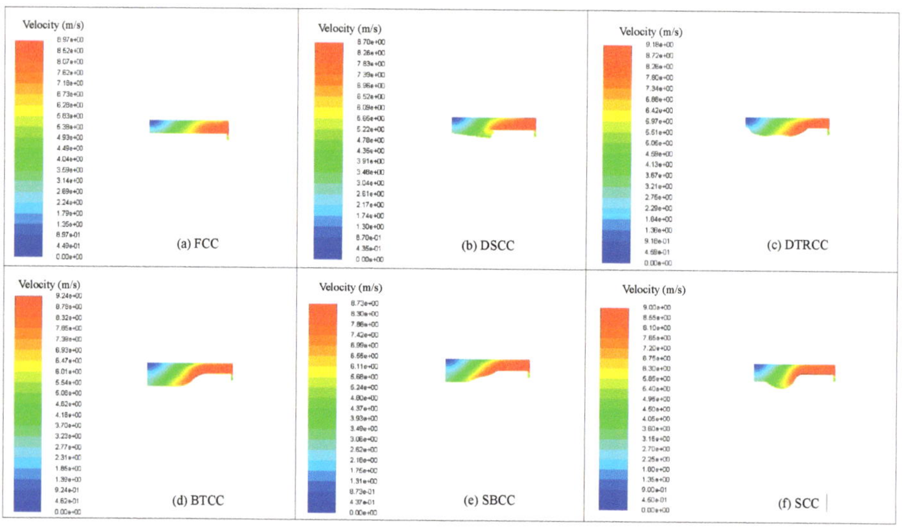

Figure 8. Velocity contours at the end of the compression stroke at 344 °CA.

3.2. *Effect of Chamber Modifications on Fuel Particle Distribution*

Figure 9 presents the fuel particle distribution with respect to velocity and temperature at 368 °CA during the injection period, which is near the top dead centre (TDC). The DSCC (2170 K) and BTCC (2140 K) showed high-temperature formation rates during the injection period compared with the FCC (2120 K). Due to poor compression rates, the DTRCC showed lower temperature formation at 1990 K compared with other chambers.

To understand fuel dispersion inside the chamber with respect to bowl shape, the temperature and velocity rate of particle traces were analysed. Bowl design signifies fuel distribution and combustion. For instance, in the DSCC and SCC, the fuel is sprayed directly into the bowl region, which restricts the fuel from spreading to the squish and crevice regions. Velocity contours also support that the DSCC and SCC have no particle motions after the bowl region. Targeting the least flow restriction benefits the air–fuel mixture because the loss in spray momentum helps distribute the fuel around the bowl region [43]. A free-flow of particle momentum was noticed with the FCC, DTRCC, BTCC, and SCC due to the absence of complex obstructions to the injected fuel. The results for these chambers indicate that the injected fuel momentum spreads towards the squish and crevice regions. The distribution rate in the squish region was greater for the FCC and DTRCC due to the absence of hard bowl obstructions. As discussed earlier, accumulated fuels in the crevice and squish regions cause improper combustion and release higher NO_x and UBHC emissions. High-density fuels such as biodiesel have more scope for forming fuel pockets near the crevice regions, making them more susceptible to uniform combustion [44].

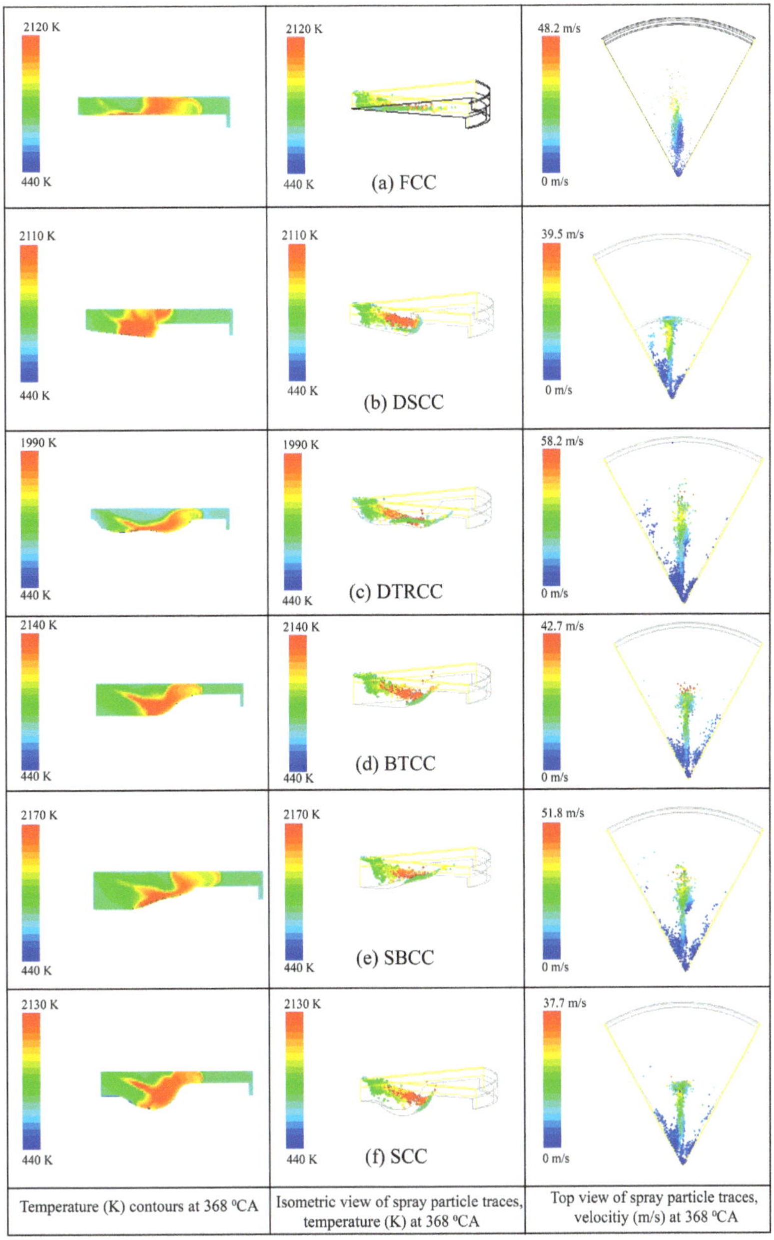

Figure 9. Injection profiles during injection at 368 °CA.

3.3. *Effect of Chamber Modifications on In-Cylinder Temperature Distribution*

Figures 10 and 11 show the in-cylinder temperature distribution for modified chambers during the middle and end of the combustion stroke at crank angles of 431 °CA and 467 °CA, respectively. As presented in Figures 10 and 11, the chambers with fewer bowl obstructions, such as the FCC and DTRCC, revealed that combustion spreads towards the cylinder wall region with respect to fuel droplet dispersion. Furthermore, because the spray mixture is more prone to hit the walls, it can cause combustion near the wall region. Therefore, there are higher chances for cylinder heat loss as the high-temperature areas are situated alongside the wall. Moreover, compared with other bowl chambers, the FCC and DTRCC showed combustion formation in the crevice regions (Figures 10 and 11). However, the bowl chambers of the DSCC, BTCC, and SCC, showed high-temperature distribution during the combustion stroke, indicating that combustion happened alongside the bowl region. A poor temperature distribution was noted in the DTRCC, where the combustion occurred near the squish region because of the shorter squish area and wider bowl chamber.

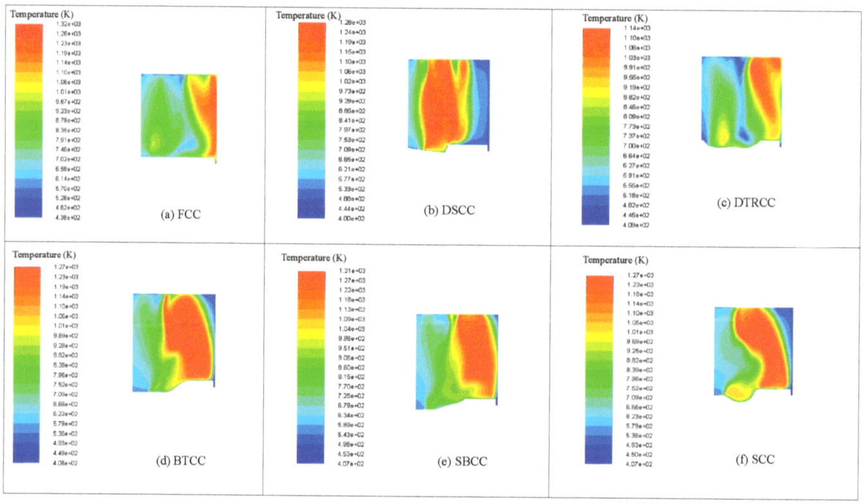

Figure 10. Temperature contours during the combustion stroke at 431 °CA.

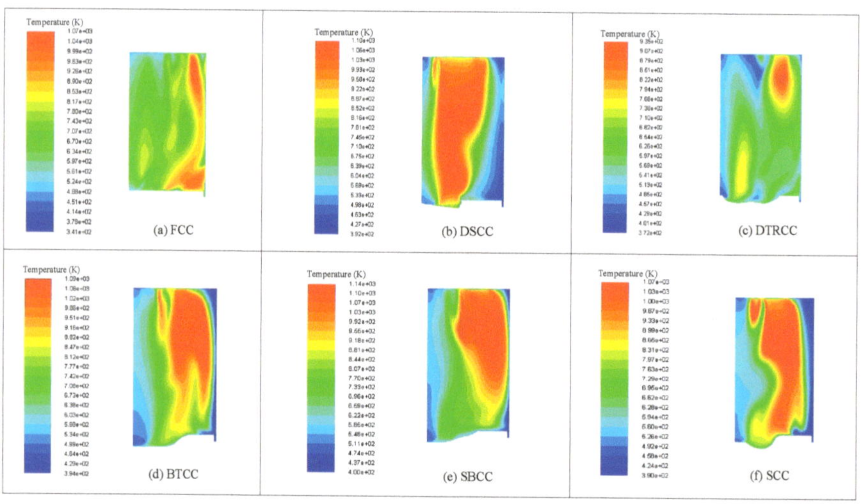

Figure 11. Temperature contours near the BDC at 431 °CA.

Fuel injection is mainly responsible for carrying the combustion near the bowl region. As shown in Figure 9, the injected fuel is dispersed into squish regions for the DTRCC, BTCC, DBCC, and SCC chambers; hence, the combustion plots (Figures 10 and 11) also show combustion at the squish regions of these chambers. Thus, it is concluded from the analysis that in order to conduct combustion with piston bowl geometries, injection type and angle should be adjusted to utilise the bowl region.

3.4. Effect of Chamber Modifications on the In-Cylinder Velocity Distribution

Figures 12 and 13 show velocity distributions at 431 °CA and 467 °CA for different modified chambers. It is observed from the results that higher velocity magnitudes are recorded on top of the piston surface as the piston moves towards the bottom dead centre BDC. Here, the velocity counters represent two main activities: turbulence and piston work [45]. The high velocities situated on the piston head for the SCC (approx. 10.5 m/s) generate significantly more power during the power stroke. In the case of the FCC, the fluid streams dispersed around the piston surface, showing higher velocities than other chambers due to the absence of a bowl region. Furthermore, the DTRCC with a broader bowl exhibited poor velocity distribution due to the formation of less turbulence inside the chamber. Higher magnitude of velocity regimes, 13.9 m/s at 431 °CA and 15 m/s at 497 °CA, were noticed for the DTRCC. However, the irregular velocity stream distribution was recorded for the DSCC due to the uneven turbulence distribution of fluid streams.

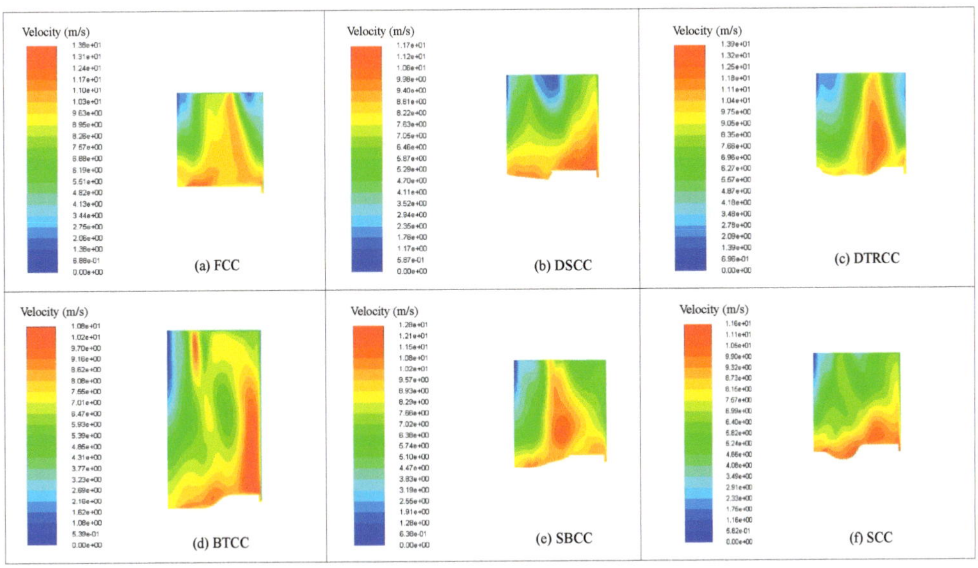

Figure 12. Velocity contours during the combustion stroke at 431 °CA.

3.5. Effect of Chamber Modifications on the Heat Release Rate

The apparent heat release rate (AHRR) for different modified chambers is illustrated in Figure 14. The three humps during the injection were mainly due to the compound HRR caused by multiple split injections. Compared with the reference FCC, all the piston bowl chambers showed high HRRs with pilot injection. This may be due to the increased turbulence rate generated by the piston bowl chambers during the injection stages. The AHRR in the DTRCC was high and more rapid than in the other chambers. However, temperature plots (Figures 10 and 11) and velocity plots (Figures 12 and 13) showed poor combustion rates and spatial distribution within the DTRCC. The main reason for these poor temperatures, velocities, and HRR was uneven and rapid combustion. Furthermore, as knocking intensity is proportional to peak HRR, the DTRCC is more susceptible to generating knocking under high turbulence conditions.

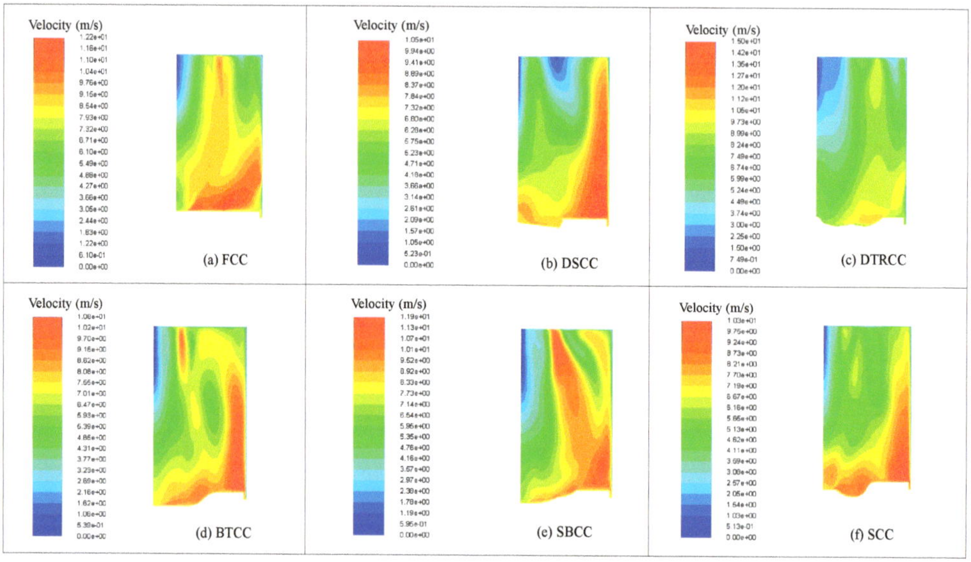

Figure 13. Velocity contours near the end of the combustion stroke at 431 °CA.

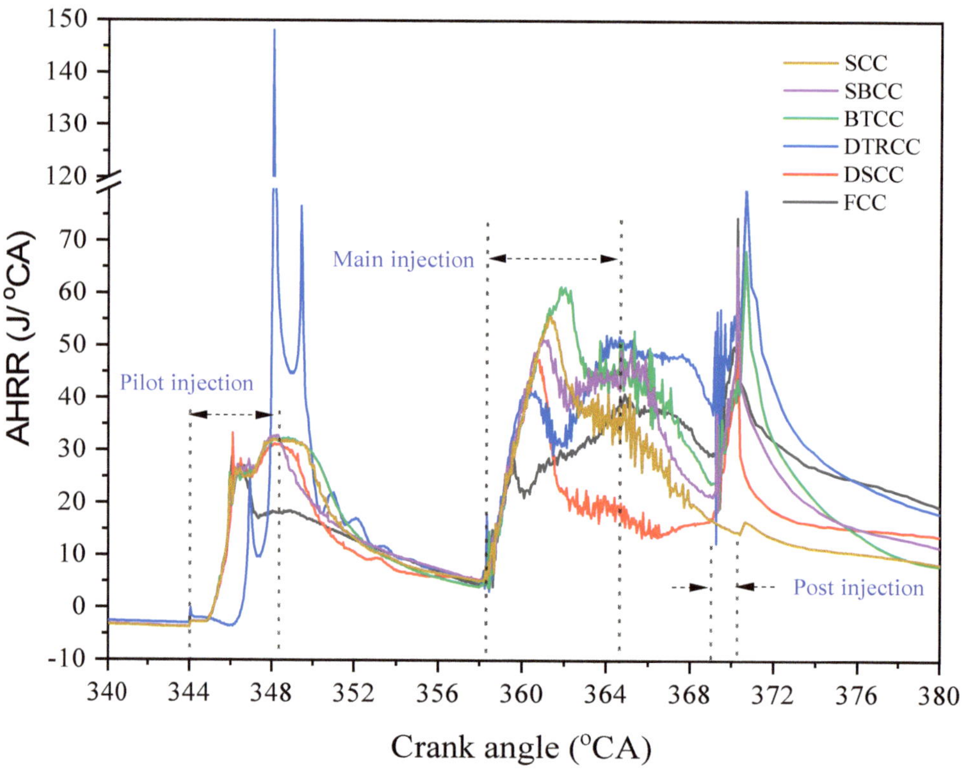

Figure 14. Heat release rate for different piston bowl geometries.

All the chambers showed high AHRRs during the main fuel injection period compared with the FCC. The DSCC and SCC gradually decreased with combustion, suggesting that better fuel combustion occurred with the DSCC and SCC than the BTCC, FCC, SBCC, and DTRCC. During the third injection, the SCC did not record significant variations in the HRR. This may be due to fuel accumulation in the deeper bowl chamber, and suggests that a third post-injection is not required in the SCC chamber model. On the other hand, all the chambers, including the FCC, reacted with the post-injection. Usually, mixture swirling occurs mainly in the bowl region, and it is highly beneficial if the chamber can utilise the squish region. In the case of the DSCC, the injected fuel split towards the bowl and squish area. Thus, the DSCC showed better combustion release rates when compared with other chambers.

3.6. Swirl Ratio and Tumble Ratio

Swirl ratio and tumble ratio greatly influence the air-fuel mixing process. Swirl and tumble flow induce the air flow field, increasing burning rate and TKE. Figure 15 shows the swirl ratio for different piston bowl geometries with respect to crank angles. It is observed that during the suction and compression stroke, swirl ratio is high, then started falling to approximately 0.0 at the end of the compression stroke (380 °CA). Again, during the combustion stroke at 382 °CA to 460 °CA, variation in swirl motions were observed for all the chambers. The DSCC showed a high swirl ratio among all the chambers because of the double swirl bowl chamber, followed by the SCC and BTCC. This clearly indicates that shorter and deeper bowl chambers have higher swirl ratios than longer and shallower bowl chambers (as seen in the SBCC, DTRCC, and FCC).

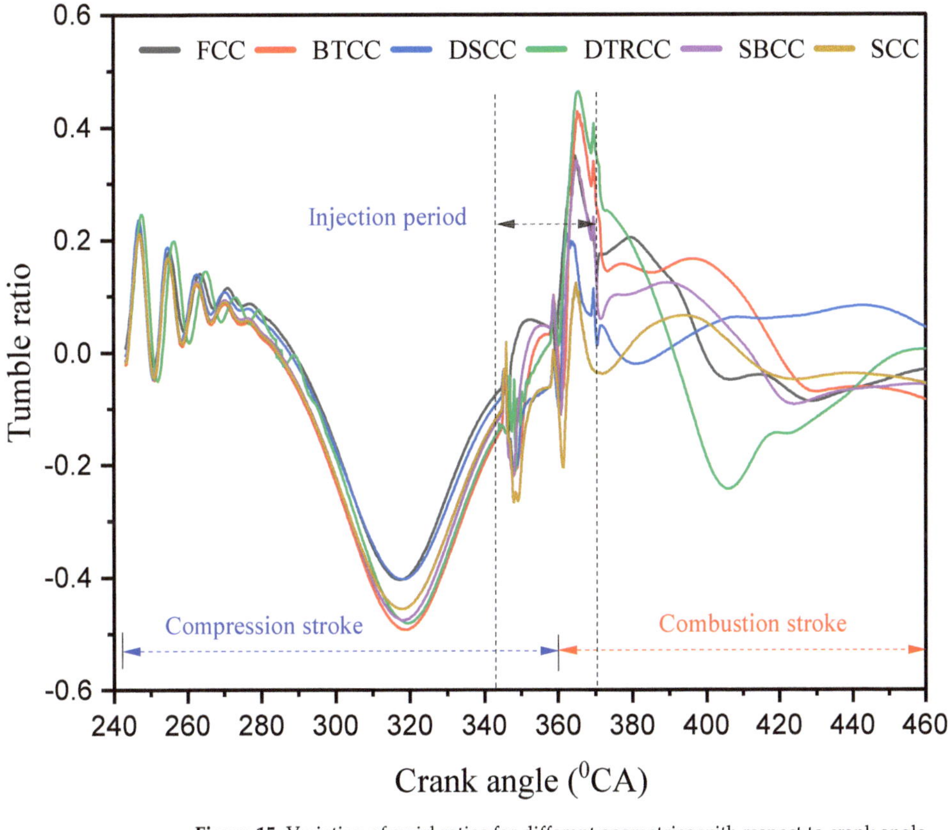

Figure 15. Variation of swirl ratios for different geometries with respect to crank angle.

Figure 16 shows the tumble ratio for different piston bowl geometries with respect to crank angles. The magnitude of the tumble and swirl ratio follows the right-hand rule; hence, the negative ranges in the graphs (Figures 15 and 16) represent the counter-clockwise direction of the flow [46]. When comparing both motions, we observed that the swirl ratio was more dominant in these chambers compared with the tumble ratio. During the suction stroke, near 243 °CA to 280 °CA, variations in tumble motions were observed; meanwhile, during the compression stroke, a negative trend in the tumble ratio was noted. Again, during the combustion stroke, variation in the tumble ratio was seen. For the DTRCC, SBCC, and FCC, a peak tumble ratio was observed over other chambers, which was the opposite of the swirl ratio results. Furthermore, a high tumble ratio was observed during the combustion stroke, whereas a high swirl ratio was observed during the suction stroke.

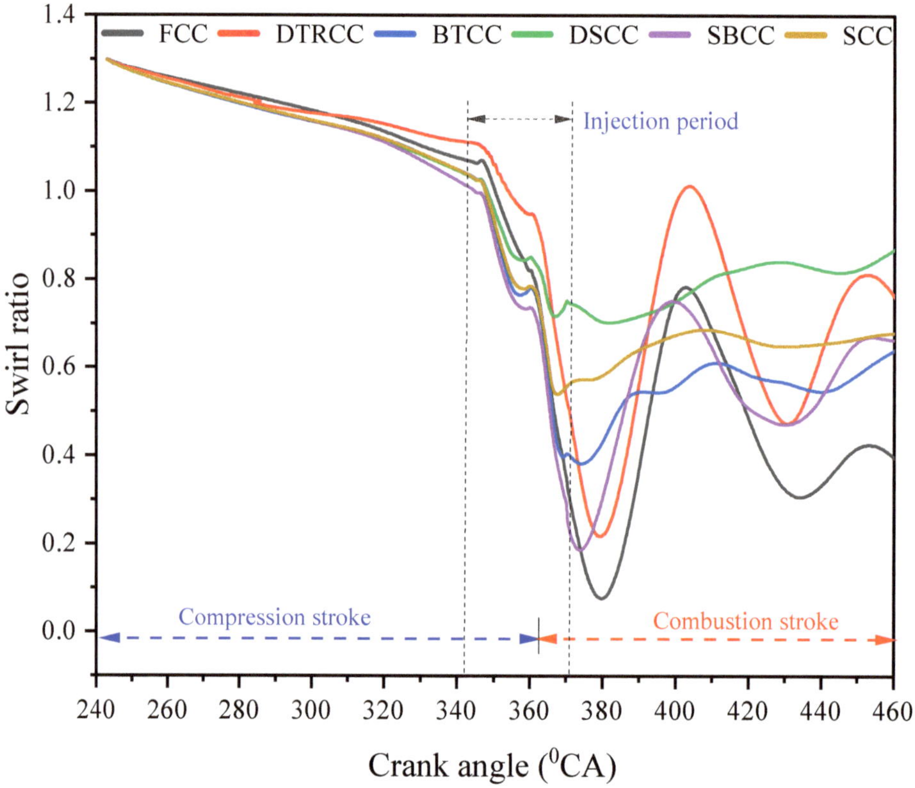

Figure 16. Variation of tumble ratios for different geometries with respect to crank angle.

4. Conclusions and Recommendations

This research paper provides the computational results of DSCC, DTRCC, BTCC, SBCC and SCC geometries and their combustion analysis with respect to the FCC. The conclusions of this study are as follows:

1. The results show that the combustion distribution pattern is quite different among the piston bowl geometries. From the examined velocity profiles, the bowl depth and length affect the spatial distribution of fluid motions inside the cylinder, which causes high-turbulence motions around the chamber.
2. The DSCC (swirl chamber) showed better combustion performance than the other chambers considering the better temperature and velocity distribution, as well as the high swirl ratio and low HRR.

3. The SCC showed better combustion rates, velocity distribution, and HRR compared with the FCC and others, whereas the DTRCC showed poor temperature formation and rapid velocity distribution, which can lead to knocking.
4. A high swirl ratio was observed during the suction stroke, and a high tumble ratio was observed during the combustion stroke, which increased the turbulence rates of the piston bowl geometries.

It is noted that this paper concentrates on the characteristics and behaviour of chamber modifications using simulation results, particularly examining the changes and impact of chamber modifications on combustion. The DSCC has a groove structure that splits the injected fuel into the squish and bowl regions. As the DSCC provided better results, it is recommended to add groove structures to bowl chambers for better combustion. This study shows that fuel injection distribution impacts combustion rates. Furthermore, injected spray plays a controlling role in causing combustion near bowl regions. Hence, one should consider adjusting injection rates and spray patterns when optimising combustion chambers. In addition, emission formation rates of modified chambers need to be tested to examine the role of chamber modifications on emission formations.

Author Contributions: Conceptualization, A.T.D. and A.K.A.; methodology, A.T.D. and A.K.A.; writing—original draft preparation, A.T.D.; visualization, A.T.D.; writing—review and editing, A.T.D., A.K.A. and M.M.K.K.; supervision, A.K.A. and M.M.K.K. All authors have read and agreed to the published version of the manuscript.

Funding: This research received no external funding.

Institutional Review Board Statement: Not applicable.

Informed Consent Statement: Not applicable.

Data Availability Statement: Not applicable.

Conflicts of Interest: The authors declare no conflict of interest.

Nomenclature

AHRR	Apparent heat release rate
BDC	Bottom dead centre
BTCC	Bathtub combustion chamber
CFD	Computational fluid dynamics
DSCC	Double swirl combustion chamber
DTRCC	Double toroidal re-entrant combustion chamber
FCC	Flat combustion chamber
HCC	Hemispherical combustion chamber
SCC	Shallow depth combustion chamber
SBCC	Stepped bowl combustion chamber
TDC	Top dead centre
TKE	Turbulent kinetic energy
TRCC	Toroidal re-entrant combustion chamber
LSCC	Lateral swirl combustion chamber
LTC	Low-temperature combustion

References

1. Kanda, T.; Hakozaki, T.; Uchimoto, T.; Hatano, J.; Kitayama, N.; Sono, H. PCCI Operation with Early Injection of Conventional Diesel Fuel. *SAE Trans.* **2005**, *114*, 584–593.
2. Praveena, V.; Martin, M.L.J.; Varuvel, E.G. Combined effects of nozzle hole variation and piston bowl geometry modification on performance characteristics of a diesel engine with energy and exergy approach. *J. Clean. Prod.* **2022**, *375*, 133946. [CrossRef]
3. Ashok, B.; Nanthagopal, K.; Saravanan, B.; Azad, K.; Patel, D.; Sudarshan, B.; Ramasamy, R.A. Study on isobutanol and Calophyllum inophyllum biodiesel as a partial replacement in CI engine applications. *Fuel* **2019**, *235*, 984–994. [CrossRef]

4. Azad, A.K.; Adhikari, J.; Halder, P.; Rasul, M.G.; Hassan, N.M.S.; Khan, M.M.K.; Naqvi, S.R.; Viswanathan, K. Performance, Emission and Combustion Characteristics of a Diesel Engine Powered by Macadamia and Grapeseed Biodiesels. *Energies* **2020**, *13*, 2748. [CrossRef]
5. Piano, A.; Roggio, S.; Millo, F.; García, A.; Micó, C.; Lewiski, F.; Pesce, F.; Vassallo, A.; Bianco, A. Numerical and optical soot characterisation through 2-color pyrometry technique for an innovative diesel piston bowl design. *Fuel* **2023**, *333*, 126347. [CrossRef]
6. Soni, D.K.; Gupta, R. Numerical analysis of flow dynamics for two piston bowl designs at different spray angles. *J. Clean. Prod.* **2017**, *149*, 723–734. [CrossRef]
7. Gafoor, C.A.; Gupta, R. Numerical investigation of piston bowl geometry and swirl ratio on emission from diesel engines. *Energy Convers. Manag.* **2015**, *101*, 541–551. [CrossRef]
8. Lawrence, K.R.; Huang, Z.; Nguyen, X.P.; Balasubramanian, D.; Gangula, V.R.; Doddipalli, R.R.; Le, V.V.; Bharathy, S.; Hoang, A.T. Exploration over combined impacts of modified piston bowl geometry and tert-butyl hydroquinone additive-included biodiesel/diesel blend on diesel engine behaviors. *Fuel* **2022**, *322*, 124206. [CrossRef]
9. Azad, A.K.; Rasul, M.; Khan, M.; Sharma, S.; Hazrat, M. Prospect of biofuels as an alternative transport fuel in Australia. *Renew. Sustain. Energy Rev.* **2015**, *43*, 331–351. [CrossRef]
10. Bhuiya, M.M.K.; Rasul, M.; Khan, M.; Ashwath, N.; Azad, A. Prospects of 2nd generation biodiesel as a sustainable fuel—Part: 1 selection of feedstocks, oil extraction techniques and conversion technologies. *Renew. Sustain. Energy Rev.* **2016**, *55*, 1109–1128. [CrossRef]
11. Bhuiya, M.M.K.; Rasul, M.; Khan, M.; Ashwath, N.; Azad, A.; Hazrat, M. Prospects of 2nd generation biodiesel as a sustainable fuel—Part 2: Properties, performance and emission characteristics. *Renew. Sustain. Energy Rev.* **2016**, *55*, 1129–1146. [CrossRef]
12. Mofijur, M.; Rasul, M.G.; Hyde, J.; Azad, A.K.; Mamat, R.; Bhuiya, M.M.K. Role of biofuel and their binary (diesel–biodiesel) and ternary (ethanol–biodiesel–diesel) blends on internal combustion engines emission reduction. *Renew. Sustain. Energy Rev.* **2016**, *53*, 265–278. [CrossRef]
13. Azad, A.; Rasul, M.; Khan, M.; Sharma, S.C.; Islam, R. Prospect of Moringa Seed Oil as a Sustainable Biodiesel Fuel in Australia: A Review. *Procedia Eng.* **2015**, *105*, 601–606. [CrossRef]
14. Azad, A.; Rasul, M.; Khan, M.; Sharma, S.C.; Mofijur, M.; Bhuiya, M. Prospects, feedstocks and challenges of biodiesel production from beauty leaf oil and castor oil: A nonedible oil sources in Australia. *Renew. Sustain. Energy Rev.* **2016**, *61*, 302–318. [CrossRef]
15. Bhuiya, M.; Rasul, M.; Khan, M.; Ashwath, N.; Azad, A.; Hazrat, M. Second Generation Biodiesel: Potential Alternative to-edible Oil-derived Biodiesel. *Energy Procedia* **2014**, *61*, 1969–1972. [CrossRef]
16. Halder, P.; Azad, K.; Shah, S.; Sarker, E. 8—Prospects and technological advancement of cellulosic bioethanol ecofuel production. In *Advances in Eco-Fuels for a Sustainable Environment*; Azad, K., Ed.; Woodhead Publishing: Cambridge, UK, 2019; pp. 211–236.
17. Uddin, S.A.; Azad, A.; Alam, M.; Ahamed, J. Performance of a Diesel Engine Run with Mustard-Kerosene Blends. *Procedia Eng.* **2015**, *105*, 698–704. [CrossRef]
18. Jaichandar, S.; Annamalai, K. Effects of open combustion chamber geometries on the performance of pongamia biodiesel in a DI diesel engine. *Fuel* **2012**, *98*, 272–279. [CrossRef]
19. Jaichandar, S.; Annamalai, K. Influences of re-entrant combustion chamber geometry on the performance of Pongamia biodiesel in a DI diesel engine. *Energy* **2012**, *44*, 633–640. [CrossRef]
20. Jaichandar, S.; Kumar, P.S.; Annamalai, K. Combined effect of injection timing and combustion chamber geometry on the performance of a biodiesel fueled diesel engine. *Energy* **2012**, *47*, 388–394. [CrossRef]
21. Channappagoudra, M.; Ramesh, K.; Manavendra, G. Comparative study of standard engine and modified engine with different piston bowl geometries operated with B20 fuel blend. *Renew. Energy* **2019**, *133*, 216–232. [CrossRef]
22. Pastor, J.V.; García, A.; Micó, C.; Lewiski, F.; Vassallo, A.; Pesce, F. Effect of a novel piston geometry on the combustion process of a light-duty compression ignition engine: An optical analysis. *Energy* **2021**, *221*, 119764. [CrossRef]
23. Li, X.; Zhou, H.; Zhao, L.M.; Su, L.; Xu, H.; Liu, F. Effect of split injections coupled with swirl on combustion performance in DI diesel engines. *Energy Convers. Manag.* **2016**, *129*, 180–188. [CrossRef]
24. Li, X.; Chen, Y.; Su, L.; Liu, F. Effects of lateral swirl combustion chamber geometries on the combustion and emission characteristics of DI diesel engines and a matching method for the combustion chamber geometry. *Fuel* **2018**, *224*, 644–660. [CrossRef]
25. Cao, L.; Bhave, A.; Su, H.; Mosbach, S.; Kraft, M.; Dris, A.; McDavid, R.M. Influence of Injection Timing and Piston Bowl Geometry on PCCI Combustion and Emissions. *SAE Int. J. Engines* **2009**, *2*, 1019–1033. [CrossRef]
26. Xu, L.; Bai, X.-S.; Li, Y.; Treacy, M.; Li, C.; Tunestål, P.; Tunér, M.; Lu, X. Effect of piston bowl geometry and compression ratio on in-cylinder combustion and engine performance in a gasoline direct-injection compression ignition engine under different injection conditions. *Appl. Energy* **2020**, *280*, 115920. [CrossRef]
27. Lee, S.; Park, S. Optimization of the piston bowl geometry and the operating conditions of a gasoline-diesel dual-fuel engine based on a compression ignition engine. *Energy* **2017**, *121*, 433–448. [CrossRef]
28. Shen, Z.; Wang, X.; Zhao, H.; Lin, B.; Shen, Y.; Yang, J. Numerical investigation of natural gas-diesel dual-fuel engine with different piston geometries and radial clearances. *Energy* **2021**, *220*, 119706. [CrossRef]
29. Han, S.; Kim, J.; Bae, C. Effect of air–fuel mixing quality on characteristics of conventional and low temperature diesel combustion. *Appl. Energy* **2014**, *119*, 454–466. [CrossRef]

30. Hao, C.; Zhang, Z.; Wang, Z.; Bai, H.; Li, Y.; Li, Y.; Lu, Z. Investigation of spray angle and combustion chamber geometry to improve combustion performance at full load on a heavy-duty diesel engine using genetic algorithm. *Energy Convers. Manag.* **2022**, *267*, 115862. [CrossRef]
31. Saito, T.; Daisho, Y.; Uchida, N.; Ikeya, N. Effects of Combustion Chamber Geometry on Diesel Combustion. *SAE Trans.* **1986**, *95*, 793–803.
32. Varun; Singh, P.; Tiwari, S.K.; Singh, R.; Kumar, N. Modification in combustion chamber geometry of CI engines for suitability of biodiesel: A review. *Renew. Sustain. Energy Rev.* **2017**, *79*, 1016–1033. [CrossRef]
33. Venu, H.; Raju, V.D.; Subramani, L. Combined effect of influence of nano additives, combustion chamber geometry and injection timing in a DI diesel engine fuelled with ternary (diesel-biodiesel-ethanol) blends. *Energy* **2019**, *174*, 386–406. [CrossRef]
34. Krishnaraj, J.; Vasanthakumar, P.; Hariharan, J.; Vinoth, T.; Karthikayan, S. Combustion Simulation and Emission Prediction of Different Combustion Chamber Geometries Using Finite Element Method. *Mater. Today Proc.* **2017**, *4*, 7903–7910. [CrossRef]
35. Bapu, B.R.; Saravanakumar, L.; Prasad, B.D. Effects of combustion chamber geometry on combustion characteristics of a DI diesel engine fueled with calophyllum inophyllum methyl ester. *J. Energy Inst.* **2017**, *90*, 82–100. [CrossRef]
36. Hazrat, M.A.; Rasul, M.G.; Mofijur, M.; Khan, M.M.K.; Djavanroodi, F.; Azad, A.K.; Bhuiya, M.M.K.; Silitonga, A. A Mini Review on the Cold Flow Properties of Biodiesel and its Blends. *Front. Energy Res.* **2020**, *8*, 598651. [CrossRef]
37. Azad, A.K.; Rasul, M.G.; Khan, M.M.K.; Sharma, S.C.; Bhuiya, M.M.K. Recent development of biodiesel combustion strategies and modelling for compression ignition engines. *Renew. Sustain. Energy Rev.* **2016**, *56*, 1068–1086. [CrossRef]
38. Varol, Y.; Oztop, H.F.; Firat, M.; Koca, A. CFD modeling of heat transfer and fluid flow inside a pent-roof type combustion chamber using dynamic model. *Int. Commun. Heat Mass Transf.* **2010**, *37*, 1366–1375. [CrossRef]
39. Han, Z.; Reitz, R.D. Turbulence modeling of internal combustion engines using RNG κ-ε models. *Combust. Sci. Technol.* **1995**, *106*, 267–295. [CrossRef]
40. Duan, X.; Zhang, S.; Liu, Y.; Li, Y.; Liu, J.; Lai, M.-C.; Deng, B. Numerical investigation the effects of the twin-spark plugs coupled with EGR on the combustion process and emissions characteristics in a lean burn natural gas SI engine. *Energy* **2020**, *206*, 118181. [CrossRef]
41. Li, J.; Yang, W.; An, H.; Maghbouli, A.; Chou, S. Effects of piston bowl geometry on combustion and emission characteristics of biodiesel fueled diesel engines. *Fuel* **2014**, *120*, 66–73. [CrossRef]
42. Banaeizadeh, A.; Afshari, A.; Schock, H.; Jaberi, F. Large-eddy simulations of turbulent flows in internal combustion engines. *Int. J. Heat Mass Transf.* **2013**, *60*, 781–796. [CrossRef]
43. Yadav, J.; Venkatesh, P.; Pischinger, S. Application of micro-genetic algorithms to optimise piston bowl geometries for heavy-duty engines running on diesel and 1-Octanol fuels. *Appl. Therm. Eng.* **2023**, *226*, 120236. [CrossRef]
44. Doppalapudi, A.; Azad, A.; Khan, M. Advanced strategies to reduce harmful nitrogen-oxide emissions from biodiesel fueled engine. *Renew. Sustain. Energy Rev.* **2023**, *174*, 113123. [CrossRef]
45. Doppalapudi, A.; Azad, A.; Khan, M. Combustion chamber modifications to improve diesel engine performance and reduce emissions: A review. *Renew. Sustain. Energy Rev.* **2021**, *152*, 111683. [CrossRef]
46. Lee, K.; Bae, C.; Kang, K. The effects of tumble and swirl flows on flame propagation in a four-valve S.I. engine. *Appl. Therm. Eng.* **2007**, *27*, 2122–2130. [CrossRef]

energies

Article

Ultra-Lean Premixed Turbulent Combustion: Challenges of RANS Modelling

Lorenzo Sforza, Suliman Abdelwahid, Tommaso Lucchini and Angelo Onorati *

ICE Group—Energy Department, Politecnico di Milano, Via R. Lambruschini 4a, 20156 Milano, Italy
* Correspondence: angelo.onorati@polimi.it

Abstract: The main challenge of improving spark ignition (SI) engines to achieve ever increasing thermal efficiencies and near-zero pollutant emissions today concerns developing turbulent combustion under homogeneous ultra-lean premixed mixtures (HULP). This continuous shift of the lean operation limit entails questions on the applicability limits of the combustion models used to date for SI engine design and optimization. In this work, an assessment of flamelet-based models, widely used in RANS SI engines simulations of premixed turbulent combustion, is carried out using an open-source 3D-CFD platform to clarify the applicability limits on HULP mixtures. Two different consolidated approaches are selected: the Coherent Flame Model (CFM) and the Flame Area Model (FAM). Both methodologies are embedded by the authors into the same numerical structure and compared against measurements over a simplified and controlled flame configuration, which is representative of engine-like conditions. The experimental steady-state flame of type "A" of the Darmstadt Turbulent Stratified Flame (TSF) burner is selected for the assessment. This configuration is characterized by flame measurements over a strong shear and mixing layer between the central high-speed CH_4-air jet and the surrounding slow air co-flow, hence, it represents an interesting controlled condition to study turbulent HULP mixtures. A comparison between computed results and experimental data on trends of mean flow velocity, turbulence, temperature and mixture stratification was carried out. This enabled us to assess that the investigated flamelet-based combustion models failed in providing accurate and reliable results when the flame approaches turbulent HULP mixture conditions, demonstrating the urgency to develop models able to fill this gap.

Keywords: ultra-lean combustion; premixed turbulent combustion; flamelet; CFD; RANS; CFM; FAM

Citation: Sforza, L.; Abdelwahid, S.; Lucchini, T.; Onorati, A. Ultra-Lean Premixed Turbulent Combustion: Challenges of RANS Modelling. *Energies* **2022**, *15*, 5947. https://doi.org/10.3390/en15165947

Academic Editors: Maria Cristina Cameretti and Roberta De Robbio

Received: 26 July 2022
Accepted: 13 August 2022
Published: 17 August 2022

1. Introduction

The rapid evolution of the transport sector towards complete environmental sustainability represents one of the major challenges that need to be overcome in the next decades. The goal of achieving a climate neutral world by mid-century is laid out in the Paris Agreement (PA), with the transport sector playing a leading role. In fact, several countries have already deployed dedicated packages of proposals to reduce greenhouse gas (GHG) emissions from the transport sector [1–4].

The spark-ignition (SI) engine represents one of the most interesting and feasible solutions in powertrain technologies to meet such ambitious targets. Currently, SI engines are under continuous improvement to achieve ever increasing thermal efficiencies as well as near-zero pollutant emissions [5–7]. In this framework, the major challenge is to operate the engine with homogeneous ultra-lean premixed (HULP) mixtures, which have the potential to minimize thermal losses, avoid soot formation and cut NO_x emissions, paving the way for a further reduction in size and costs of the after-treatment system [8,9]. Recently, some innovative solutions, such as the prechamber turbulent jet ignition (TJI), have shown reduced combustion duration and increased lean operation limits when applied to HULP mixtures [10–12].

This continuous shift in the lean operation limit towards extremely high relative air/fuel ratios (e.g., $\lambda > 2$ for methane–air mixtures [13]) entails questions on the applicability limits of the numerical models used for SI engine design and optimization. In this context, 3D-CFD RANS approaches are extensively used, thanks to their compromise between limited computational efforts and accuracy of results [14,15]. The modeling of the turbulent premixed combustion usually relies on the *flamelet* assumption [16–21]. This hypothesis provides a great advantage from the modelling point of view because the reacting problem can be fully separated from the flame–turbulence interaction prediction. Concerning this last topic, several strategies are commonly applied for SI engines simulations: the Coherent Flame Model (CFM) proposed by Marble [16] and then developed by several authors [18,22], Weller's Flame Area Model (FAM) [19,23] and the G-equation model from Peters [20,24].

However, a suitable assessment on HULP mixtures from the above flamelet-based approaches need to be carried out in simpler experimental configurations than SI engines. This is essential to detach the turbulent flame phenomenon from transient effects and/or external influences (e.g., the wall temperature effect) characterizing an internal combustion engine. Burners designed to stabilize premixed turbulent steady-state flames are suited for this goal, and from the literature, we know that several configurations are available. The turbulent stratified flame (TSF) burner of TU Darmstadt, characterised by three concentric pipes operating at ambient conditions, allows us to study several premixed flame types with unlimited optical access (open flame) and under controlled conditions [25]. The Cambridge stratified swirl burner (SwB) provides a more complex configuration. Its open-flame design, characterized by a central bluff-body together with a variable swirling flow, tries to emulate the flow complexity and flame stabilization methods seen in practical combustion systems [26,27].

In this work, an assessment of flamelet-based models used in RANS 3D-CFD simulations for premixed turbulent combustion is carried out to clarify their applicability limits for HULP mixtures. Two different consolidated approaches are selected: the Coherent Flame Model (CFM) and the Flame Area Model (FAM). To ensure a consistent comparison, both models are embedded into the same numerical structure used for SI engine simulations and implemented by the authors in the LibICE code. LibICE is a set of solvers and libraries based on the OpenFOAM® technology and recently applied for the simulation of the spark-ignition combustion process in gasoline and natural gas engines [28,29].

The model assessment is carried out on the steady-state open flame type "A" operated on the TSF burner (TSF-A) [25]. Despite the TSF-A configuration having been designed for flames interacting with a limited fuel stratification without flow shear, the actual experimental flame undergoes a relative air–fuel ratio varying from $\lambda \approx 1.1$ to infinite (pure air). This gradual stratification is also characterized by a significant turbulence intensity level u' induced by the interaction between the central high-speed premixed CH_4-air flow ($\approx$10 m/s) with the surrounding slow air co-flow ($\approx$0.1 m/s). Therefore, through a numerical–experimental comparison in terms of mean flow velocity, turbulence, temperature and mixture stratification, the TSF-A configuration lets us clarify the limit of any RANS flamelet-based approach in predicting the existence of a flame front with respect to local levels of λ and u'.

2. Approaches for Premixed Turbulent Combustion Modelling

2.1. The Coherent Flame Model

According to the available literature [18,22], the general architecture of the Coherent Flame Model (CFM) is characterised by a normalised progress variable c equation [22]:

$$\frac{\partial \rho \tilde{c}}{\partial t} + \nabla \cdot (\rho \tilde{\mathbf{U}} \tilde{c}) - \nabla \cdot (\mu_t \nabla \tilde{c}) = \rho_u \tilde{S}_u \Sigma \tag{1}$$

which simulates the flame propagation process, coupled with the following transport equation for Σ [18,22]:

$$\frac{\partial \rho \tilde{\Sigma}}{\partial t} + \nabla \cdot (\rho \tilde{\mathbf{U}} \tilde{\Sigma}) - \nabla \cdot \left(\mu_t \nabla \left[\frac{(\rho \Sigma)}{\rho} \right] \right) + (\nabla \cdot \tilde{\mathbf{U}}) \rho \Sigma = S \rho \Sigma - D \rho \Sigma^2 \tag{2}$$

to model the flame–turbulence interaction.

In this work, the production S and destruction D terms of Equation (2) are modelled according to a general approach. First, the Choi–Huh strategy [22] of different closures for S and D between laminar and turbulent stages is selected. The applied formulations are:

- Laminar stage

$$\begin{cases} S_{lam} = 2\, \rho_u / \rho_b \, S_u / r_k \\ D_{lam} = 0 \end{cases} \tag{3}$$

- Turbulent stage

$$\begin{cases} S_{turb} = \Gamma_{\overline{K}} \, u' / l_{tc} \\ D_{turb} = \left(S_u + C\, k^{1/2} \right) / [b(1-b)] \end{cases} \tag{4}$$

For the laminar stage, the expression proposed in [22] is employed, with the early flame kernel radius r_k computed according to the 0-D model used in [29]. For fully turbulent conditions, D_{turb} formula comes from the CFM-2b model version reported by Duclos [18], while the production term S_{turb} is achieved by introducing the ITNFS model [18,30] on the Choi–Huh CFM-2 source term [22]. Then, Equations (3) and (4) are blended into a single general expression to model the laminar-to-turbulent flame transition:

$$\begin{cases} S = \alpha \left[(1-f)\, S_{lam} + f\, S_{turb} \right] + A_{ik}\, \partial \tilde{U}_k / \partial x_i \\ D = \beta \left[(1-f)\, D_{lam} + f\, D_{turb} \right] \end{cases} \tag{5}$$

where α and β are model constants that need a suitable calibration.

The averaging parameter $0 \leq f \leq 1$ is used to model the flame evolution from laminar ($f = 0$) to turbulent ($f \to 1$) features, and it is computed according to [29]. The term $A_{ik}\, \partial \tilde{U}_k / \partial x_i$ is introduced to consider the Σ production from the flame stretch caused by mean flow gradients [18,31]. In case of isotropic turbulence, for the sake of simplicity, the same constant value can be assumed for each A_{ik} term after a proper calibration.

Simplified ignition approaches [29], as well as more detailed methodologies [32–34], can be included into the numerical model structure implemented by the authors. However, the ignition model details are not discussed here because the present work is focused on steady-state flames.

The correlation proposed by Gulder [35] is selected for the unstrained laminar flame speed S_u estimation. Even though this semi-empirical correlation is restricted by the validity limits of pressure, temperature and equivalence ratio, its application to open flames (ambient conditions) in the context of HULP mixtures could artificially extend the lean operation limit. Figure 1 shows a comparison between Gulder's correlation and the tabulation approach based on a GRI-Mech 3.0 kinetic mechanism [36,37] for the laminar flame speed estimation at ambient conditions of CH_4-air mixtures. As can be observed, close to the flammability limit ($\phi < 0.4$), the correlation proposed by Gulder overestimates the S_u value with respect to tabulation strategy, yielding an artificial extension of the lean operation limit.

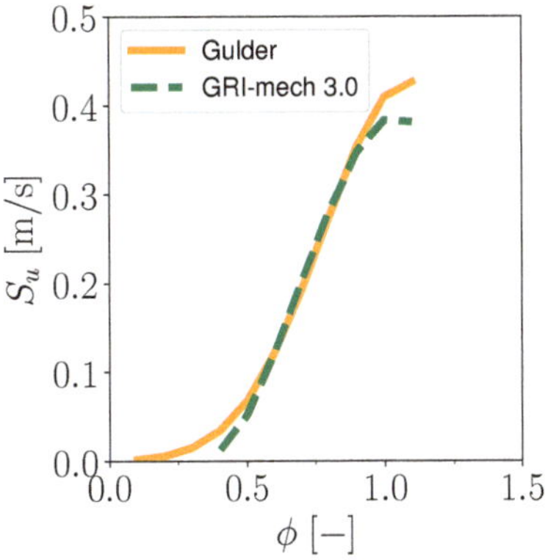

Figure 1. Estimation of the laminar flame speed S_u value as a function of the fresh CH_4-air mixture equivalence ratio ϕ at ambient conditions (p = 1.01325 bar; T = 300 K): comparison between Gulder's correlation and a tabulation approach with the GRI-mech 3.0 mechanism.

The tabulated kinetics approach proposed in [36] is chosen to compute the chemical composition of the reacting fuel–air mixture. Similar to [29], only the chemical composition of burned and unburned gases is retrieved from the lookup table at equilibrium conditions because the reaction rate from premixed combustion comes from Equation (1). Then, the actual composition of the reacting mixture is achieved by weighting the previous two compositions on the normalized progress variable c. The final structure of the implemented model is represented in Figure 2, where the red dashed box is disabled when the CFM is used.

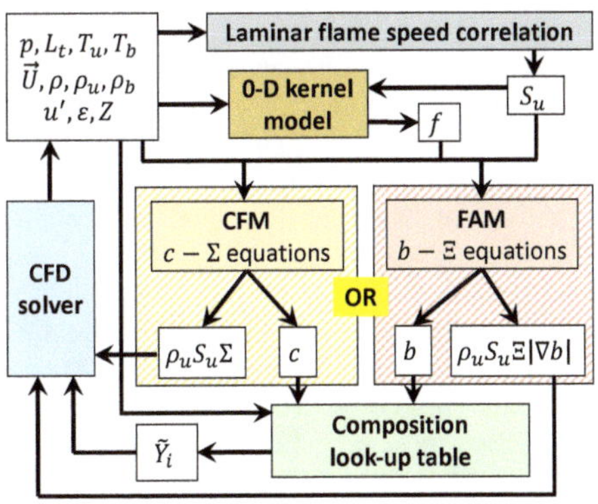

Figure 2. Schematic of the CFM and FAM combustion models embedded into the same numerical structure.

2.2. The Flame Area Model

The Flame Area Model proposed by Weller [19,23] is based on the resolution of a regress variable b equation [19,23]:

$$\frac{\partial \rho \tilde{b}}{\partial t} + \nabla \cdot (\rho \tilde{\mathbf{U}} \tilde{b}) - \nabla \cdot (\mu_t \nabla \tilde{b}) = \rho_u \tilde{S}_u \tilde{\Xi} |\nabla \tilde{b}| + \dot{\omega}_{ign} \tag{6}$$

while the flame–turbulence interaction is modelled by solving a transport equation for the flame wrinkle parameter Ξ [19]:

$$\frac{\partial \rho \tilde{\Xi}}{\partial t} + \rho \widehat{\mathbf{U}}_s \cdot \nabla \tilde{\Xi} = \rho G \Xi - \rho R (\Xi - 1) \tag{7}$$

In Equation (7), coefficients G and R are estimated as [19]

$$G = R\,\frac{\Xi_a^* - 1}{\Xi_a^*} \quad \text{and} \quad R = \frac{0.28}{\tau_\eta}\frac{\Xi_a^*}{\Xi_a^* - 1} \tag{8}$$

in which Ξ_a^* is modelled according to [29]

$$\Xi_a^* = 1 + f\left(\Xi_{eq}^* - 1\right) \tag{9}$$

to mimic the transition from laminar ($\Xi_a^* = 1$) to fully turbulent conditions ($\Xi_a^* = \Xi_{eq}^*$) through the averaging parameter f. More details about f computation can be found in [29]. In this work, Ξ_{eq}^* is modelled according to Gulder [38]:

$$\Xi_{eq}^* = 1 + \Xi_{coef}\sqrt{\frac{u'}{S_u}\mathcal{R}_\eta} \tag{10}$$

where $\mathcal{R}_\eta$ is the Kolmogorov Reynolds number. The interpolation of many sets of experimental data suggested that Ξ_{coef} should be 0.62 [38], but fine-tuning of this parameter is allowed.

Regarding the prediction of both the laminar flame speed and the mixture chemical composition, the same methodologies described in Section 2.1 were used. The model schematic is shown by Figure 2, where the yellow dashed box is disabled when the FAM is used.

3. Experimental Configuration and Numerical Setup

3.1. The Darmstadt Stratified Burner: Features and Investigated Cases

The experimental rig of the Darmstadt TSF burner is characterised by three staged co-flow concentric tubes [25]. The central tube is the pilot from which hot burned products are released to ignite the surrounding fresh mixture. The pilot wall is made of ceramic material, but it is not perfectly insulated [39]. However, as a first step in this work, it is assumed adiabatic, leaving the study of the heat losses to a future analysis. The other two annular tubes (called slots 1 and 2) supply a fresh and turbulent fuel–air mixture. The burner is placed inside a 600 mm-wide air co-flow to shield the flame from the surrounding environment. Further details about the applied experimental measurement techniques can be found in [25].

Between the available flame configurations [25], the CH_4-air flame of type A is selected. In particular, the isothermal flow case A-i2 and the reactive case A-r are investigated in this work; their main features are summarised in Table 1. The isothermal non-reacting case A-i2 is used to evaluate the capability of the chosen numerical methodology in predicting the flow field features where the combustion will occur, assessing the mesh quality performance. Then, the reacting case A-r is selected to clarify the applicability limits of the selected flamelet-based approaches (CFM and FAM) on HULP mixtures. In fact, the interaction between the high-speed lean mixture from slot 2 (10 m/s, $\phi = 0.6$) and the external slow air co-flow (0.1 m/s) generates a region where the combined effects of shear and mixture stratification can be used to clarify the limit of flamelet-based approaches towards turbulent HULP mixtures.

Table 1. Flow parameters of the two investigated conditions.

Case	ϕ_{pilot}	ϕ_{slot1}	ϕ_{slot2}	u_{pilot}	u_{slot1}	u_{slot2}	Re_{pilot}	Re_{slot1}	Re_{slot2}	u_{coflow}
A-i2	0	0	0	10	10	10	9800	13,800	13,300	0.1
A-r	0.9	0.9	0.6	1	10	10	980	13,800	13,300	0.1

Equivalence ratios ϕ (–); calculated bulk velocities from unburned gas properties u (m/s); Reynolds numbers Re (–) (based on the bulk velocity and hydraulic diameter).

3.2. Numerical Setup

The flamelet-based approaches investigated in this study are implemented by the authors in the LibICE code and embedded into the same numerical structure used for SI engine simulations. LibICE is a set of solvers and libraries based on the OpenFOAM® technology, which is an open-source platform involving a 3-D finite volume discretization. The conservation equations of mass, momentum, chemical species and energy are solved with the RANS approach using the standard k-ϵ model for turbulence.

Concerning the adopted computational domain, a 3-D axisymmetric analysis of the Darmstadt TSF burner is considered. This choice is motivated by the symmetrical features of the burner geometry, which is characterized by three concentric tubes. Therefore, a 3-D wedge domain of 5 degrees is considered, as depicted in Figure 3. The flow effects on the third dimension (z-axis, namely the tangential direction) are therefore considered, thanks to a suitable boundary condition applied on the wedge sides.

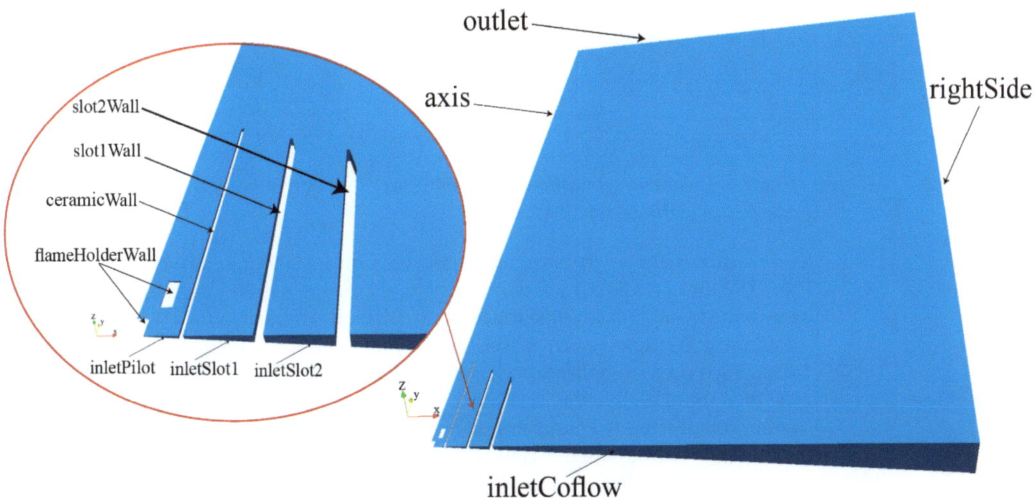

Figure 3. 3-D axisymmetric wedge domain for the numerical investigation of the Darmstadt TSF burner.

The jet injection direction (pipes axis) is defined as y coordinate, whereas the radial direction is defined as r (or x).Because experimental measurements were performed until 200 mm downstream from the pilot exit section [25], whose axial position is defined as y = 0 mm (see Figure 4), the computational domain was extended until y = 300 mm to avoid possible numerical effects of the outlet boundary on the region of interest. In the radial direction, the domain was extended up to 250 mm (Figure 3, the right side), while to simulate the effect of the flame holder, a mesh was generated up to 50 mm upstream from the pilot exit section.

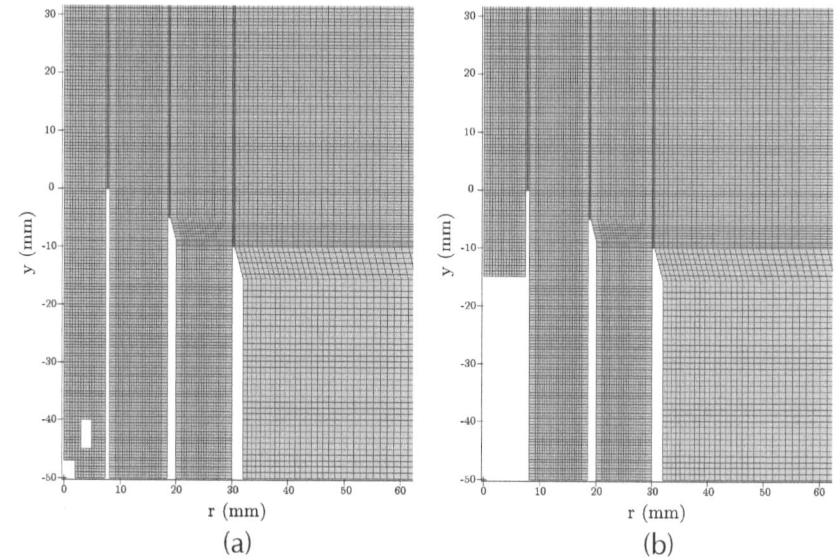

Figure 4. The computational mesh in the vicinity of the burner exit region for: (**a**) the non−reacting case A-i2; and (**b**) the reacting case A-r.

Figure 4 shows the structured grid of cubic hexahedra employed for the numerical investigation. An average cell dimension of 0.5 mm is used inside the burner and where the flame is stabilized, while the mesh is gradually coarsened until 1 mm close to the side boundary. At pipes walls, the cell side is calibrated to achieve y+ values greater than 30, allowing a correct application of wall functions for the k-ϵ turbulence model. To optimize the total mesh dimension in terms of cells number, a sensitivity analysis is carried out on both non-reacting A-i2 and reacting A-r cases. In light of this investigation, available in Appendix A, a mesh of 80k cells is selected for all numerical simulations.

Similar to [40], the flame holder geometry is not included in reacting case A-r simulations; hence, for the sake of simplicity, only 15 mm of the domain upstream from the pilot exit section are considered, as shown in Figure 4b.

Concerning boundaries, no-slip adiabatic conditions are used at walls, whereas slip conditions are employed for the lateral confinement of the air co-flow. A fully-turbulent flow velocity profile is imposed at slot 1 and 2 inlet because the considered computational domain is shorter than the actual pipe length. A uniform velocity profile of 0.1 m/s is applied, instead, at the inlet of the coflow. For the reacting case A-r only, a fully burned mixture (c = 1) and a uniform bulk flow velocity of about 10 m/s are imposed at the pilot inlet because no experimental measurements are available inside the pilot pipe.

The values of the constants belonging to the investigated flamelet-based models are specified in Table 2, where all parameters are chosen after a suitable fine calibration. Concerning the prediction of chemical composition, the lookup table used for this purpose is generated by using the GRI-Mech 3.0 kinetic model [37].

Table 2. Model constants used in this work.

Constant	CFM	FAM
α	2.0	-
l_{tc}	1.26×10^{-3}	-
A_{ik}	1.5	-
β	1.2	-
C	0.7	-
Ξ_{coef}	-	0.1

CFM: CFM model (Section 2.1); FAM: FAM model (Section 2.2). For the sake of clarity, α and β coefficients belong to Equation (5).

4. Results and Discussions

4.1. Non-Reacting Case A-i2

The non-reacting case A-i2 is basically used to assess the numerical setup and the mesh quality before advancing to the reactive cases. No combustion modelling is required because of the nature of the case (pure air flows out from all pipes, see Table 1). The numerical–experimental comparison is carried out only at three different axial distances from the pilot exit section because measurements on condition A-i2 are available only at y = 1, 50 and 100 mm, as shown by Figure 5.

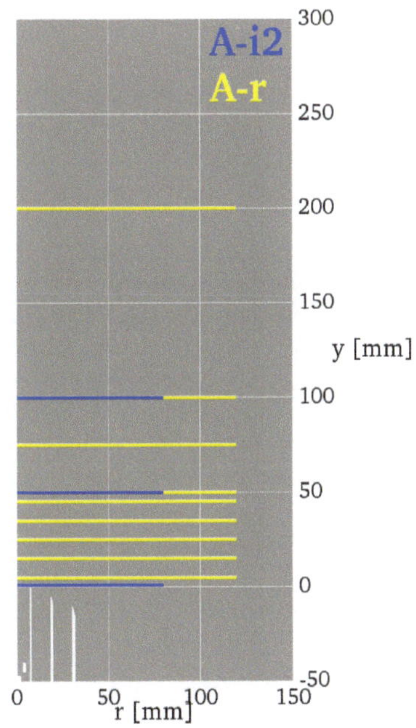

Figure 5. Axial locations where experimental measurements of conditions A-i2 (blue) and A-r (yellow) of Table 1 are available.

In Figure 6, the computed axial U_y and radial U_r velocity distributions, as well as the predicted turbulence intensity u' trend, are compared with the experimental measurements. At the lowest axial position (y = 1 mm), namely close to the burner exit, the trend shows three hump-shaped axial velocity profiles because of the wake regions generated above the concentric pipe walls by the three high-velocity outflow jets. Further downstream, the velocity humps smear out due to the flow mixing. The numerical results of axial velocity show a satisfactory agreement with experimental measurements despite a slight under-prediction at y = 1 mm in the wake of the flame holder. In addition, the agreement between computed and measured radial velocity distributions is rather satisfactory, in particular at y = 50 mm and y = 100 mm.

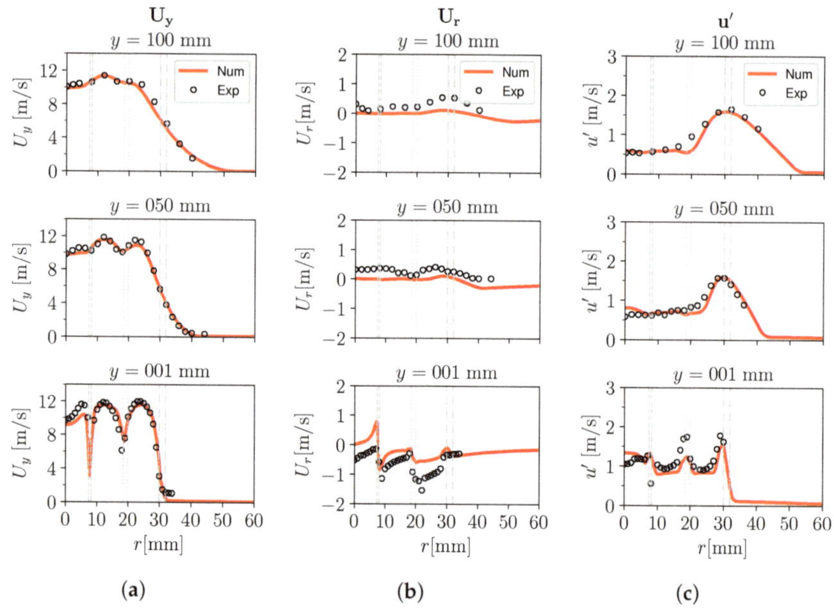

Figure 6. Case A-i2: radial distributions of mean axial velocity U_y (**a**); mean radial velocity U_r (**b**); and mean turbulence intensity u' (**c**); computed at various axial positions y in comparison with experimental data. Hereafter, following the increase of r, the vertical grey dashed, dotted and dashed-dotted lines correspond to the radial positions of the inner and outer walls of the pilot, slot 1 and slot 2, respectively.

Concerning turbulence intensity, as expected three peaks are found at y = 1 mm due to the presence of mixing shear layers in the wake regions above the concentric pipes walls. Further downstream from the burner, the flow seems mixed; hence only one peak appears due to the velocity gradient between the high-velocity jet flows and the slower co-flow. As can be also observed, the k-ϵ turbulence model performs well, especially far from the burner exit (y = 50 mm and y = 100 mm) where the flow velocity gradients are smoother (compare to left and central columns of Figure 6), producing anyway a reasonable prediction of the complex flow field at y = 1 mm. The good agreement between the numerical results and experimental measurements on the non-reacting case A-i2 indicates that the numerical setup in terms of mesh quality, way of imposing the boundary conditions and turbulence model are suitable to numerically study the TSF burner and advance to the combustion case.

4.2. Reacting Case A-r

4.2.1. Flame Regimes

With the purpose to verify the applicability limits of the flamelet concept, an assessment on the combustion regimes experienced by the TSF burner flame at various positions after the pilot exit section was performed on the investigated condition A-r. However, as confirmed by an a priori analysis based on Figure 6, it is expected that at high axial positions y, the flame will interact with a broad shear layer generated by the surrounding slow co-flow of air, hence with a region characterized also by a strong mixture stratification.Therefore, to provide a deeper understanding of this phenomenon, at each investigated axial distance from pilot the flame regime was evaluated at several positions inside the flame brush, especially at low c values where there is an high probability of detecting fresh, as well as lean, mixture conditions.

Figure 7 shows a comparison on the the Borghi–Peters diagram [20,41] between the combustion regimes achieved with CFM and FAM combustion models, represented in red and blue colours, respectively, at all axial locations where measurements of reacting scalars are available. Three different c values were selected for the investigation. The first is $c = 0.5$ (Figure 7 bottom row), where the probability to find fresh or burnt mixture is the same. The other two are $c = 0.05$ and 0.01 (Figure 7 central and top row) because the regions are very close to the "cold-tail" of the flame brush. Axial positions up to $y = 45$ mm are shown in Figure 7a column, while the furthest locations are represented in Figure 7b column. A graphical representation of the investigated flame brushes and the related c iso-values is provided by Figure 8.

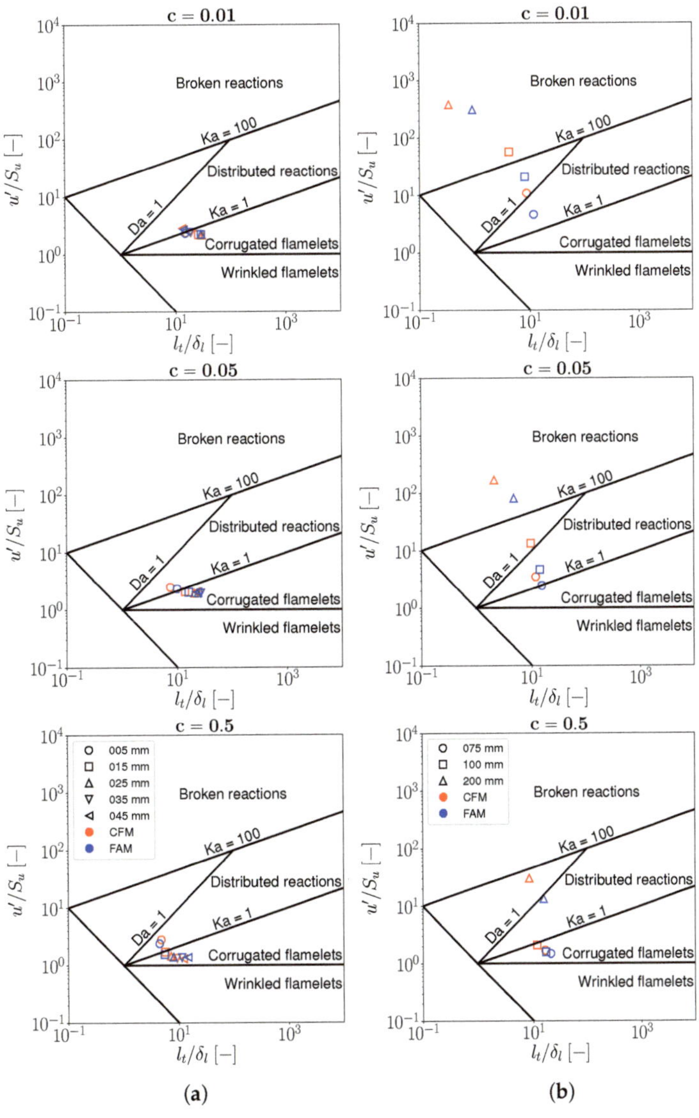

Figure 7. Case A-r: numerical flame regimes achieved with CFM (red) and FAM (blue) at different axial positions y, evaluated at three values of normalized progress variable: $c = 0.5$, $c = 0.05$ and $c = 0.01$, from bottom to top line. Column (**a**): 5 mm $\leq y \leq$ 45 mm; column (**b**): 75 mm $\leq y \leq$ 200 mm.

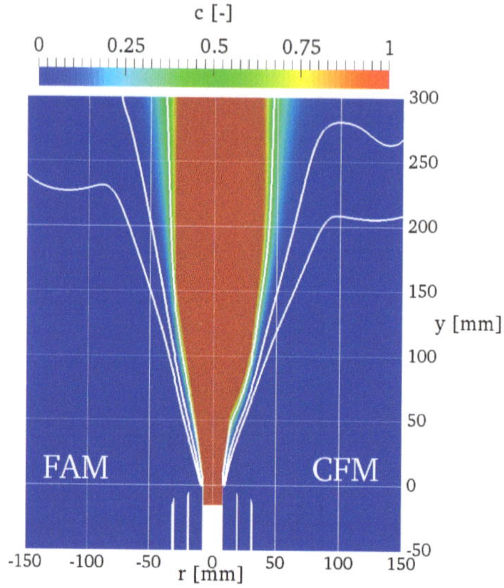

Figure 8. Case A-r: comparison between normalized progress variable c fields achieved with FAM (left half) and CFM (right half) models, including iso−lines (solid white) of $c = 0.5$, $c = 0.05$ and $c = 0.01$ (from burner axis towards increasing radii, respectively).

As can be observed in Figure 7a column, at all axial positions closer to the pilot exit section a *corrugated flamelet* can be detected through the whole flame brush, with a slight shift towards the *distributed reaction* region as the c value decreases. A small exception is represented by the $y = 5$ mm position, where the trend seems opposite; however, its flame regimes at all c values are very close to a unitary Karlovitz number. Therefore, up to $y = 45$ mm, the flamelet assumption can be considered valid over the whole flame brush for both adopted combustion models.

Different conclusions can be drawn for the furthest axial locations investigated by Figure 7b column. As can be seen, at each y position the flame regime tends to move from the *corrugated* flamelet towards the *broken reaction* condition as the "cold-tail" of the flame brush is approached (from bottom to top row). This behaviour is observed with both combustion models, with the CFM always more shifted towards the broken reaction condition. A sensitivity with respect to y value can be also noticed. In particular, the $y = 200$ mm position represents the worst case in terms of flamelet assumption validity. Here, both CFM and FAM models predict a *broken reaction* combustion regime over at least half of the flame brush ($0.01 \leq c \leq 0.5$). Therefore, from $y = 75$ mm onward, the flamelet assumption cannot be considered valid over the whole flame brush for both adopted combustion models. This limit affects higher and higher values of c as much as the axial distance from the pilot increases.

4.2.2. Numerical–Experimental Comparison

A comparison between the available experimental measurements and the numerical results achieved with both CFM and FAM models was carried out to clarify the impact of what we observed in Section 4.2.1 about flame regimes.

Figure 9 compares numerical and experimental radial trends of mean temperature T ((a) column) and mixture fraction Z ((b) column) for axial positions up to $y = 45$ mm. Here, the validity of the flamelet assumption for numerical results was verified (Figure 7a column), and a graphical representation of computed T and Z fields, including the c iso-values used for the flame regime investigation, is provided by Figure 10. As can be observed from Figure 9a column, all numerical approaches seem to provide a rather good estimation of the flame brush position and evolution in terms of T distribution. This is true despite two issues: an overestimation of the temperature above the pilot pipe (r < 7 mm), especially close to the exit section (y = 5 mm), and the tendency to slightly underestimate the flame position as the y value increases (y > 25 mm). Both aspects can be explained by the adiabatic condition applied to the pilot wall, which is not fully consistent with the measured heat losses [39,40,42]. In fact, because the fine-tuning of numerical models coefficients was performed close to the pilot wall (y = 5 mm), a slight underestimation of the flame speed has to be expected where heat loss effects become negligible (y > 25 mm).

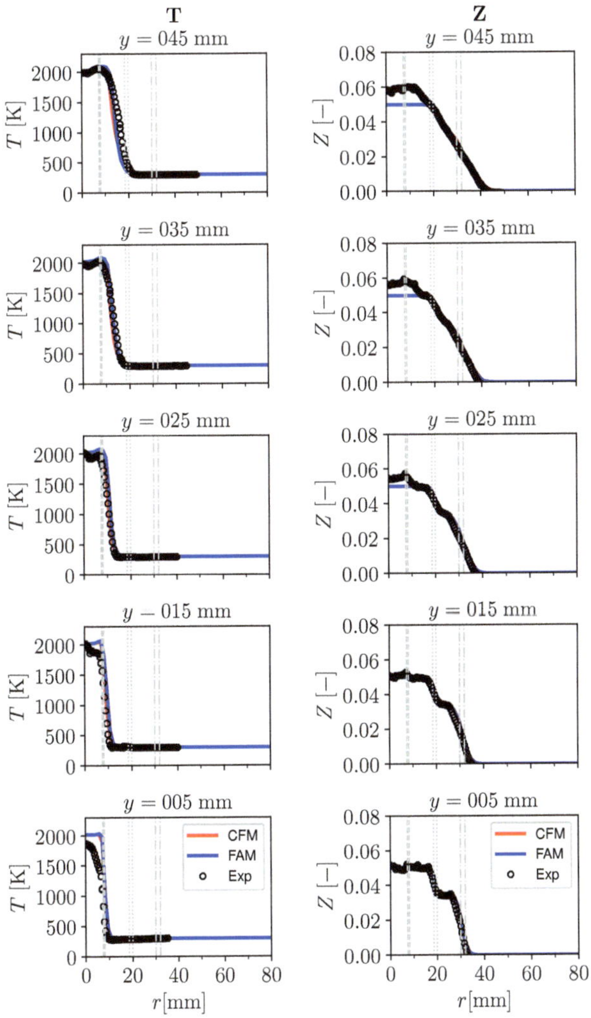

(a) (b)

Figure 9. Case A-r: radial profiles of mean temperature T (**a**); and mean mixture fraction Z (**b**); computed with different models at axial positions 5 mm $\leq y \leq$ 45 mm in comparison with experimental data.

The satisfactory behaviour of both CFM and FAM models is also confirmed by the radial evolution of Z (Figure 9b column), which is almost perfectly captured. Nevertheless, it is worth mentioning that nonphysical values of Z were measured close to the burner axis from y = 25 mm and above, causing an overestimation of its real value. This can be attributed to improper background corrections at low number densities of CO and CH_4 [42].

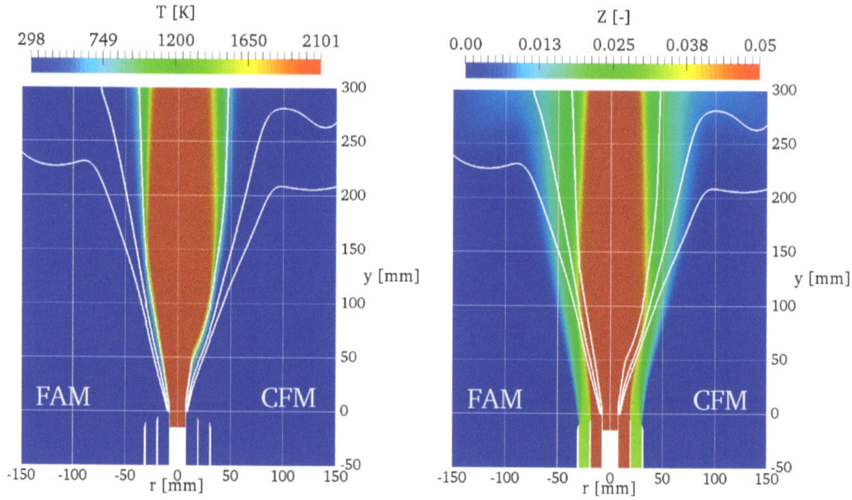

(**a**) (**b**)

Figure 10. Case A-r: comparison between mean temperature T (**a**); and mean mixture fraction Z (**b**); fields achieved with FAM and CFM models, including iso-lines (solid white) of $c = 0.5$, $c = 0.05$ and $c = 0.01$ (from burner axis towards increasing radii, respectively).

Focusing, instead, on the farthest region from the pilot exit section, different observations can be made.Figure 11 compares numerical and experimental radial trends of mean temperature T ((a) column) and mixture fraction Z ((b) column) for axial positions $y \geq 75$ mm. Here, the flamelet assumption cannot be considered valid over the whole flame brush for both CFM and FAM models (Figure 7b column). A graphical representation of computed T and Z fields is still provided by Figure 10. By looking at Figure 11a column, a significant difference between computed and measured results in terms of T distribution can be observed. All numerical flame brushes are narrower than the experimental one, with a difference that increases with the y value. This phenomenon can be explained by the limitation of the flamelet assumption on which both CFM and FAM approaches rely when approaching turbulent HULP mixtures (increasingly low Z values and $u' > 0.5$ m/s). In particular, because this limitation involves higher c values as the axial distance from the pilot increases, a growing underestimation of the measured T close to the "cold-tail" of the flame brush is observed as y increases (Figure 11a column). This misprediction of the combustion process also affects the computed Z distribution. In fact, as shown by Figure 11b column, increasing under-predictions of the experimental findings are observed, especially at high radial values.

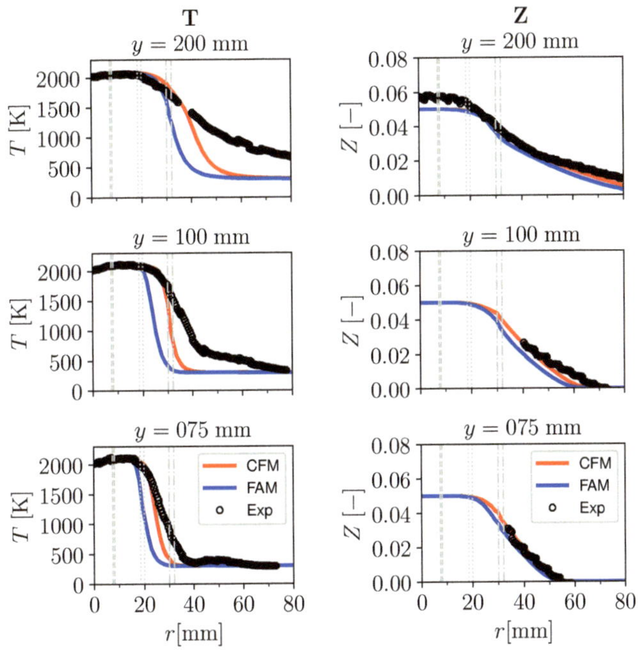

(a) (b)

Figure 11. Case A-r: radial profiles of mean temperature T (**a**); and mean mixture fraction Z (**b**); computed with different models at axial positions 75 mm $\leq y \leq$ 200 mm in comparison with experimental data.

To further support the thesis that limitations of the flamelet assumption cause significant discrepancies between numerical results and experimental measurements when the flame approaches turbulent HULP mixtures (Figure 11), a possible misalignment between the mixing and reaction layers can be neglected, as demonstrated by a previous study on the TSF A-r condition [43].

The effects produced by the limitation of the flamelet assumption on both CFM and FAM models, observed when $y \geq 75$ mm, also impact the flow velocity and turbulence prediction. Figure 12 compares numerical and experimental radial trends of mean axial U_y ((a) column) and radial U_r ((b) column) velocities, as well as the turbulence intensity u' ((c) column), at three different axial positions: $y = 50$, 75 and 200 mm (from bottom to top row). As can be observed, a satisfactory numerical–experimental agreement is achieved when the combustion phenomenon is properly captured (Figure 12, bottom row). In fact, near $y = 50$ mm the flamelet assumption is still valid over the whole flame brush (Figure 7a column). Nevertheless, the adiabatic assumption imposed at the pilot wall still has some consequences inside the flame core at this distance. The first is the U_y over-prediction close to the burner axis (Figure 12a column, bottom row). The second is the underestimation of the U_r imposed by the flame to the mixture above slot 2 (Figure 12b column, bottom row, 20 mm < r < 30 mm). Concerning the furthest axial locations $y = 75$ and 200 mm, from the central and top rows of Figure 12 significant discrepancies are observed between numerical results and experimental measurements for all investigated magnitudes. This is consistent with the observed limitation of the flamelet assumption on both CFM and FAM models in these regions.

A final observation can be made by comparing CFM and FAM model behaviour. By looking at Figures 11 and 12, it can noticed how the flame predicted by the FAM approach seems more limited by the validity range of the flamelet assumption than the CFM strategy.

In fact, if the flame surface density model is able to predict a wider and more shifted flame brush towards the air co-flow, this seems not true for the flame area approach.

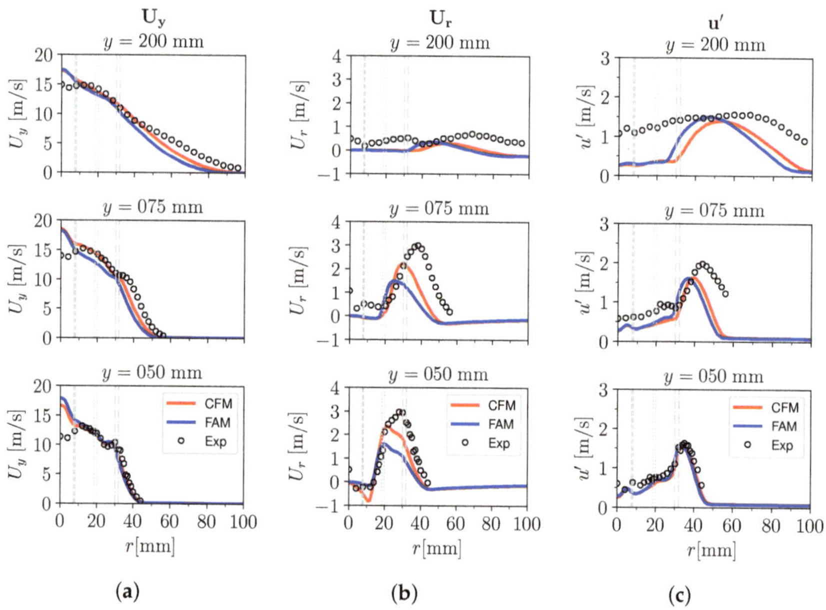

Figure 12. Case A-r: radial profiles of mean axial velocity U_y (**a**); mean radial velocity U_r (**b**); and mean turbulence intensity u' (**c**); computed with different models at axial positions y = 50, 75 and 200 mm in comparison with experimental data.

4.2.3. The Combustion Effects: Case A-i2 vs. Case A-r

Despite the limits observed when approaching turbulent HULP mixtures, this section provides a deeper insight into the capabilities of the selected modelling approaches (CFM and FAM) in predicting the combustion effects on the flow field of the TSF burner. A comparison between the isothermal (A-i2) and the reacting (A-r) configurations of condition A is carried out.

Figure 13 shows the effects of the combustion event on the radial profiles of axial velocity U_y ((a) column), radial velocity U_r ((b) column) and turbulence intensity u' ((c) column). Here, the experimental findings are plotted, together with all achieved numerical results, to assess the models capabilities in predicting the measured trends. At both available axial positions, all numerical approaches seem able to estimate the combustion effects on the flow field of case A, such as the wider radial distribution of U_y ((a) column), the radial acceleration of the fresh mixture ((b) column) and the shift towards higher radii of the shear layer with the air co-flow ((c) column). Analyzing each axial position, at y = 50 mm the results are rather nice, and small discrepancies can be observed. As already discussed, they are caused by the adiabatic assumption adopted in this study on the pilot wall (see Section 4.2.2). For what concerns the other axial distance (y = 100 mm), the investigated combustion models fail in providing accurate and reliable results. Only a rough estimation of the reacting process is achieved due to the limitation of the flamelet assumption detected at this position and the related mispredictions on the "cold-tail" of the flame brush (see Sections 4.2.1 and 4.2.2). Therefore, this demonstrates how tricky this limitation of flamelet-based models on turbulent HULP mixtures is,particularly when modelling more complex configurations (e.g., SI engines). In fact, in such cases, the presence of many additional uncertainties affecting the numerical results, as well as the apparently satisfactory numerical trend, make the detection of this source of error almost impossible .

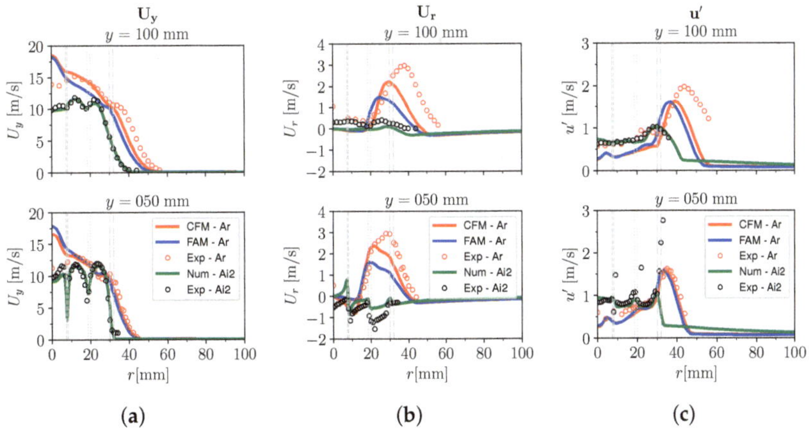

Figure 13. Radial distributions of mean axial velocity U_y (**a**); mean radial velocity U_r (**b**); and mean turbulence intensity u' (**c**); computed at y = 50 and 100 mm from the pilot exit section, in comparison with the experimental data, for both the non-reacting (A-i2) and reacting (A-r) cases of condition A. For case A-i2, the single non-reacting result is included (Figure 6), while for case A-r both CFM and FAM models are considered (Figure 12).

5. Conclusions

An assessment of two different flamelet-based models, the Coherent Flame Model (CFM) and the Flame Area Model (FAM), is carried out to clarify their applicability limits on homogeneous ultra-lean premixed (HULP) mixtures. The experimental steady-state, open flame type "A" of the Darmstadt TSF burner, is selected for the assessment.

In light of the achieved results, it can be stated that both CFM and FAM approaches are able to predict the combustion evolution with reasonable accuracy only when "classical" flamelet configurations are detected. This was observed in the vicinity of the burner exit (y $\leq$ 45 mm) where the flamelet assumption is fully valid because an homogeneous and shear-less CH_4-air mixture, with a $\lambda \approx 1.1$, is burned.

On the other hand, when typical conditions of turbulent HULP mixtures are observed, the investigated models failed to provide accurate and reliable results because the combustion regime is significantly outside the flamelet applicability limits ($Ka > 100$). Only a rough estimation of the reacting process is achieved in such conditions, as observed at the furthest axial locations (y $\geq$ 75 mm). In fact, as y increases, turbulent HULP mixtures conditions ($\lambda \rightarrow \infty$ and $u' = 1.5 \div 2$ m/s) gradually affect the experimental flame features. As a consequence, the flamelet assumption starts to fail. This happens, first, at the "cold-tail" of the flame brush ($c \approx 0.01$), then higher c values are affected as y increases. Accordingly, all numerical predictions of the measured flame position become increasingly worse.

However, if more complex experimental rigs are investigated (e.g., SI engines) instead of simplified flame configurations (e.g., TSF flame A), the aforementioned numerical–experimental discrepancies could be ascribed to other uncertainties. This explains why such numerical models can also be applied to novel SI engine configurations operated with increasingly lean mixtures.

Nevertheless, considering that the current pathway of SI engine design and optimization is characterized by a continuous shift of the lean operation limit towards extremely high relative air/fuel ratios (turbulent HULP mixtures), these flamelet-based models need to be extended or modified to significantly improve 3D-CFD simulations of such extreme combustion modes.

Author Contributions: Conceptualization, L.S. and T.L.; methodology, L.S. and T.L.; software, L.S. and S.A.; validation, S.A. and L.S.; formal analysis, L.S. and S.A.; resources, T.L. and A.O.; data curation, L.S. and S.A.; writing—original draft preparation, L.S. and S.A.; writing—review and editing, L.S., T.L. and A.O.; supervision, T.L.; project administration, A.O. All authors have read and agreed to the published version of the manuscript.

Funding: This research received no external funding.

Institutional Review Board Statement: Not applicable.

Informed Consent Statement: Not applicable.

Data Availability Statement: The original numerical results presented in this study are openly available in OSF at https://osf.io/wndus/, accessed on 9 August 2022.

Acknowledgments: The authors thank A. Dreizler (TU Darmstadt, Germany) for providing the experimental measurements of the Darmstadt TSF burner.

Conflicts of Interest: The authors declare no conflict of interest.

Nomenclature

The following nomenclature is used in this manuscript:

A_{ik}	tensor describing flame orientation factors
b	regress variable (unburned mass fraction)
c	normalised progress variable (burned mass fraction)
C	CFM model constant for destruction term
C_S	ignition strength parameter
d_{ign}	ignition diameter
D	flame surface density destruction term
D_{lam}	flame surface density destruction term for laminar-only regime
D_{turb}	flame surface density destruction term for fully-turbulent regime
f	parameter for laminar-to-turbulent flame transition (laminar: $f = 0$; turbulent: $f \to 1$)
G	coefficient of flame wrinkling generation term
k	turbulent kinetic energy
l_t	integral length scale
l_{tc}	length scale (introduced for dimensional reasons and set according to [22])
l_δ	flame inner layer thickness
$\mathbf{n}$	flame surface normal direction towards fresh mixture
p	pressure
p_{ref}	reference pressure
r_{ign}	ignition radius
r_k	early flame kernel radius
R	coefficient of flame wrinkling reduction term
R_η	Kolmogorov Reynolds number
S	flame surface density production term
S_{kernel}	kernel flame surface imposed by the ignition process
S_{lam}	flame surface density production term for laminar-only regime
S_{turb}	flame surface density production term for fully-turbulent regime
S_u	unstrained laminar flame speed
$S_{u,ref}$	unstrained laminar flame speed at reference conditions (T_{ref},p_{ref})
S_Ξ	coefficient for flame wrinkling distribution across the flame
t	time
T	temperature
T_{ref}	reference temperature
u'	turbulence intensity
$\mathbf{U}$	flow velocity vector
U_i	flow velocity component
$\widehat{\mathbf{U}}_s$	surface-filtered flame velocity
V_{cell}	volume of the cell including the ignition position

W	coefficient for reference unstrained laminar flame speed computation
x_i	Cartesian coordinate
$Y_{b,i}$	chemical species mass fraction in burned mixture
Y_c	not normalised progress variable
Y_i	chemical species mass fraction
$Y_{u,i}$	chemical species mass fraction in unburned mixture
Z	mixture mass fraction
α	coefficient of CFM model production term S; temperature ratio exponent for unstrained laminar flame speed computation; alignment angle between progress variable ∇Y_c (or ∇c) and mixture fraction ∇Z gradients
β	coefficient of CFM model destruction term D; pressure ratio exponent for unstrained laminar flame speed computation
$\Gamma_{\overline{K}}$	ITNFS (Intermittent Turbulent Flame Stretch) model function, according to [18,30]
δ_l	laminar flame thickness
Δt_{ign}	ignition duration
ε	turbulent kinetic energy dissipation rate
η	Kolmogorov length scale; Equivalence ratio exponent for reference unstrained laminar flame speed computation
μ_t	turbulent dynamic viscosity
μ_u	dynamic viscosity of unburned mixture
ξ	exponent for reference unstrained laminar flame speed computation
Ξ	flame wrinkling
Ξ_a	algebraic model of flame wrinkling distribution
Ξ_a^*	algebraic model of flame wrinkling distribution at equilibrium conditions
Ξ_{coef}	calibration coefficient of algebraic flame wrinkling
Ξ_{eq}^*	algebraic model of flame wrinkling distribution at fully turbulent equilibrium conditions
ρ	mixture density
ρ_u	unburned mixture density
Σ	flame surface density
Σ_k	flame surface density of the mesh cell including the ignition position
τ_η	Kolmogorov time scale
ϕ	equivalence ratio
$\dot{\omega}_{ign}$	ignition source term
∂	partial derivative
∇	nabla operator

Appendix A. Mesh Sensitivity Analysis

The optimization of the computational grid is carried out on both non-reacting (A-i2) and reacting (A-r) investigated cases, to minimize the computational costs without grid-biased numerical results. The mean axial U_y and radial U_r velocities, as well as the mixture fraction Z, are selected as reference parameters to understand when numerical results are independent from mesh refinement. Three different mesh dimensions are tested: 20k, 80k and 160k cells.

As it can be evinced from Figures A1 and A2, in which also the non-reacting u' distributions are reported (Figure A1 (c) column), the convergence of all numerical results is achieved at a mesh dimension of 80k cells. Therefore, this mesh type is chosen as the best trade-off between computational time and accuracy of results. A detail of this mesh near burner walls can be observed in Figure 4 and it spans from cubic cells of about 0.5 mm in the most refined regions to cells of 1 mm side far from the burning area, as mentioned in Section 3.2.

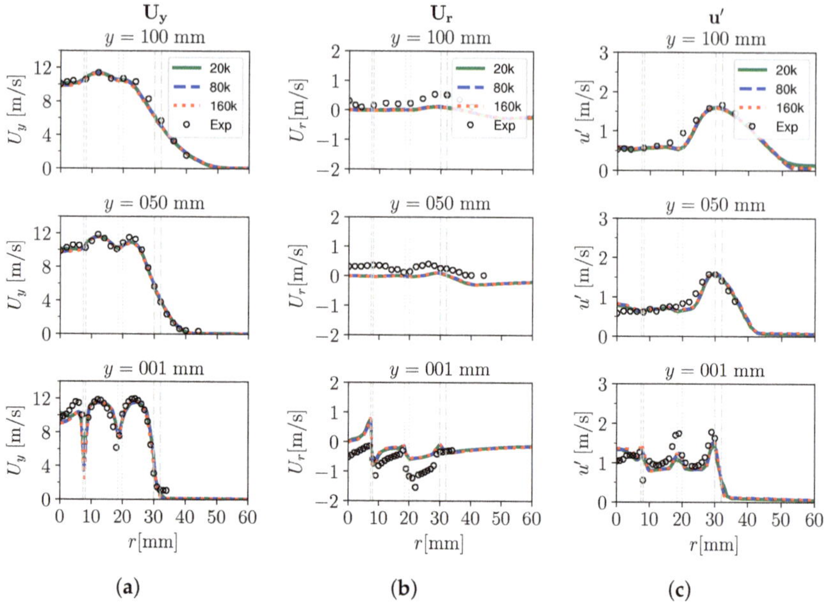

Figure A1. Case A-i2: radial distributions of mean axial velocity U_y (**a**), mean radial velocity U_r (**b**) and mean turbulence intensity u' (**c**) computed at various axial positions (y = 1, 50 and 100 mm) with different mesh resolutions (20k, 80k and 160k cells) in comparison with experimental data.

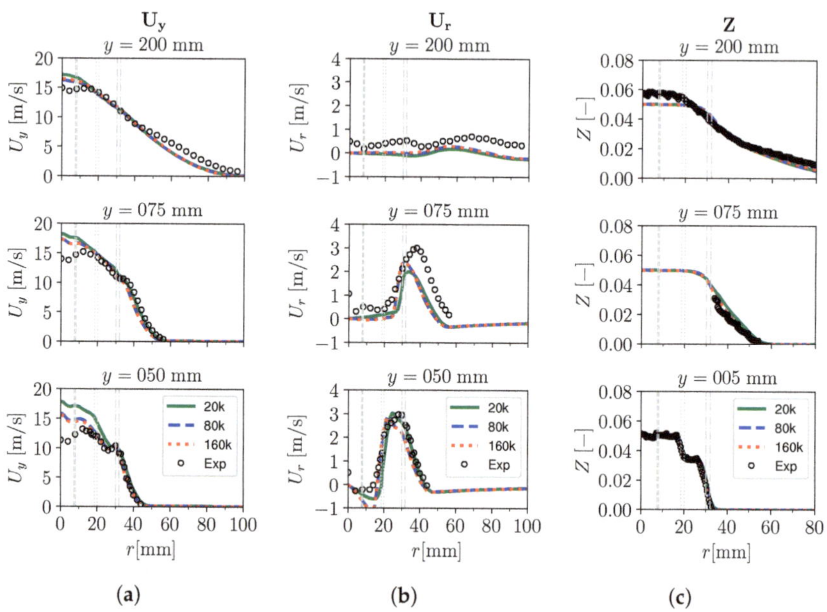

Figure A2. Case A-r: radial distributions of mean axial velocity U_y (**a**), mean radial velocity U_r (**b**) and mean mixture fraction Z (**c**) computed at various axial positions (y = 50, 75 and 200 mm) with different mesh resolutions (20k, 80k and 160k cells) in comparison with experimental data.

References

1. The European Commission. *The European Green Deal*; The European Commission: Brussels, Belgium, 2019.
2. The European Commission. *'Fit for 55': Delivering the EU's 2030 Climate Target on the Way to Climate Neutrality*; The European Commission: Brussels, Belgium, 2021.
3. The European Commission. *Regulation of the European Parliament and of the Council amending Regulation (EU) 2019/631 as Regards Strengthening the CO_2 Emission Performance Standards for New Passenger Cars and New LIGHT commercial Vehicles in Line with the Union's Increased Climate Ambition*; The European Commission: Brussels, Belgium, 2021.
4. US Department of Energy. *The United States' Nationally Determined Contribution*; US Department of Energy: Washington, DC, USA, 2021.
5. Guido, C.; Napolitano, P.; Alfuso, S.; Corsetti, C.; Beatrice, C. How engine design improvement impacts on particle emissions from an HD SI natural gas engine. *Energy* **2021**, *231*, 120748. [CrossRef]
6. Liu, Z.; Sun, P.; Du, Y.; Yu, X.; Dong, W.; Zhou, J. Improvement of combustion and emission by combined combustion of ethanol premix and gasoline direct injection in SI engine. *Fuel* **2021**, *292*, 120403.
7. Solomon, A.; Battiston, P.; Sczomak, D. Lean-Stratified Combustion System with Miller Cycle for Downsized Boosted Application—Part I. In Proceedings of the SAE WCX Digital Summit, Virtual, 12–15 April 2021. [CrossRef]
8. Attard, W.; Parsons, P. A normally aspirated spark initiated combustion system capable of high load, high efficiency and near Zero NOx emissions in a modern vehicle powertrain. *SAE Int. J. Engines* **2010**, *3*, 269–287. [CrossRef]
9. Toulson, E.; Watson, H.; Attard, W. *Gas Assisted Jet Ignition of Ultra-Lean LPG in a Spark Ignition Engine*; SAE Technical Papers; SAE International: Warrendale, PA, USA, 2009.
10. Shah, A.; Tunestal, P.; Johansson, B. *Effect of Pre-Chamber Volume and Nozzle Diameter on Pre-Chamber Ignition in Heavy Duty Natural Gas Engines*; SAE Technical Papers; SAE International: Warrendale, PA, USA, 2009.
11. Toulson, E.; Huisjen, A.; Chen, X.; Squibb, C.; Zhu, G.; Schock, H.; Attard, W. Visualization of Propane and Natural Gas Spark Ignition and Turbulent Jet Ignition Combustion. *SAE Int. J. Engines* **2012**, *5*, 1821–1835. [CrossRef]
12. Zhu, S.; Akehurst, S.; Lewis, A.; Yuan, H. A review of the pre-chamber ignition system applied on future low-carbon spark ignition engines. *Renew. Sustain. Energy Rev.* **2022**, *154*, 111872. [CrossRef]
13. Zhou, L.; Liu, P.; Zhong, L.; Feng, Z.; Wei, H. Experimental observation of lean flammability limits using turbulent jet ignition with auxiliary hydrogen and methane in pre-chamber. *Fuel* **2021**, *305*, 121570. [CrossRef]
14. Kim, J.; Scarcelli, R.; Som, S.; Shah, A.; Biruduganti, M.; Longman, D. Numerical investigation of a fueled pre-chamber spark-ignition natural gas engine. *Int. J. Engine Res.* **2021**. [CrossRef]
15. Kim, J.; Scarcelli, R.; Som, S.; Shah, A.; Biruduganti, M.; Longman, D. Assessment of turbulent combustion models for simulating prechamber ignition in a natural gas engine. *J. Eng. Gas Turbines Power* **2021**, *143*, 091004. [CrossRef]
16. Marble, F.E.; Broadwell, J.E. *The Coherent Flame Model for Turbulent Chemical Reactions*; Technical Report; TRW Defense and Space Systems Group: New York, NY, USA, 1977.
17. Boudier, P.; Henriot, S.; Poinsot, T.; Baritaud, T. A model for turbulent flame ignition and propagation in spark ignition engines. *Symp. Int. Combust.* **1992**, *24*, 503–510. [CrossRef]
18. Duclos, J.; Veynante, D.; Poinsot, T. A comparison of flamelet models for premixed turbulent combustion. *Combust. Flame* **1993**, *95*, 101–117. [CrossRef]
19. Weller, H.G.; Tabor, G.; Gosman, A.D.; Fureby, C. Application of a Flame Wrinkling LES Combustion Model to a Turbulent Mixing Layer. In Proceedings of the Twenty-Seventh Symposium on Combustion, Boulder, CO, USA, 2–7 August 1998; pp. 899–907.
20. Peters, N. *Turbulent Combustion*; Cambridge University Press: Cambridge, UK, 2000.
21. Herrmann, M. Refined Level Set Grid Method for Tracking Interfaces. In *Annual Research Briefs 2005*; Center for Turbulence Research: Stanford, CA, USA, 2005; pp. 3–18.
22. Choi, C.R.; Huh, K.Y. Development of a Coherent Flamelet Model for a Spark-Ignited Turbulent Premixed Flame in a Closed Vessel. *Combust. Flame* **1998**, *114*, 336–348. [CrossRef]
23. Weller, H.G. *The Development of a New Flame Area Combustion Model Using Conditional Averaging*; Thermo-Fluids Section Report TF 9307; Imperial College of Science, Technology and Medicine: London, UK, 1993.
24. Ewald, J.; Peters, N. A Level Set Based Flamelet Model for the Prediction of Combustion in Spark-Ignition Engines. In Proceedings of the 15th International Multidimensional Engine Modeling User's Group Meeting, Detroit, MI, USA, 20 April 2005.
25. Seffrin, F.; Fuest, F.; Geyer, D.; Dreizler, A. Flow field studies of a new series of turbulent premixed stratified flames. *Combust. Flame* **2010**, *157*, 384–396. [CrossRef]
26. Sweeney, M.S.; Hochgreb, S.; Dunn, M.J.; Barlow, R.S. The structure of turbulent stratified and premixed methane/air flames I: Non-swirling flows. *Combust. Flame* **2012**, *159*, 2896–2911. [CrossRef]
27. Sweeney, M.S.; Hochgreb, S.; Dunn, M.J.; Barlow, R.S. The structure of turbulent stratified and premixed methane/air flames II: Swirling flows. *Combust. Flame* **2012**, *159*, 2912–2929. [CrossRef]
28. Lucchini, T.; D'Errico, G.; Paredi, D.; Sforza, L.; Onorati, A. *CFD Modeling of Gas Exchange, Fuel-Air Mixing and Combustion in Gasoline Direct-Injection Engines*; SAE Technical Papers; SAE International: Warrendale, PA, USA, 2019.
29. Sforza, L.; Lucchini, T.; Gianetti, G.; D'Errico, G. *Development and Validation of SI Combustion Models for Natural-Gas Heavy-Duty Engines*; SAE Technical Papers; SAE International: Warrendale, PA, USA, 2019. [CrossRef]

30. Meneveau, C.; Poinsot, T. Stretching and quenching of flamelets in premixed turbulent combustion. *Combust. Flame* **1991**, *86*, 311–332. [CrossRef]
31. Prasad, R.; Gore, J. An evaluation of flame surface density models for turbulent premixed jet flames. *Combust. Flame* **1999**, *116*, 1–14. [CrossRef]
32. Lucchini, T.; Cornolti, L.; Montenegro, G.; D'Errico, G.; Fiocco, M.; Teraji, A.; Shiraishi, T. *A Comprehensive Model to Predict the Initial Stage of Combustion in SI Engines*; SAE Paper 2013-01-1087; SAE International: Warrendale, PA, USA, 2013.
33. Zhu, X.; Sforza, L.; Ranadive, T.; Zhang, A.; Lee, S.Y.; Naber, J.; Lucchini, T.; Onorati, A.; Anbarasu, M.; Zeng, Y. Experimental and Numerical Study of Flame Kernel Formation Processes of Propane-Air Mixture in a Pressurized Combustion Vessel. *SAE Int. J. Engines* **2016**, *9*, 1494–1511. [CrossRef]
34. Sforza, L.; Lucchini, T.; Onorati, A.; Zhu, X.; Lee, S. Modeling Ignition and Premixed Combustion Including Flame Stretch Effects. In *Proceedings of the WCX™ 17: SAE World Congress Experience*; SAE International: Warrendale, PA, USA, 2017. [CrossRef]
35. Gulder, O.L. *Correlations of Laminar Combustion Data for Alternative S.I. Engine Fuels*; SAE Technical Paper; SAE International: Warrendale, PA, USA, 1984. [CrossRef]
36. Lucchini, T.; D'Errico, G.; Onorati, A.; Frassoldati, A.; annd Gilles Hardy, A.S. Modeling Non-Premixed Combustion Using Tabulated Kinetics and Different Flame Structure Assumptions. *SAE Int. J. Engines* **2017**, *2*, 593–607. [CrossRef]
37. Smith, G.P.; Golden, D.M.; Frenklach, M.; Moriarty, N.W.; Eiteneer, B.; Goldenberg, M.; Bowman, C.T.; Hanson, R.K.; Song, S.; Gardiner, J.; et al. GRI-Mech 3.0. Available online: http://www.me.berkeley.edu/gri_mech/ (accessed on 25 October 2019).
38. Gulder, O.L. Turbulent Premixed Flame Propagation Models for Different Combustion Regimes. In Proceedings of the the Twenty-Third Syposium on Combustion (International), Orleans, France, 22–27 July 1990; pp. 743–750.
39. Kuenne, G.; Euler, M.; Ketelheun, A.; Avdić, A.; Dreizler, A.; Janicka, J. A Numerical Study of the Flame Stabilization Mechanism Being Determined by Chemical Reaction Rates Submitted to Heat Transfer Processes. *Z. Phys. Chem.* **2015**, *229*, 643–662. [CrossRef]
40. Mercier, R.; Auzillon, P.; Moureau, V.; Darabiha, N.; Gicquel, O.; Veynante, D.; Fiorina, B. LES Modeling of the Impact of Heat Losses and Differential Diffusion on Turbulent Stratified Flame Propagation: Application to the TU Darmstadt Stratified Flame. *Flow Turbul. Combust.* **2014**, *93*, 349–381. [CrossRef]
41. Borghi, R. On the Structure and Morphology of Turbulent Premixed Flames. In *Recent Advances in the Aerospace Sciences: In Honor of Luigi Crocco on His Seventy-Fifth Birthday*; Casci, C.; Bruno, C., Eds.; Springer: Boston, MA, USA, 1985; pp. 117–138. [CrossRef]
42. Kuenne, G.; Seffrin, F.; Fuest, F.; Stahler, T.; Ketelheun, A.; Geyer, D.; Janicka, J.; Dreizler, A. Experimental and numerical analysis of a lean premixed stratified burner using 1D Raman/Rayleigh scattering and large eddy simulation. *Combust. Flame* **2012**, *159*, 2669–2689. Special Issue on Turbulent Combustion. [CrossRef]
43. Sforza, L. Development of a Comprehensive CFD Model for Spark-Ignition Engine Combustion. Ph.D. Thesis, Politecnico di Milano, Milan, Italy, 2018.

energies

MDPI

Article

Numerical Modeling and Simulation of a Spark-Ignition Engine Fueled with Ammonia-Hydrogen Blends

Gabriele D'Antuono, Davide Lanni *, Enzo Galloni and Gustavo Fontana *

Department of Civil and Mechanical Engineering, University of Cassino and Southern Latium, 03043 Cassino, Italy

* Correspondence: davide.lanni@unicas.it (D.L.); fontana@unicas.it (G.F.)

Abstract: Carbon-free fuels, in particular ammonia and hydrogen, could play a significant role in the decarbonization of the mobility sector. In this work, the authors assessed the operation of a light-duty spark-ignition engine fueled with an ammonia–hydrogen blend (85% ammonia and 15% hydrogen by volume) using a 1D predictive model. Three-dimensional computations have been used in order to verify the reliability of the 1D model. The addition of hydrogen to the air–fuel mixture allows the operating capacity of the engine to be extended with respect to neat ammonia fueling. The engine can be properly regulated between 1500 rpm and 3000 rpm. Its operating range reduces as engine speed increases, and it cannot run at 6000 rpm. This is due to different engine operating constraints being exceeded. The maximum engine torque is about 240 Nm and is reached at 1500 rpm. The engine efficiency ranges between 42% and 19%, and the specific fuel consumption varies from about 350 g/kWh to about 750 g/kWh. The results provide both performances and operating ranges of the engine allowing us to define optimized engine maps obtained by means of a constrained optimization.

Keywords: ammonia; hydrogen; carbon-free fuels; e-fuels; SI engines; downsizing

Citation: D'Antuono, G.; Lanni, D.; Galloni, E.; Fontana, G. Numerical Modeling and Simulation of a Spark-Ignition Engine Fueled with Ammonia-Hydrogen Blends. *Energies* **2023**, *16*, 2543. https://doi.org/10.3390/en16062543

Academic Editors: Roberta De Robbio and Maria Cristina Cameretti

Received: 10 February 2023
Revised: 2 March 2023
Accepted: 6 March 2023
Published: 8 March 2023

1. Introduction

Today, in terms of the pollutant emissions of the Internal Combustion Engines (ICEs) complying with the current regulations, we can affirm that these engines have won the battle against those emissions that are considered to be harmful for the human health. However, the emissions of greenhouse gases, deriving from the combustion of traditional fossil fuels, still require significant attention.

In order to drastically reduce CO_2 emissions, the only effective solution is to find new carbon-free fuels.

As carbon-free fuels, ammonia [1–3] and hydrogen [4–6] represent good choices as fuels for ICEs. Furthermore ammonia can be used as a raw material to produce hydrogen [7,8].

The combustion characteristics of ammonia are very different from those of hydrogen and gasoline, and in Table 1, a brief summary of these is reported.

Table 1. Ammonia properties vs. gasoline and hydrogen properties [2,9,10].

	Ammonia	Gasoline	Hydrogen
Storage	Liquid (300 K–11 bar)	Liquid (300 K–1 bar)	Compressed (300 K, 700 bar)
LHV (MJ/kg)	18.8	44.5	120
Stoichiometric air to fuel ratio (-)	6.05	14.8	34.3
Heat released by a volume unit of stoichiometric air-fuel mixture (MJ/m^3)	2.81	3.46	2.88
Laminar burn velocity (m/s) at ambient condition	0.015	0.58	3.51
Octane number	130	92–98	>130
Auto-ignition temperature (K)	930	503	503
Explosion limit (% volume ratio)	16–28	1.4–1.6	4.5–75
Density (kg/m^3)	0.703	740	0.082

Ammonia clearly shows the worst combustion characteristics compared to hydrogen and gasoline: its laminar flame speed is 57% less than that of gasoline and 98% less than hydrogen; ammonia's LHV is much smaller than the LHVs of both gasoline and hydrogen. This suggests that ammonia is better to use in blends [11]. However, at present, neat ammonia, as a fuel for ICEs, is also under study.

In [12], the authors focused their attention on the performance of a spark-ignition engine fueled with neat ammonia. By means of 1D-3D analyses, they investigated the influence of boost pressure, load conditions and spark timing on the behavior of the engine running with ammonia. In the end, they found out that ammonia drastically reduces the typical operation map of a light-duty SI engine. However, ammonia could be effectively used in engines running at low–medium speeds, preferably featuring high pressure ratios.

Scharl et al. [13] investigated the mixture formation and spray combustion on a rapid compression–expansion machine. Combining 1D analyses and optical investigations, they found out that ammonia flames do not appear stable after ignition. Considering ammonia spray, the authors reported that close to the nozzle, the temperatures are low and the equivalence ratio drops below 1. Cui et al. [14] experimentally analyzed the effect of a pre-chamber with a jet flame in order to improve the ignition phase of ammonia and its combustion. They found that increasing the pre-chamber volume leads to an increase in jet duration and ignition energy.

Ammonia can also be added to conventional fuels such as diesel or gasoline. El Fattah et al. [15] analyzed the emissions of a spark-ignition engine powered by water ammonia–gasoline blends. In particular, using this blend, they obtained a higher engine thermal efficiency, but also a higher amount of NO_x.

In [16], Zhang et al. experimentally investigated the performance of a two-stroke low-speed engine powered by a dual-fuel direct injection mixture composed by ammonia and diesel. Varying the ammonia's amount, they discovered that diesel injection timing helps the regulation of the ignition and early stage of combustion. The authors also pointed out that this solution is important for the reduction in ammonia's combustion duration, and, consequently, the indicated thermal efficiency increases.

Nowadays, ammonia–hydrogen blends are objects of study. In particular, in [17], a numerical study on the combustion characteristics of a spark-ignition engine fueled by ammonia and hydrogen which is obtained from the reformation of the exhaust gas is presented. The authors found that by adding hydrogen, up to 10% by volume, the in-cylinder pressure increased, and the combustion duration decreased, leading to an improvement in combustion and thermal efficiencies.

In [18], Lhuillier et al. experimentally investigated the combustion of ammonia–hydrogen blends in spark-ignition engines. They considered a constant engine speed equal to 1500 rpm, an intake pressure between 0.1 and 0.12, an equivalence ratio ranging from 0.6 to 1.2, and a fraction of hydrogen in the fuel mixture varying from 0 to 0.15 by mass. They found out that adding hydrogen also makes the engine run in lean operations with a COV_{IMEP} less than 5%. In this way, hydrogen allows an improvement in stability and performance as well, since the combustion duration decreases at a high rate until the H_2 percentage is less than 20% (volume basis); over this threshold, the decrease slows down.

Frigo et al. [19] experimentally investigated a spark-ignition engine, used as a range extender for electric vehicles, powered by ammonia and hydrogen where hydrogen is obtained by ammonia. After defining the type of injector and its timings, they considered either full-load operations or half-load conditions, with the engine running from 2500 rpm to 5000 rpm and stoichiometric mixtures. The authors showed the amount of hydrogen necessary to increase the combustion velocity mainly depends on load and less on engine speed, and, in particular, they considered a percentage of hydrogen of 6% (volume basis) at full load and 10% at partial load.

In [20], there is a numerical study on the ammonia–hydrogen blends' combustion characteristics compared to methane, methanol, and ethanol combustion. Using a 1D approach, the authors observed several engine conditions by varying the equivalence ratio (from 0.7 to 1.3), the hydrogen mole fraction (from 0 to 100%), and the compression ratio. They found out that the laminar burning velocity needs an amount of 40–70% of hydrogen in order to reach the same level of hydrocarbon fuels unless the compression ratio increases. The ignition energy shows an opposite trend to laminar burning velocity. The authors also found that due to an incomplete oxidation, the combustion efficiency decreases if the amount of hydrogen and the equivalence ratio increase.

Pyrc et al. [21] experimentally investigated an SI engine characterized by a variable pressure ratio and modified it in order to work as a dual-fuel engine fueled with ammonia and hydrogen. Considering two different pressure ratios (8 and 10) and an amount of the energy shared by hydrogen from 0 to 70%, the results show that with 12% of energy shared by hydrogen, the instability related to the ignition process disappears. In particular, with a compression ratio equal to 8, the authors observed that ignition delay decreases two times, while for a compression ratio equal to 10, it decreases almost three times; naturally, the specific energy consumption decreases as the compression ratio increases.

Due to the low engine speeds that characterize the heavy-duty engines, ammonia could effectively be fired in applications such as marine engines, as reported in [22,23]. In addition, in [24], the authors theoretically investigated the behavior of a diesel marine engine firing with ammonia–hydrogen blends and performing an HCCI combustion mode able to reduce the NO_X formation.

The state of the art demonstrates that engines fueled by ammonia–hydrogen are interesting due to their low environmental impact and good operating capacity.

Thus, in this paper, the operation of a light-duty spark-ignition engine fueled with an ammonia–hydrogen blend was investigated by means of a 1D numerical approach.

The blend was composed by 85% of ammonia and 15% of hydrogen by volume. It was chosen considering a previous literature analysis of these blends. In particular, a low amount of hydrogen, such as the one chosen in this work (15% by volume), could be directly produced on-board vehicles from an ammonia cracking process.

The analyses were performed taking into account the operating constraints of the engine. In particular, limits on the in-cylinder pressure, exhaust gas temperature, spark advance, knock intensity, and fraction of fuel burnt were considered. To the best of the authors' knowledge, previous studies do not show complete operating maps of a light-duty SI PFI engine fueled with the selected ammonia–hydrogen mixture. Therefore, the aim of this work was to assess both the performance and operating range of the engine, in order to obtain optimized engine maps which guaranteed its proper operation with the analyzed fuel blend.

2. Materials and Methods

With the aim of investigating the possible utilization of neat ammonia as a fuel for ICEs, the fueling change of a turbocharged gasoline engine, designed for light-duty vehicles, was analyzed (Section 2.1).

Engine performances were evaluated by using a numerical 1D model able to simulate the engine performances for different fuels (Section 2.2). The dependability of the 1D results strongly depended on a proper prediction of the in-cylinder pressure curves. The predicted in-cylinder pressure curves were compared to those obtained by a 3D numerical approach. The latter was based on a detailed chemical–kinetic mechanism able to predict the ammonia–hydrogen–air mixture oxidation (Section 2.3).

2.1. Engine and Fuel Blend

The engine under study is a PFI turbo-charged spark-ignition engine. It was designed to be fueled by regular gasoline. Some engine characteristics are shown in Table 2 which reports the nominal performances of the engine running with gasoline as designed by the manufacturer.

Table 2. Engine main characteristics for gasoline fueling.

Model	4 Cylinders, 4 Valves/cyl
Displacement (cm^3)	1368
Bore/Stroke/Con. Rod (mm)	72/84/129
Compression Ratio (-)	9.8
Combustion Chamber Shape	Pent Roof
Max Power (ISO Conditions) (kW)	110 @ 5500 rpm
Max Torque (ISO Conditions) (Nm)	230 @ 3000 rpm
Turbocharger group	IHI RHF3

In this work, the fuel system was modified in order to inject both ammonia and hydrogen fuels in the intake runners. In this way, the engine fired with a premixed hydrogen–ammonia–air mixture. In the following, the fuel was a blend of 85% ammonia and 15% hydrogen by volume. The main properties of the considered blend are summarized in Table 3.

Table 3. Fuel blend properties (ammonia 85%–hydrogen 15% by volume).

	Blend
LHV (MJ/kg)	33.98
Stoichiometric air to fuel ratio (-)	6.63
Heat released by a volume unit of stoichiometric air–fuel mixture (MJ/m^3)	2.82
Laminar burn velocity (m/s) in ambient condition *	0.088
Density (kg/m^3)	0.69

* Data calculated by Equation (1).

2.2. 1D Numerical Approach

A 1D model approach was used in order to evaluate the operation of the engine firing with ammonia–hydrogen blends. The model was already validated considering the engine running with gasoline [25] and different gasoline–alcohol blends [26]. The average error with respect to experimental data was less than 5% for fuel consumption, exhaust gas temperature, and maximum in-cylinder pressure. The same model was also used to predict the performance of the same engine fueled by hydrogen [6] or pure ammonia [12].

The numerical approach used for the combustion modeling decouples turbulence and chemical effects. Implementing the chemical properties of the ammonia–hydrogen–air mixture (laminar flame speed and induction time), since the kind of fuel does not affect the fluid dynamic behavior in a significant manner, the approach was still considered valid to predict the combustion process of the ammonia–hydrogen blend.

In Figure 1, the whole engine layout is presented. Briefly, the turbocharger was modeled by using the operation maps provided by the manufacturer, while in-pipe flow was estimated using a finite volume approach. The inlet air filter, which affects the performance and emissions of both spark-ignition engines [27] and compression ignition engines [28,29], was modeled as a volume. The friction losses were also considered according to data provided by the manufacturer.

The wastegate valve operation was simulated, allowing a portion of the exhaust gases to bypass the turbine. This was needed to control the boost pressure. The heat transfer rate through the cylinder walls was estimated using the well-known Woshni correlation. The friction losses were calculated as a function of the engine speed.

The injector was modeled considering a mixture with properties that nominally represent the fuel blend characteristics.

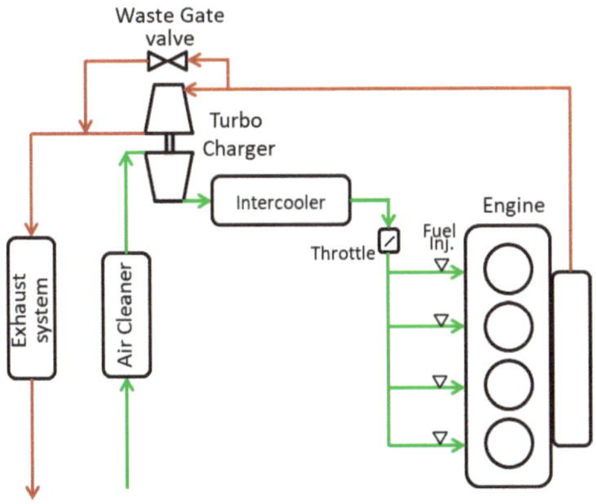

Figure 1. Schematic layout of the engine [20,27].

The burning rate for homogeneous charge spark-ignition engines in this model was predicted using the Keck and Tabaczynski model [30,31]. The calculated turbulent flame speed depended on both laminar flame speed and turbulent flow indices according to an entrainment model. Following the hierarchical approach described in [32], the mean turbulent length scale and the mean turbulent intensity were imposed, exploiting the flow field details provided by 3D calculations.

The thermochemical properties of the ammonia-hydrogen blend were utilized, while the laminar flame speed (S_L) was computed by means of the correlation presented in [33].

$$S_{L,NH3,H2} = S_{LO_2} * T_n^{\alpha_2} * P_n^{\beta_2} \tag{1}$$

$$\alpha_2 = a_{2,P_3} + b_{2,P_3} * \lambda + c_{2,P_3} * \lambda^2 + d_{2,P_3} * \lambda^3 \tag{2}$$

$$\beta_2 = a_{3,P_3} + b_{3,P_3} * \lambda + c_{3,P_3} * \lambda^2 \tag{3}$$

$$S_{LO_2} = a_1 * b_{1,P_4}^{\lambda} * \lambda^{c_{1,P_4}} \tag{4}$$

$$a_1 = a_{10} + (1 - a_{10}) * \left(1 - e^{-a_{11}*xH_2 - a_{12}*x^2H_2 - a_{13}*x^3H_2}\right) \tag{5}$$

where T_n and P_n are the normalized temperature and pressure with respect to the reference conditions of T and P, respectively, equal to 300 K and 0.1 MPa.

$$T_n = T_u / T_0 \tag{6}$$

$$P_n = P_u / P_0 \tag{7}$$

The coefficients a, b, c, and d are tabulated as a function of temperature and pressure in [33]. λ is the air to fuel ratio.

Figure 2 compares the calculated laminar flame speed of the hydrogen-enriched ammonia to those of pure ammonia and gasoline. It is worth noting that with a small amount of hydrogen, the laminar flame speed increased by 47.8% at an equivalence ratio equal to 1.

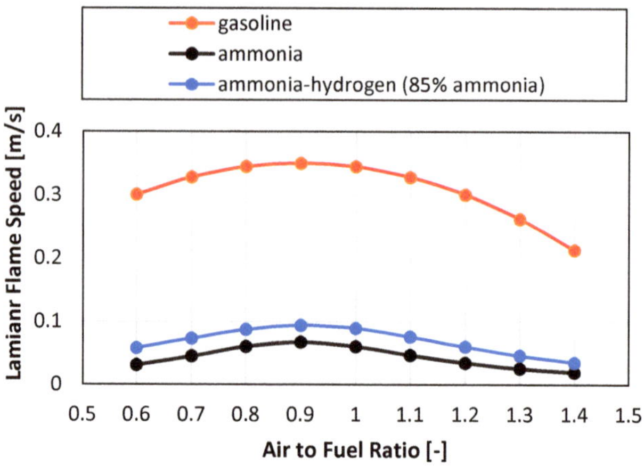

Figure 2. Laminar flame speed comparison in ambient condition. Data were calculated according to Equations (1)–(5) for ammonia and hydrogen–ammonia blend, according to [34] for gasoline.

The knock occurrence was detected considering the induction time of the air fuel mixture. It was assumed that knock occurs when $\int_0^t dt/\tau_{id}(t) = 1$ [35]. The ignition delays were provided by detailed chemical kinetic calculations previously performed in the Cantera environment [36]. In particular, the Nakamura mechanism [37], based on 38 species and 232 reactions, was solved in a "Ideal Gas Reactor" component. Several levels of pressure and temperature like those characterizing the engine operation were considered. The calculated data were tabulated, and the resulting database was implemented in the 1D model.

When knock is detected, its intensity is evaluated as described in the following equation:

$$KI = (1 - x_b)(cr - 1)\sqrt{1 - \frac{\theta_k}{\theta_{ref}}\frac{N}{N_{ref}}} \quad (8)$$

where x_b is the mass of burned fuel, cr is the compression ratio, θ_k is the knock onset crank angle and N the engine speed, and θ_{ref} is the maximum crank angle for which knock is still audible, set to 50 CAD. N_{ref} is a tuning parameter set as equal to 3500 rpm. It was assumed that tolerable knock corresponds to a KI level less than or equal to 0.5, as described in [12,25].

2.3. Three-Dimensional Numerical Approach

A 3D computational model, previously built by means of the commercial code AVL Fire, was used to reproduce some 1D results. Briefly, a first-order (Euler) implicit differencing scheme was used for time discretization, while second-order schemes were used for both continuity and momentum equations. An upwind scheme was used for both turbulence and energy equations. Turbulent effects were modeled using the well-known k-ε approach. This model was widely used in the past to reproduce the behavior of the same engine running with gasoline, obtaining very reliable results [32]. It was adapted to the ammonia–hydrogen engine operation referring to the Nakamura mechanism previously described. Thus, different chemical species were considered during the simulations, and the fuel oxidation process was simulated by means of a chemical–kinetic approach. The engine mesh, reported in Figure 3, reproduces only half geometry due to the symmetry of the engine geometry. The minimum number of the grid cells is about 50,000 (firing TDC), while the maximum number is approximately 350,000 (BDC during the intake stroke). Boundary and initial conditions were set by utilizing the computations provided by the 1D model previously depicted. In the inlet runner, a premixed air–fuel mixture was considered.

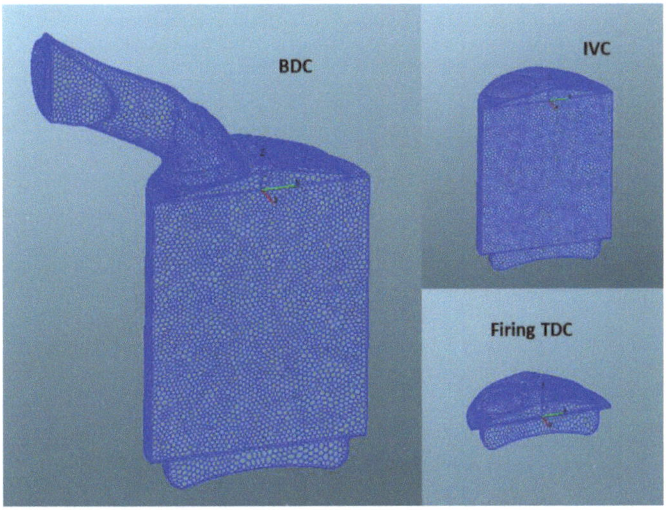

Figure 3. Three-dimensional engine mesh.

3. Results

The model described in Section 2.2 was used in order to evaluate the operation of the engine considering fueling with an ammonia–hydrogen blend. In particular, as already specified above, a blend of 85% ammonia and 15% hydrogen by volume was used.

Steady-state engine operating points in stoichiometric conditions were investigated. First, 3D computations were performed in order to investigate some operating points and to compare the 3D results with those obtained from the 1D computations (Section 3.1). In a second step, different analyses were carried out by using the 1D model in order to investigate both the performances and the operation limits of the engine fueled with the ammonia–hydrogen mixture.

The 1D simulation results are presented in the following. Low load operation was investigated considering no boost conditions and different values of the throttle opening (Section 3.2). Medium and high load operation was evaluated by varying the boost pressure at wide-open throttle (WOT) (Section 3.3).

In the end, taking into account several constraints which guarantee a proper operation of the engine (Table 4), the engine operating map is built in Section 3.4.

Table 4. Operating constraints.

Peak pressure (bar)	≤100
Exhaust gas temperature (K)	≤900
Spark advance (°bTDC)	≤110
Knock intensity (-)	≤0.5
Burned fuel fraction (%)	≥98

The details of each analysis are described in the respective subsection.

3.1. 1D–3D Comparisons

The 3D approach was used to reproduce the WOT operating points presented in Table 5. The spark advance was chosen by means of the 1D simulation results aiming toward the maximum engine torque. It is worth noting that the spark advance values, as discussed in [12] and [18], were bigger than the ones typical for the engine running with gasoline but were much lower than the values obtained for pure ammonia. As reported in Table 5, when increasing the engine speed, the spark advance moves far away from the top dead center (TDC).

Table 5. Three-dimensional simulated points. WOT.

Engine Speed	1500	3000
Spark advance (°bTDC)	54.4	64.3
Equivalence ratio (-)	1	1
Boost pressure (bar)	No boost	No boost

Figure 4 shows both in-cylinder pressure and heat release curves for each investigated case. The best agreement between the models was reached at 3000 rpm with a difference in pressure peak of about 0.47%, while at 1500 rpm, the difference was of about 4.9%. The agreement between the two models seemed to be good. These results confirm that the 1D model effectively predicted the behavior of the analyzed engine. Thus, it was implemented for the following analyses involving the fuel blend under study.

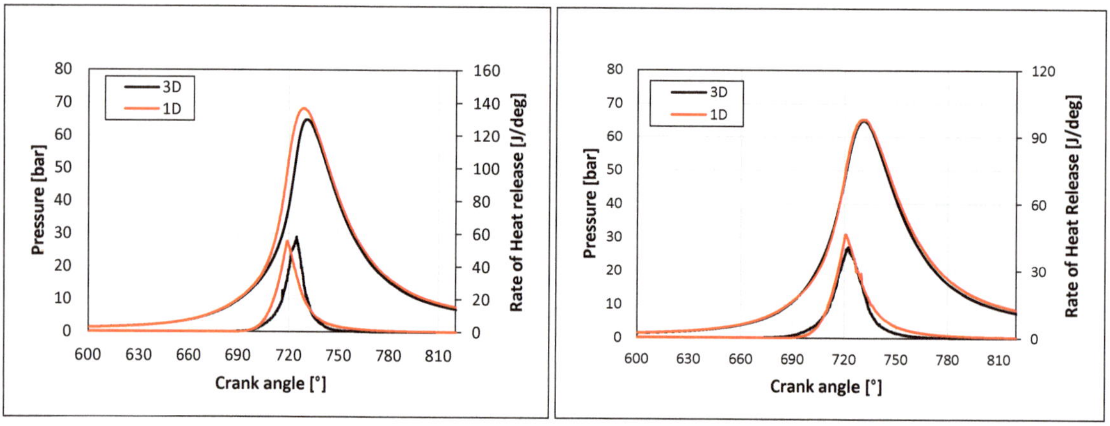

Figure 4. In-cylinder pressure and rate of heat release. The 1500 (**left**)–3000 (**right**) rpm, no boost, WOT.

3.2. Low Load Operation

In order to assess the engine behavior at partial load, a parametrical analysis was carried out by varying the throttle opening until a minimum torque of about 10 Nm was reached. The whole analysis was performed considering different engine speeds (from 1500 to 6000 rpm), no boost condition, an overall equivalence ratio equal to 1, and spark timing set to the maximum torque value.

As an example, Figure 5 shows the engine efficiency calculated at 3000 rpm when the load changes. Of course, due to the increasing pumping losses, when reducing the engine load by throttling, the efficiency decreased.

Furthermore, the typical increase in the angular combustion duration (CA 0–90) for increasing engine speeds (Figure 6, right) became critical for the lowest loads at medium regimes and for almost all the loads at high regimes. In fact, in the latter cases, the minimum allowable burned fuel fraction (98%) was not reached (Figure 6, left). As a consequence, the operating range of the engine significantly reduced. Partial load conditions were sustainable for the whole considered range at 1500 and 2000 rpm, with a burned fuel fraction at minimum load, respectively, equal to 98.8% and 98.9%. At 3000 and 4000 rpm, the entire map was not exploitable since the lowest loads did not respect the imposed constraint. At 5000 rpm, only WOT and very low throttling conditions guaranteed a proper engine operation, while at 6000 rpm, the 98% burned fuel fraction was not reachable. It is worth underlining that, in this partial load analysis, all the other constraints were respected; therefore, they are not reported.

Engine running at 3000 rpm, no boost.

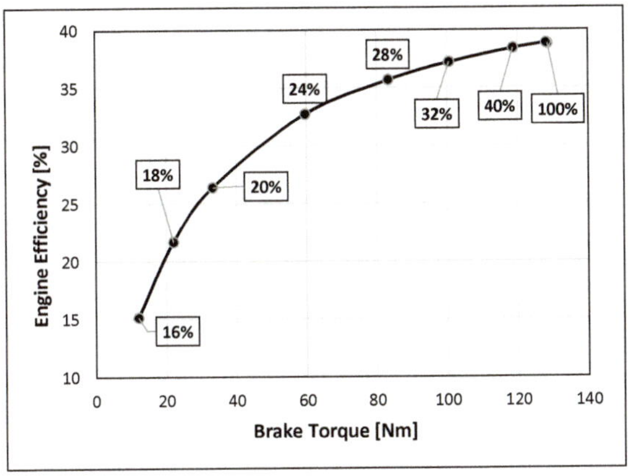

Figure 5. Engine efficiency as a function of engine torque for different throttle opening values.

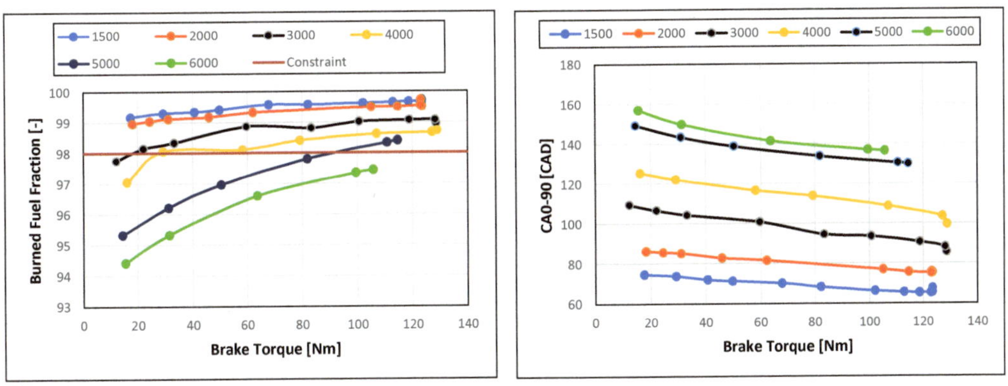

Figure 6. Burned fuel fraction (**left**) and CA 0–90 (**right**) for different regimes. No boost.

3.3. Medium and High Load Operation

In this subsection, the behavior of the engine at WOT was evaluated considering different engine speeds, from 1500 to 6000, and different boost pressure (BP) levels, up to 2 bar. Stoichiometric mixtures were always considered.

Figure 7 (left) shows the calculated engine torque delivered by the engine when the spark timing was set to the maximum torque knock limited value (Figure 7, middle). Of course, the engine torque increased when the boost pressure level grew. The optimal knock limited spark advance followed the trend of the combustion duration (Figure 7, right) which increased for increasing engine speeds and decreased when the boost pressure increased. It is worth noting that the combustion durations, presented in Figure 7 (right) for the speed of 1500 rpm, agreed with those measured in [18].

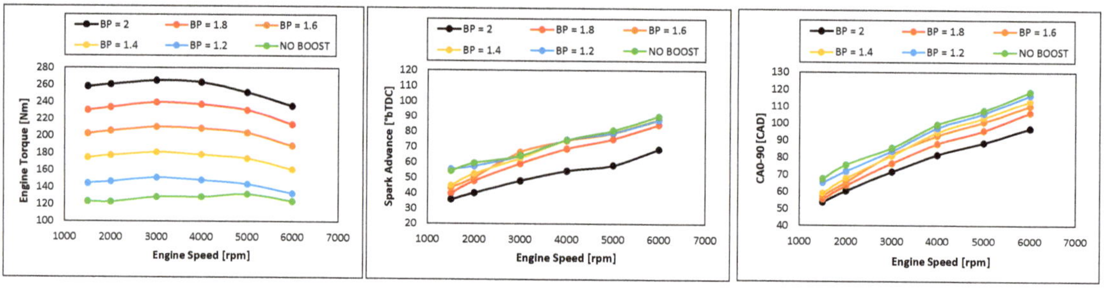

Figure 7. Engine torque (**left**) calculated at maximum torque knock limited spark advance (**middle**) and relative combustion duration (**right**) at different boost pressure levels.

Unfortunately, considering the operating constraints reported in Table 4, not all the points shown in Figure 7 were acceptable (Figure 8). The engine running with the optimal spark advance reached an unacceptable peak pressure for boost greater than or equal to 1.6 at any rotational speed. In addition, it reached unacceptable exhaust gas temperatures when it ran above 5000 rpm at any boost pressure levels. The exhaust gas temperature only did not limit the engine operation below 3000 rpm, while in the range of 3000–5000 rpm, it determined a maximum allowable boost pressure.

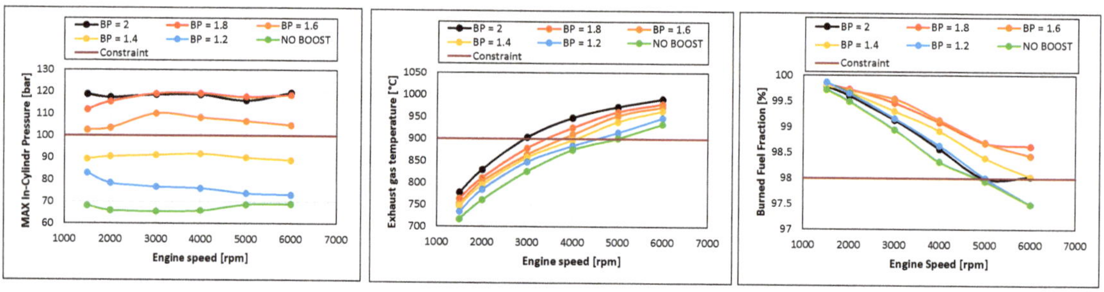

Figure 8. Maximum in-cylinder pressure (**left**), exhaust gas temperature (**middle**) and burned fuel fraction (**right**) calculated for the engine running at the maximum torque knock limited spark advance.

Finally, above 5000 rpm and at low boost pressure, the engine also showed unacceptable unburned fuel fractions at exhaust.

3.4. Engine Maps

The results reported in previous sections show that engine operation is strongly limited by the constraints reported in Table 4.

However, for some operating points, by varying the spark advance with respect to the optimal value, which guarantees the maximum torque, it is possible to extend the operating range of the engine. As an example, at 3000 rpm and a boost level equal to 1.6, and by delaying the spark time, the maximum pressure decreased until it became acceptable. Naturally, both exhaust gas temperature and unburned fuel fraction increased, but they still remained within allowable limits (Figure 9). This result was achieved and only lost about 3% of engine efficiency (Table 6).

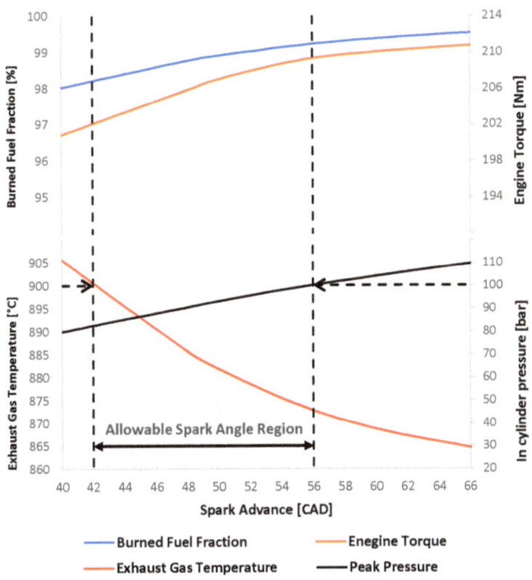

Figure 9. In-cylinder pressure, burned fuel fraction, exhaust gas temperature and engine torque versus spark advance. Engine speed: 3000 rpm, boost pressure: 1.6 bar. For this engine operating point, the maximum torque spark advance is 66.9 crank angle degrees, and the knock intensity is always zero. The allowable spark advance region is determined by constraints on the peak pressure and exhaust gas temperature.

Table 6. Engine recalibration at 3000 rpm, boost 1.6.

	Maximum Torque Operating Point	Recalibrated Operating Point
Max. In-Cylinder Pressure (bar)	101	89.9
Engine Torque (Nm)	212	210
Engine Power (kW)	66.5	65.9
Engine Efficiency (%)	41.4	40.0
Spark advance (°bTDC)	66.7	55.0
Exhaust Gas Temperature (°C)	869	874
Burned Fuel Fraction (%)	99.5	99.2

Thus, it is possible to identify an operating area where all parameters are within acceptable limits for many engine operating points that have been previously excluded.

The results of this constrained optimization are shown in Figure 10, where various engine parameters are represented as a function of engine speed and torque.

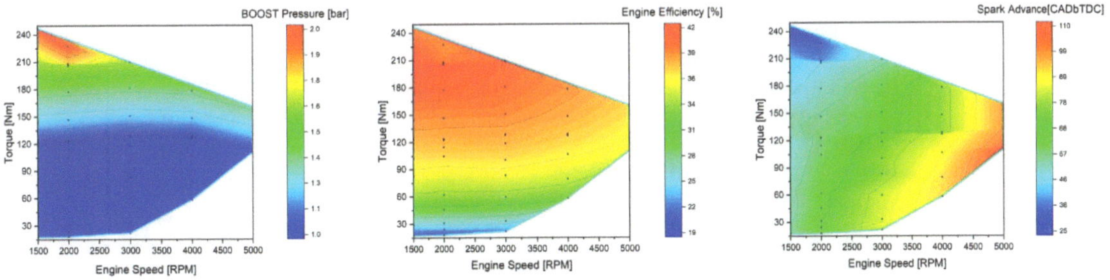

Figure 10. Boost level, engine efficiency, and spark advance as a function of engine speed and torque.

The feasible region was large enough at low engine speeds where good engine operation was possible. Between 1500 rpm and 3000 rpm, the engine could be regulated, delivering a torque ranging from a minimum value around 15 Nm to a maximum value greater than 200 Nm. The maximum engine torque was about 240 Nm; it was delivered at 1500 rpm, where a boost pressure equal to 2 bar could be achieved. This boost level could be kept only up to 2000 rpm.

Above 3000 rpm, the engine could run with an average maximum boost pressure of 1.4 bar, and the maximum delivered toque decreased almost linearly with the engine speed. In addition, the minimum load achievable by the engine decreased significantly as the engine speed increased. The combination of these two effects reduced the operating capacity of the engine as the speed increased. It could not run at 6000 rpm.

The engine efficiency was higher than 33% for almost all points; it never fell below about 19%. The maximum efficiency was very high: it was about 42% when the engine ran at low speed and the load was slightly less than the allowable maximum. Correspondingly, the specific consumption turned out to vary from about 350 g/kWh to about 750 g/kWh.

Spark advances are always very high. They ranged from a minimum of 25 CAD at 1500 rpm, maximum load, to a maximum of 110 CAD at 5000 rpm, low load.

4. Conclusions

In this paper, the authors investigated the behavior of a turbocharged spark-ignition light-duty engine powered by an ammonia–hydrogen blend (15% of hydrogen by volume) considering a port fuel injection. The analysis was carried out using a predictive 1D model. Furthermore, 3D computations were performed in order to verify the reliability of the results obtained by the 1D model.

The aim of the work was to assess both the performance and operating range of the engine in order to obtain optimized engine maps which guaranteed its proper operation with the chosen ammonia–hydrogen mixture.

The main results of the analysis can be summarized as:

- The small amount of hydrogen (by mass) improved the combustion behavior with respect to pure ammonia fueling, leading to reduced combustion durations and extending the operating range of the engine.
- Partial load operation became critical for the lowest loads at medium regimes and for almost all the loads at high regimes, due to unacceptable values of the burned fuel fraction. Low regimes (1500 and 2000 rpm) were sustainable for the whole considered throttling range. Medium regimes (3000 and 4000 rpm) were not sustainable at the lowest loads. At high regimes (5000 and 6000 rpm), a proper engine operation was not reachable, except for WOT and very low throttling conditions at 5000 rpm.
- At medium and high loads, the engine running with the optimal spark advance reached unacceptable values of the different constraints in various conditions, especially at the highest regimes (5000 and 6000 rpm) and at high boost levels (BP $\geq$ 1.6 bar). However, for some operating points, a specific tuning of the spark advance allowed all the operating constraints to be complied with (peak pressure, exhaust gas temperature, spark advance, knock intensity, and burned fuel fraction).
- Between 1500 rpm and 3000 rpm, the engine could be properly regulated. The engine could not reach 6000 rpm.
- At 1500 and 2000 rpm, the engine could reach a boost pressure equal to 2, while for the other regimes, the maximum feasible boost pressure was about 1.4.
- The maximum engine torque (about 240 Nm) was delivered at 1500 rpm. The engine efficiency ranged between 42% and 19%. Correspondingly, the specific consumption varied from about 350 g/kWh to about 750 g/kWh.

Author Contributions: Conceptualization, D.L. and E.G.; methodology, D.L., E.G. and G.D.; formal analysis, G.D., D.L. and E.G.; investigation, G.D.; data curation, G.D. and D.L.; writing—original draft preparation, G.D., D.L. and E.G.; writing—review and editing, G.D., D.L., E.G. and G.F.; supervision, E.G. and G.F. All authors have read and agreed to the published version of the manuscript.

Funding: This research received no external funding.

Data Availability Statement: Not applicable.

Conflicts of Interest: The authors declare no conflict of interest.

Nomenclature

1D	One-dimensional
3D	Three-dimensional
BDC	Bottom dead center
BP	Boost pressure
bTDC	Before top dead center
CA0–90	0–90% mass fraction fuel burn duration
CAD	Crank angle degree
COV_{IMEP}	Coefficient of variation of IMEP
HCCI	Homogeneous charge compression ignition
ICE	Internal combustion engine
IMEP	Indicated mean effective pressure
IVC	Intake valve closing
KI	Knock intensity
LHV	Lower heating value
NO_x	Nitrogen oxides
PFI	Port fuel injection
SI	Spark-ignition
TDC	Top dead center
WOT	Wide open throttle

References

1. Chiong, M.C.; Chong, C.T.; Ng, J.H.; Mashruk, S.; Chong, W.W.; Samiran, N.A.; Mong, G.R.; Valera-Medina, A. Advancements of combustion technologies in the ammonia-fuelled engines. *Energy Convers. Manag.* **2021**, *244*, 114460. [CrossRef]
2. Valera-Medina, H.A.; Xiao, M.; Owen-Jones, M.; David, W.I.; Bowen, P.J. Ammonia for power. *Prog. Energy Combust. Sci.* **2018**, *69*, 63–102. [CrossRef]
3. Tornatore, C.; Marchitto, L.; Sabia; de Joannon, M. Ammonia as Green Fuel in Internal Combustion Engines: State-of-the-Art and Future Perspectives. *Front. Mech. Eng.* **2022**, *8*, 944201. [CrossRef]
4. Ji, C.; Wang, S. Effect of hydrogen addition on combustion and emissions performance of a spark ignition gasoline engine at lean conditions. *Int. J. Hydrogen Energy* **2009**, *34*, 7823–7834. [CrossRef]
5. Fayaz, H.; Saidur, R.; Razali, N.; Anuar, F.; Saleman, A.; Islam, M. An overview of hydrogen as a vehicle fuel. *Renew. Sustain. Energy Rev.* **2012**, *6*, 5511–5528. [CrossRef]
6. Galloni, E.; Lanni, D.; Fontana, G.; D'Antuono, G.; Stabile, S. Performance estimation of a downsized SI engine running with hydrogen. *Energies* **2022**, *15*, 4744. [CrossRef]
7. Comotti, M.; Frigo, S. Hydrogen generation system for ammonia–hydrogen fuelled internal combustion engines. *Int. J. Hydrogen Energy* **2015**, *40*, 10673–10686. [CrossRef]
8. Kojima, Y.; Yamaguchi, M. Ammonia as a hydrogen energy carrier. *Int. J. Hydrogen Energy* **2022**, *47*, 22832–22839. [CrossRef]
9. CMørch, S.; Bjerre, A.; Gøttrup, M.; Sorenson, S.C.; Schramm, J. Ammonia/hydrogen mixtures in an SI-engine: Engine performance and analysis of a proposed fuel system. *Fuel* **2011**, *90*, 854–864. [CrossRef]
10. Onorati, A.; Payri, R.; Vaglieco, B.M.; Agarwal, A.K.; Bae, C.; Bruneaux, G.; Canakci, M.; Gavaises, M.; Günthner, M.; Hasse, C.; et al. The role of hydrogen for future internal combustion engines. *Int. J. Engine Res.* **2022**, *23*, 529–540. [CrossRef]
11. Xu, X.; Liu, E.; Zhu, N.; Liu, F.; Qian, F. Review of the Current Status of Ammonia-Blended Hydrogen Fuel Engine Development. *Energies* **2022**, *15*, 1023. [CrossRef]
12. Lanni, D.; Galloni, E.; Fontana, G.; D'Antuono, G. Assessment of the Operation of an SI Engine Fueled with Ammonia. *Energies* **2022**, *15*, 8583. [CrossRef]
13. Scharl, V.; Lackovic, T.; Sattelmayer, T. Characterization of ammonia spray combustion and mixture formation under high-pressure, direct injection conditions. *Fuel* **2023**, *333*, 126454. [CrossRef]
14. Cui, Z.; Tian, J.; Zhang, X.; Yin, S.; Long, W.; Song, H. Experimental Study of the Effects of Pre-Chamber Geometry on the Combustion Characteristics of an Ammonia/Air Pre-Mixture Ignited by a Jet Flame. *Processes* **2022**, *10*, 2102. [CrossRef]
15. El Fattah, S.F.A.; Ezzat, M.F.; Mourad, M.A.; Elkelawy, M.; Youssef, I.M. Experimental Investigation of the Performance and Exhaust Emissions of a Spark-Ignition Engine Operating with Different Proportional Blends of Gasoline and Water Ammonia Solution. *J. Eng. Res.* **2022**, *5*, 38–45.

16. Zhang, Z.; Long, W.; Dong, P.; Tian, H.; Tian, J.; Li, B.; Wang, Y. Performance characteristics of a two-stroke low speed engine applying ammonia/diesel dual direct injection strategy. *Fuel* **2023**, *332*, 126086. [CrossRef]
17. Zhang, H.; Li, G.; Long, Y.; Zhang, Z.; Wei, W.; Zhou, M.; Belal, B.Y. Numerical study on combustion and emission characteristics of a spark-ignition ammonia engine added with hydrogen-rich gas from exhaust-fuel reforming. *Fuel* **2023**, *332*, 125939. [CrossRef]
18. Lhuillier, C.; Brequigny; Contino, F.; Mounaïm-Rousselle, C. Experimental study on ammonia/hydrogen/air combustion in spark ignition engine conditions. *Fuel* **2020**, *269*, 117448. [CrossRef]
19. Frigo, S.; Gentili, R.; Doveri, N. Ammonia plus hydrogen as fuel in a SI engine: Experimental results. *SAE Tech. Pap.* **2012**. [CrossRef]
20. Wang, D.; Ji, C.; Wang, S.; Yang, J.; Wang, Z. Numerical study of the premixed ammonia-hydrogen combustion under engine-relevant conditions. *Int. J. Hydrogen Energy* **2021**, *46*, 2667–2683. [CrossRef]
21. Pyrc, M.; Gruca, M.; Tutak, W.; Jamrozik, A. Assessment of the co-combustion process of ammonia with hydrogen in a research VCR piston engine. *Int. J. Hydrogen Energy* **2022**, *48*, 2821–2834. [CrossRef]
22. Rodríguez, C.G.; Lamas, M.I.; Rodríguez, J.D.; Abbas, A. Possibilities of Ammonia as Both Fuel and NOx Reductant in Marine Engines: A Numerical Study. *J. Mar. Sci. Eng.* **2022**, *10*, 43. [CrossRef]
23. Liu, L.; Tan, F.; Wu, Z.; Wang, Y.; Liu, H. Comparison of the combustion and emission characteristics of NH_3/NH_4NO_2 and NH3/H2 in a two-stroke low speed marine engine. *Int. J. Hydrogen Energy* **2022**, *47*, 17778–17787. [CrossRef]
24. Wang, Y.; Zhou, X.; Liu, L. Theoretical investigation of the combustion performance of ammonia/hydrogen mixtures on a marine diesel engine. *Int. J. Hydrogen Energy* **2021**, *46*, 14805–14812. [CrossRef]
25. Lanni, D.; Galloni, E.; Fontana, G. Numerical analysis of the effects of port water injection in a downsized SI engine at partial and full load operation. *Appl. Therm. Eng.* **2022**, *205*, 118060. [CrossRef]
26. Scala, F.; Galloni, E.; Fontana, G. Numerical analysis of a downsized spark-ignition engine fueled by butanol/gasoline blends at part-load operation. *Appl. Therm. Eng.* **2016**, *102*, 383–390. [CrossRef]
27. Abdullah, N.R.; Shahruddin, N.S.; Mamat, A.M.I.; Kasolang, S.; Zulkifli, A.; Mamat, R. Effects of Air Intake Pressure to the Fuel Economy and Exhaust Emissions on a Small SI Engine. *Procedia Eng.* **2013**, *68*, 278–284. [CrossRef]
28. Dziubak, T.; Karczewski, M. Experimental Study of the Effect of Air Filter Pressure Drop on Internal Combustion Engine Performance. *Energies* **2022**, *15*, 3285. [CrossRef]
29. Dziubak, T.; Karczewski, M. Experimental Studies of the Effect of Air Filter Pressure Drop on the Composition and Emission Changes of a Compression Ignition Internal Combustion Engine. *Energies* **2022**, *15*, 4815. [CrossRef]
30. Metghalchi, M.; Keck, J.C. Burning velocities of mixtures of air with methanol, isooctane, and indolene at high pressure and temperature. *Combust. Flame* **1982**, *48*, 191–210. [CrossRef]
31. Metghalchi, M.; Keck, J.C. Laminar burning velocity of propane-air mixtures at high temperature and pressure. *Combust. Flame* **1980**, *38*, 143–154. [CrossRef]
32. Bozza, F.; Fontana, G.; Galloni, E.; Torella, E. 3D-1D analyses of the turbulent flow field, burning speed and knock occurrence in a turbocharged SI engine. *SAE Trans.* **2007**, *116*, 1495–1507.
33. Goldmann, A.; Dinkelacker, F. Approximation of laminar flame characteristics on premixed ammonia/hydrogen/nitrogen/air mixtures at elevated temperatures and pressures. *Fuel* **2018**, *224*, 366–378. [CrossRef]
34. Heywood, J.B. *Combustion Engine Fundamentals*, 1st ed.; McGraw-Hill Education: New York, NY, USA, 1988; Volume 25, pp. 1117–1128.
35. Livengood, J.; Wu, P.C. Correlation of Autoignition Phenomena in Internal Combustion Engines and Rapid Compression Machines. In Proceedings of the 5th Symposium (International) on Combustion, Pittsburg, PA, USA, 30 September–3 October 1955.
36. Goodwin, D.G.; Moffat, H.K.; Speth, R.L. *Cantera: An Object-Oriented Software Toolkit for Chemical Kinetics, Thermodynamics, and Transport Processes, Version 2.2. 1.*; Cantera Developers: Warrenville, IL, USA, 2017.
37. Nakamura, H.; Hasegawa, S.; Tezuka, T. Kinetic modeling of ammonia/air weak flames in a micro flow reactor with a controlled temperature profile. *Combust. Flame* **2017**, *185*, 16–27. [CrossRef]

energies

MDPI

Article

Linear Model of a Turboshaft Aero-Engine Including Components Degradation for Control-Oriented Applications

Teresa Castiglione [1], Diego Perrone [1], Luciano Strafella [2], Antonio Ficarella [2] and Sergio Bova [1,*]

[1] Department of Mechanical Energy and Management Engineering, University of Calabria, 87036 Rende, Italy; teresa.castiglione@unical.it (T.C.)
[2] Department of Engineering for Innovation, University of Salento, 73100 Lecce, Italy
* Correspondence: sergio.bova@unical.it

Abstract: The engine fuel control system plays a crucial role in engine performance and fuel economy. Fuel control, in traditional engine control systems, is carried out by means of sensor-based control methods, which correct the fuel flow rate through correlations or scheduled parameters in order to reduce the error between a measured parameter and its desired value. In the presence of component degradation, however, the relationship between the engine measurable parameters and performance may lead to an increase in the control error. In this research, linear models for advanced control systems and for direct fuel control in the presence of components degradation are proposed, with the main objective being to directly predict and correct fuel consumption in the presence of degradation instead of adopting measurable parameters. Two techniques were adopted for model linearization: Small Perturbation and System Identification. Results showed that both models are characterized by high accuracy in predicting the output engine variables, with the mean errors between model prediction and data below 1%. The maximum errors, recorded for shaft power, were about 6% for Small Perturbation and lower than 3% for System Identification. A simple correlation between engine performance and components degradation was also demonstrated; in particular, the achieved results allow one to conclude that the Small Perturbation approach is the best candidate for controller development when a prediction of components degradation is included.

Keywords: turbo-shaft engine model; gas turbine linear models; compressor degradation; small perturbation; system identification

Citation: Castiglione, T.; Perrone, D.; Strafella, L.; Ficarella, A.; Bova, S. Linear Model of a Turboshaft Aero-Engine Including Components Degradation for Control-Oriented Applications. *Energies* **2023**, *16*, 2634. https://doi.org/10.3390/en16062634

Academic Editors: Roberta De Robbio, Maria Cristina Cameretti and Andrzej Teodorczyk

Received: 30 December 2022
Revised: 2 March 2023
Accepted: 9 March 2023
Published: 10 March 2023

1. Introduction

The aviation industry is committed to reducing emissions and complying with the even more restrictive regulations aimed at reaching the target of net-zero carbon emissions by 2050 [1]. New technologies are being developed to mitigate the impact of the aviation sector on the global climate. To this aim, new ultra-high efficiency and ultra-low emissions aero-propulsion systems, which include, for instance, Ultra High Bypass Ratio (UHBR) engines and Hybrid Electric Propulsion (HEP) [2–4], are being considered by engine manufacturers. These options, however, require major design changes in aircrafts, and are only viable as long-term solutions. In the short-term, strategies for optimizing flight routes for minimum emissions, for improving maintenance schedules and other operational measures, are being adopted. In addition, the use of alternative fuels, the so-called sustainable aviation fuels (SAF), which encompass bio and synthetic fuels, are currently the most viable option in achieving future emission goals and contributing to lowering the dependency on fossil fuels [5]. However, if improvements in the engine components design and development and the use of alternative fuels are necessary to guarantee the achievement of such ambitious targets, the development of advanced engine control systems to satisfy the requirements of the more complex engine architectures operating with the new fuels safely and efficiently will be a fundamental contribution [6,7].

Traditionally, the control system has the function of providing the desired thrust, and guaranteeing that the engine operates within safe boundaries. The fuel flow rate, therefore, needs to be tuned in order to achieve the desired power, and to satisfy the various control modes simultaneously. To fulfil these requirements, min-max control strategies are usually adopted. Moreover, by means of gain scheduling, the controller parameters are tuned at different operating points in such a way that the desired performance characteristics can be achieved over the whole flight envelope [8]. The control of thrust, inlet temperature of high-pressure turbine and surge margins is usually made in an indirect way, by means of measurable parameters such as shaft speeds and pressure [9,10]; thus, a significant number of sensors are installed in the engine. This approach requires a complex architecture [11], and is characterized by a heavy computational burden that arises from multiple iterative control loops [12]. To overcome these limitations and to improve engine performance, modern control methodologies are being developed.

Novel control concepts, known as model-based control (MBC), aim at achieving the direct control of unmeasurable performance parameters through on-line model outputs instead of adopting feedback sensor data [13,14]. Among the various approaches, great attention is being paid to Model Predictive Control (MPC), which is attracting significant attention in all those fields where increasing energy efficiency is required [15], and in particular, to aircraft engines [16–18]. Montazeri et al. [6] developed a turbofan engine control system based on an MPC approach. They first developed a thermodynamic non-linear engine model, which was linearized for model predictive controller application. They demonstrated that the controller satisfies all of the engine constraints, and confirmed successful software and hardware implementation of the control algorithm in real-time. Pang et al. [19] developed a non-linear state-space model for MPC, which, differently from [6], applied an extended Kalman filter every sampling instant, and was based on a quadratic constrained problem rather than a higher-order nonlinear problem. Simulation results demonstrated that this approach allows one to achieve smaller control errors than the standard model predictive control algorithm does, but also that the proposed controller is more time-saving than the standard model predictive control (about 61%). In this context, the development of reliable control-oriented on-board models plays a fundamental role.

In this regard, several approaches to on-board control-oriented engine modeling were investigated for MPC algorithms. These can be classified into non-linear and linear approaches. Among the non-linear approaches, real-time component-level modeling (CLM) uses the physical laws governing the system to derive the mathematical model of the aero-engine. Several models based on CLM were developed over years, and can be found in [20–22]. A full description, including the dynamic equations of all the engine components, can be found in Montazeri et al. [6]. However, owing to the complex involved dynamics, the development of a thorough and precise mathematical model of an aero-engine is generally very challenging. Therefore, alternative approaches, based on data-driven modeling methods, are gaining popularity in the aero-engine community [23]. In particular, artificial intelligence algorithms are currently widely utilized for engine real-time modeling and simulations [24,25]. However, these approaches usually rely on a large amount of data training, and the computational complexity is usually high. Moreover, from a control point of view, the implementation of a non-linear model always results in a non-linear constrained problem to be optimized [17,26]. On the contrary, the adoption of linear real-time models of the engine generates a quadratic constrained problem, which allows one to achieve an optimal solution with less computational effort in comparison to non-linear formulations [27–29]. In addition, a large number of control methods are based on linear control theory. It is, therefore, convenient to deal with linear models, provided that they are sufficiently accurate. These linear models are often obtained by linearization procedures applied to non-linear models.

Several approaches to develop linear models can be found in the literature. The most common approach for the derivation of a linearized model around an equilibrium condition is based on small input and state perturbations [30]. These models adequately capture the system dynamics near the operating condition, and are valid in a neighborhood of an operating point in certain flight conditions. However, they deviate in accuracy during transients of the system away from equilibrium. To overcome these limitations, Chung et al. [31] presented an analytical method to linearize the engine equations in each time step, and extracted a precise linear model to be used for controller design. When large amounts of experimental data are available instead of a non-linear thermodynamic model, System Identification can be adopted to derive a linear model [32,33]. A comprehensive study about the accuracy of different numerical methods adopted to perform the linearization of the thermodynamic non-linear model of a turbofan engine can be found in Montazeri et al. [34]. The results showed that the linear models obtained from system identification and central difference perturbation methods have higher percentages of compliance compared to the others. However, it is worth pointing out that in the analyzed cases, the on-board modelling and linearization of gas turbine aero-engines was developed for nominal engines, which represent new or clean engine performance.

Control-oriented on-board engine modeling cannot ignore engine components degradation. In fact, due to the normal aging in gas turbine aero-engines with increasing flight cycles, the engine performance deviates from its nominal state; in particular, unmeasured safety-critical parameters such as thrust, surge margin and turbine entry temperature are strongly affected by the engine health status. Therefore, on-board clean engine models cannot predict the control variables of degraded engines correctly, with adverse effects on controller reliability. Degraded engines on-board modeling is, therefore, strongly motivated. A few studies regarding the inclusion of degradation effects in engine models can be found in the literature; however, the majority of this research focused on the adoption of non-linear engine models and require a great amount of data, which burden the on-line computation. Models based on using a Kalman filter for the estimation of engine degradation parameters, aimed at tuning the piecewise linear model to match the degraded engine, can be found in [35–37]. Recently, Wei et al. [7] proposed a novel Hybrid Wiener Model (HWM), which takes into account the degradation effects using post-flight data. Sun et al. [8] included the degradation effects in a non-linear component-level model by means of degradation parameters such as efficiency and flow capacity indexes, and proposed an approach for controller adjustment to deal with degradation implications by developing an industrial min-max control strategy.

In the present study, the possibility of including the effects of engine components degradation directly into the linear model coefficients was investigated. The novel contribution relies on the estimation of the linear model coefficients as a function of components degradation, in order to extrapolate, if any, a simple correlation law. This would allow the achievement of 'real-time' corrections for direct fuel control and improved engine response in the presence of degradation. Therefore, the control objective of the proposed model is to predict the specific fuel consumption directly instead of adopting measurable parameters. In particular, two linearization techniques were evaluated: small perturbation and system identification state-space models. In both cases, the study considered only compressor degradation, evaluated in terms of compressor efficiency and flow coefficient, and analyzed the matrices of the linear model for different degradation values estimated through the monitoring system. The research was developed in the framework of an EU-funded wider project, whose final goal is to define engine controllers that take into account the natural degradation of all engine components simultaneously.

2. Materials and Methods

In order to develop an accurate control-oriented engine linear model that takes into account component degradation, a comprehensive physics-based thermodynamic model was built using the component-level modeling approach. The non-linear model results were adopted for linearization purposes. In particular, two methodologies were proposed for model linearization: Small Perturbations [38,39] and System Identification [40–42] state-space models. Both approaches were applied to the engine model, with inclusion of a certain level of components degradation, and the linear coefficients as a function of component degradation were estimated in order to evaluate the possibility of extrapolating a correlation law, if any.

In the following subsections, the non-linear model and the adopted linearization approaches are discussed.

2.1. Engine Thermodynamic Model and Test Conditions

A rotorcraft with a suitable take-off weight for urban air mobility was studied in this paper. The rotorcraft was equipped with a two-spool turboshaft engine, the PW200 Pratt & Whitney Canada, which is developed specifically for helicopter applications. The propulsion system included a free turbine (Low Pressure Turbine (LPT)), connected to the rotor shaft with a nominal speed of 6000 rpm, and a single stage centrifugal compressor driven by a single stage turbine (High Pressure Turbine (HPT)). The specifications for the PW200 are reported in Table 1.

Table 1. PW200 technical specifications.

Description	Value
Power	200–400 [kW]
Weight	110 [kg]
Pressure ratio	8:1
Turbine Inlet temperature	1173.5 [K]
SFC (Specific Fuel Consumption)	0.426–0.33 [kg/kWh]
Compressor Configuration	1 centrifugal
Turbine configuration	HPT, LPT

Due to a lack of experimental data for the selected engine, detailed simulations with GSP software (Gas Turbine Simulation Program) [43] were performed in the present study. The GSP is a tool developed at the Aerospace Department of Delft Technical University, which allows for both transient and steady-state simulations of almost all types of gas turbines (turboshaft, turboprop, turbofan, single and multi-shafts, etc.) by properly establishing a specific arrangement of engine components, and through careful model calibration.

The arrangement of engine components adopted in the present research, which reflects the PW200 configuration, is presented in Figure 1.

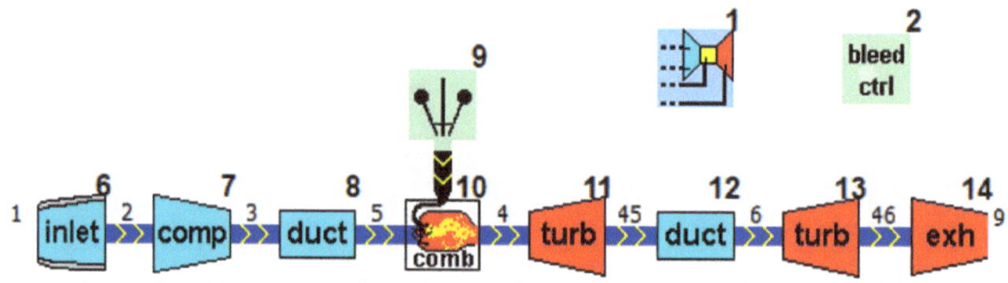

Figure 1. Model of the turboshaft in the GSP.

The model links an inlet (block 6), a compressor (block 7), a combustion chamber (block 10), a high-pressure turbine (block 11) and a low-pressure turbine (block 13) with its exhaust (block 14) connected to the output shaft. Block 1 represents the "manual case control", which changes the desired power, the altitude and the Mach number according to the design of the experiments. Block 2 is the controller of the bleeding air, and block 9 sets the specification of the fuel. Two additional components named "duct" were included in the scheme, and were utilized in transient models to take into account dynamics and volumetric effects, as well as to perform mass balance calculations according to the ICV-Intercomponent Volume method [44,45]. They have no effect on the engine's performance and operational characteristics during steady-state conditions.

As for model calibration, several steps were required:

- The design point, whose main parameters are reported in Table 2, was assumed to be sea-level static (SLS) and at international standard atmospheric (ISA) conditions (temperature = 288.25 K, pressure = 1.013 bar, altitude = 0 m, Mach = 0);
- The default compressor and turbine maps were scaled according to a set of input parameters taken from the database of the reference engine, i.e., intake pressure ratio, design rotor speed, design shaft speed;
- An iteration was run in the GSP by adjusting the design air and fuel flow rates injected into the combustion chamber, in order to obtain the desired power output and SFC. The iteration allowed for an accuracy of 1% in the prediction of take-off power with respect to the design take-off power of the reference engine;
- Using block 1, which allowed for manual case control, the off-design engine variables when changing the operating conditions (desired power, Mach number, altitude) were obtained.

Table 2. PW200 design points.

Description	Value
Power (POW)	295 [kW]
Intake Pressure Ratio (PR)	0.988
Air Flow Rate (W_a)	2 [kg/s]
Compressor pressure ratio (β_c)	7.17
Combustion Efficiency (η_b)	0.985
Fuel Flow Rate (W_f)	0.0315 [kg/s]
Compressor Rotor Speed (N)	40,891 [rpm]
Compressor Efficiency (η_c)	0.825
LPT Rotor Speed (N_{LP})	6000 [rpm]
Turbine Efficiency (η_t)	0.88
Spool Mechanical Efficiency (η_m)	0.99
Mach, Altitude	0, 0

This research examined two mission profiles reported in Figure 2, a simple delivery trip (Mission A for "Transport"), and a search-and-rescue (SAR) mission in a mountain environment (Mission B for "Mountain Rescue"). These missions were also used to generate several points for engine failure, as shown in Table 3.

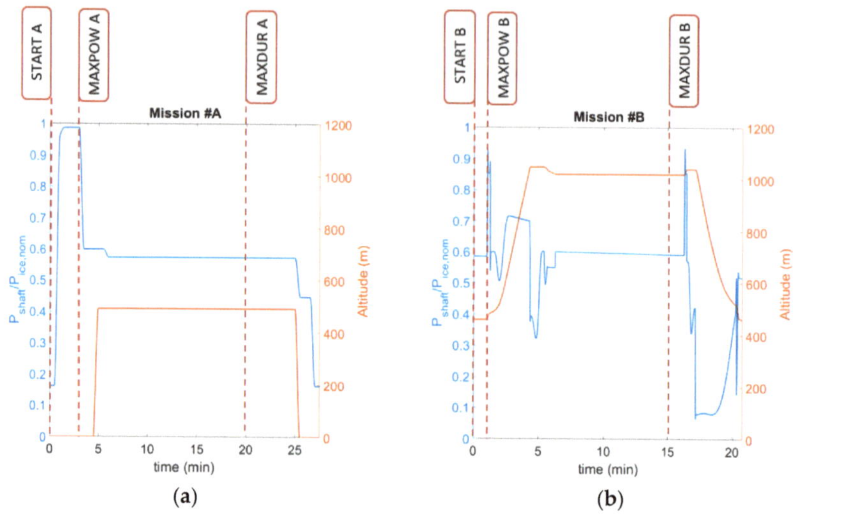

Figure 2. Reference missions. (**a**) Mission A for "Transport"; (**b**) Mission B for "Mountain Rescue".

Table 3. Simulated working points.

Description	Speed [m/s]	Altitude [m]	Power [KW]	Low-Pressure Turbine Speed [rpm]
START-A	30.6	0	48	6000
START-B	0	1150	173	6000
MAXPOW-A	30.6	0	291	6000
MAXPOW-B	3.96	1154	273	6000
MAXDUR-A	30.6	492	169	6000
MAXDUR-B	0	2550	176	6000

However, the purpose of the present study relied on the achievement of a linear model of the engine around a working point, which is representative of an equilibrium condition. In particular, the selected operating condition is indicated in Figure 2 as MAXDUR-A, and is characterized by a power lever angle (PLA) of about 50%, altitude h = 50 m and Mach = 0.09. In order to predict the dynamic behavior of the system, the engine equilibrium point must be altered by enforcing small changes to the PLA by applying step, ramp and pulse signals. In this case, variations of ±10% with respect to the steady state value were applied to the PLA according to the sequence displayed in Figure 3; transient simulations, which include dynamic phenomena such as spool inertia, heat transfer in turbomachinery and volumetric effects, were carried out by means of the GSP. The time series data sets obtained from simulations included compressor pressure ratio, compressor rotational speed, inlet and outlet turbine temperature, fuel flow rate, specific fuel consumption, engine torque and power.

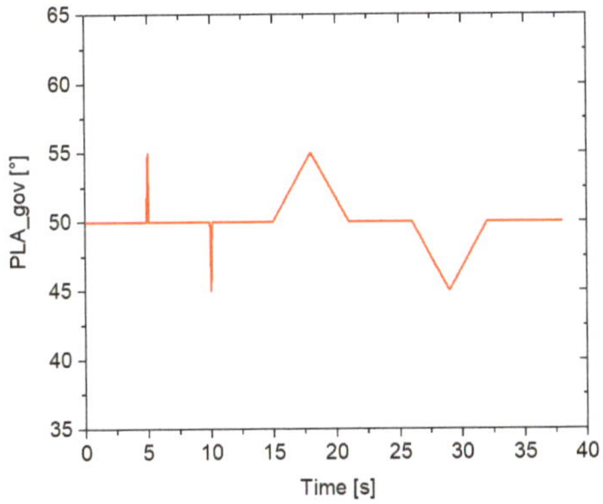

Figure 3. PLA signal.

In the current simulations, the machinery degradation was accounted for. During its life cycle, in fact, an engine undergoes several issues due to over-temperatures, erosion of the compressor and turbine blades, fouling, bird strikes, etc., which contribute to increased machine degradation; as a direct consequence, engine performance reduces. In the literature, these mechanisms cause performance losses approaching 10% in the process [45]. In order to predict the engine performance degradation, the exhaust gas temperature (EGT), which reflects the performance degradation of the aeroengine, is usually monitored and used for engine control, fault diagnosis and maintenance decisions. The GSP allows one to illustrate the engine's deterioration in a variety of ways. The most common method is to conduct a simulation with a continuous degree of degradation, such as continuous reductions in compressor efficiency and mass flow rate. In the current simulations, step changes, represented as a modification of the compressor isentropic efficiency and mass flow parameter, were investigated in order to make the analysis as generic and feasible as possible. The step changes included reductions in compressor efficiency and mass flow parameters amounting to 2%, 4%, 8%, 12% and 15% with respect to healthy conditions. These deterioration levels were still compatible with safe operation of the propeller, as reported in [45,46].

2.2. Linear Model

The primary problem in controller design is to develop a linear model, at the considered operating point of the engine, which can be presented in the following form:

$$\begin{cases} \dot{x} = Ax + Bu \\ y = Cx + Du \end{cases} \tag{1}$$

where *x(t)* is the state of the model, *u(t)* is the input variable vector and *y(t)* is the output vector. *A*, *B*, *C* and *D* are the linearization matrices. In this formulation, the heat-transfer process can be neglected, owing to the fact that the thermal inertia of the engine components is usually much smaller than that of the rotors [47,48]. Therefore, only the engine shafts dynamics can be taken into account. In addition, in the analyzed engine configuration, the low-pressure turbine shaft is connected to the rotor, which is kept at a constant speed. Hence, the only considered dynamic is the high-pressure turbine shaft rotational speed, N_{HP}. In summary, the state of the system, *x(t)*, is given by the high-pressure turbine shaft speed (N_{HP}); the input vector, *u(t)*, is given by the fuel flow rate (*Wf*), which, from a controller point of view, is the manipulating variable; and the output vector, *y(t)*, is defined according to physical and structural limits that must be satisfied in the controller design to protect the engine components. These limits include over-pressurization, over-temperature and over-speed, which are controlled by means of some of the following engine performance parameters [34]: the total pressure at the compressor exit (PT_3), the total temperature at the HPT exit (TT_{45}), the HP shaft speed, N_{HP}, and the shaft power. Therefore, the linear model had 1 input, 4 outputs and 1 state variable, which are fully listed in Table 4.

Table 4. State variables, control inputs and outputs of the desired engine.

State Variables, x	Output Variables, y	Input Variables, u
HP shaft speed, N_{HP} [rpm]	HPC exit total pressure, PT_3 [bar] HPT exit total temperature, TT_{45} [K] HP shaft speed, N_{HP} [rpm] Shaft Power, *Pow* [kW]	Fuel flow, W_f [kg/s]

The approaches proposed for the turboshaft engine dynamic model linearization are discussed in the next subsections.

2.2.1. Perturbation Method

The general form of engine non-linear equations, which include fast and slow dynamics, can be written as follows [31]:

$$\dot{z} = g(x, z,\ u) \tag{2}$$

$$\dot{x} = f(x, z,\ u) \tag{3}$$

where z denotes the fast states (mainly flow dynamics), while x denotes the slow states (shaft dynamics, etc.). However, as already mentioned, if it is assumed that the fast dynamics can be neglected and only the high-pressure turbine shaft dynamics is included, only Equation (3) can be considered for linearization purposes. To this aim, an arbitrary position $\overline{x},\ \overline{z}_0,\ \overline{u}$ is selected, and perturbations (δx,δu) from a linearization point are applied:

$$\dot{x} = f(\overline{x} + \delta x, \overline{z}_0,\ \overline{u} + \delta u) \tag{4}$$

By differentiating Equation (4) and neglecting the higher order terms, the following equation is obtained:

$$\dot{x} \approx f(\overline{x}, \overline{z_0},\ \overline{u}) + \left.\frac{\partial f}{\partial x}\right|_{\overline{x},\overline{z}_0,\overline{u}} \delta x + \left.\frac{\partial f}{\partial u}\right|_{\overline{x},\overline{z}_0,\overline{u}} \delta u \tag{5}$$

As linearization is often achieved about an equilibrium condition where $f(\overline{x}, \overline{z_0},\ \overline{u}) = 0$, Equation (5) can be simplified to the following:

$$\dot{x} = A\delta x + B\delta u \tag{6}$$

where $A = [\partial f/\partial x]$ and $B = [\partial f/\partial u]$.

Similarly, the non-linear equations of the output variables are written as follows:

$$y = h(x, z,\ u) \tag{7}$$

A Taylor approximation is applied to the engine model, Equation (7), at the equilibrium point ($\overline{x},\ \overline{z}_0,\ \overline{u}$), and the retaining constant and first-order terms yield the following state variable model:

$$y \approx h(\overline{x}, \overline{z_0},\ \overline{u}) + \left.\frac{\partial h}{\partial x}\right|_{\overline{x},\overline{z}_0,\overline{u}} \delta x + \left.\frac{\partial h}{\partial u}\right|_{\overline{x},\overline{z}_0,\overline{u}} \delta u = C\delta x + D\delta u \tag{8}$$

where $C = [\partial h/\partial x]$ and $D = [\partial h/\partial u]$ are the system matrices with appropriate dimensions.

The non-linear engine model in the current case included only the conservation of the angular momentum of the high-pressure turbine shaft. The non-linear equation that describes the transient process is given by the following [49]:

$$\dot{N}_{HP} = \frac{\Delta Pow_{HP}}{I_{HP} N_{HP} \left(\frac{2\pi}{60}\right)^2} \tag{9}$$

In the above equation, $\dot{N}_{HP}$ is the angular acceleration of the high-pressure turbine shaft, ΔPow_{HP} is the difference between the power produced by the turbine and the power required by the compressor, and I_{HP} is the moment of inertia of the moving parts of the spool. ΔPow_{HP} is obtained by modeling the compressor and turbine with the help of the machinery maps, which allows the evaluation of pressure, flow rate and efficiency while varying the rotational speed of the components. The correlation between ΔPow_{HP} and fuel flow was obtained by modeling the burner. In fact, fuel flow is a direct input to the burner; inside the combustor, the air and fuel are mixed and burned to produce high temperature gas, which drives the turbine and produces torque. However, the function f, which correlates ΔPow_{HP} to shaft speed and fuel flow, is not explicitly defined in Equation (9). Therefore, the state-space representation of the engine dynamic system can be written as follows:

$$\begin{aligned} \dot{N}_{HP} &= \frac{1}{I_{HP}} \times f\left(N_{HP}, W_f\right) \\ y_i &= h\left(N_{HP}, W_f\right) \end{aligned} \tag{10}$$

By applying the small perturbations in the form of Equation (6), the expansion of Equation (10) can be written as follows [34]:

$$\begin{aligned} \Delta\dot{N}_{HP} &= \frac{1}{I_{HP}}\left(\frac{\partial f}{\partial N_{HP}}\Delta N_{HP} + \frac{\partial f}{\partial W_f}\Delta W_f\right) \\ \Delta y_i &= \frac{\partial h_i}{\partial N_{HP}}\Delta N_{HP} + \frac{\partial h_i}{\partial W_f}\Delta W_f \end{aligned} \tag{11}$$

By a comparison between Equations (6), (8) and (11), it can be observed that the matrices A, B, C and D contain the partial derivatives, and are therefore expressed as follows:

$$\begin{aligned} A &= \left[\frac{\partial f}{\partial N_{HP}}\right] \\ B &= \left[\frac{\partial f}{\partial W_f}\right] \\ C &= \left[\frac{\partial y_1}{\partial N_{HP}}\ \frac{\partial y_2}{\partial N_{HP}}\ \frac{\partial y_3}{\partial N_{HP}}\ \frac{\partial y_4}{\partial N_{HP}}\right] \\ D &= \left[\frac{\partial y_1}{\partial W_f}\ \frac{\partial y_2}{\partial W_f}\ \frac{\partial y_3}{\partial W_f}\ \frac{\partial y_4}{\partial W_f}\right] \end{aligned} \tag{12}$$

The values of the partial derivatives were obtained by deviating the input in Equation (11) from its steady value and by collecting, by means of the GSP, the corresponding output and state variables. Their variations, i.e., Δy_i, ΔN_{HP}, ΔW_f, were obtained at each sampling time. Similarly, $\dot{N}_{HP}$ was obtained from finite differences. Therefore, the partial derivatives in Equation (12), which represent the unknown parameters of the system, were obtained through a linear regression over the collected data.

Each linear model was valid only in a neighborhood of an operating point in certain flight conditions (Mach, altitude, weather conditions, etc.). Therefore, in order to develop a controller capable of covering all flight conditions, this procedure had to be repeated for different operating points, and a set of linearized models was collected; hence, piecewise control was implemented.

2.2.2. System Identification

An alternative way to obtain a linear model is based on the adoption of System Identification [40–42]. This methodology is based on a "black-box" approach, in which the mathematical model of a dynamic system is based on observed data. In practice, no physical details of the non-linear model are needed, only the input/output data of the dynamic system.

System identification requires several steps, which can be summarized as follows:

- Attaining input/output data of the dynamic system;
- Selecting a suitable linear model structure from among discrete/continuous transfer functions or state-space forms;
- Fitting a suitable linear model to data and considered model structure;
- Verifying whether the model is good enough to represent the system.

As for the input/output data (a), these could be collected by means of experimental data or through a valid simulation tool. Concerning the linear model structure (b), the state-space form is usually preferable for the design of the controller. The linear model fitting (c) is usually available in various computational tools; in this case, it was carried out using the system identification toolbox of MATLAB [49]. Finally, a number of validation tests were available from the literature (d); these included correlation tests, one-step ahead prediction, model-predicted output, estimation, and test data.

From the literature [34], it was expected that the system identification approach would provide better results in terms of percentages of compliance to the experimental data with respect to the perturbation approach. However, since the linearization procedure was the result of a best fitting procedure, the drawback of the SI was that it relied on the loss of a physical meaning of the matrices elements, which had a relevant impact on the development of the controller, as will be discussed below.

3. Results and Discussion

In this section, a comparison between the proposed approaches for system linearization to a case study is presented. The case study was based on a simple delivery trip, in which the operating equilibrium conditions were characterized by a power lever angle (PLA) of about 50%, altitude h = 50 m and Mach = 0.09 (Figure 2). The dynamic behavior of the system was predicted in GSP software for different degradation levels of the compressor through the application of step, ramp and pulse signals to the PLA (Figure 3).

For each perturbation of the PLA around the equilibrium condition, the small perturbation and system identification approaches were applied to linearize the engine model. For both approaches, the input/output data and the state variables, obtained from simulations carried out by means of GSP software, were used for the estimation of the matrix coefficients of the linear model during the dynamic maneuvers. For the perturbation method, the matrix coefficients were obtained through a linear regression, whereas the system identification method obtained these by means of the MATLAB toolbox. Both methods allowed for prediction of the output engine variables with different levels of compressor degradation.

In order to assess the accuracy of linearization, the maximum error percentage, *MaxErr*, the mean error percentage, *MeanErr* and the coefficient of determination, R^2, were defined as follows:

$$
\begin{aligned}
MaxErr &= \frac{max(|Y_{GSP} - Y_{LM}|)}{max(|Y_{GSP}|)} \times 100 \\
MeanErr &= \frac{\frac{\sum_{i=1}^{N}(|Y_{GSP}(i) - Y_{LM}(i)|)}{N}}{max(|Y_{GSP}|)} \times 100 \\
R^2 &= 1 - \frac{\sum_{i=1}^{N}(Y_{GSP}(i) - Y_{LM}(i))^2}{\sum_{i=1}^{N}(Y_{GSP}(i))^2 - \frac{\left(\sum_{i=1}^{N} Y_{LM}(i)\right)^2}{N}}
\end{aligned}
\qquad (13)
$$

In Equation (13), Y_{GSP} is the output signal from the non-linear GSP model, Y_{LM} is the output signal from the linear model, and N represents the number of sampling times.

3.1. Linearized Turboshaft Model with Perturbation Method

Table 5 summarizes the comparison between the linear model and the GSP data for the output variables for the different levels of compressor degradation and flow coefficient, for the engine operating at the point previously indicated as MaxDurA (Table 3). It can be observed that the linear model predicts the output variables with high accuracy. The results show that the mean error was below 1% for all of the output variables, and that the maximum error, which was lower than 6%, was recorded for shaft power. These values, which can be considered excellent according to the literature [34], allowed us to conclude that the linear model obtained through the perturbation method provides optimal predictive capabilities.

Table 5. Linear model validation.

OUTPUTVAR. / DEGRAD. LEVEL	PT_3 [bar]			TT_{45} [K]			Power [kW]		
	Mean Err. %	Max Err. %	R^2	Mean Err. %	Max Err. %	R^2	Mean Err. %	Max Err. %	R^2
Clean	0.08	1.36	0.994	0.06	3.46	0.968	0.23	5.81	0.987
2%	0.07	1.39	0.995	0.06	3.40	0.969	0.23	5.86	0.987
4%	0.07	1.48	0.994	0.06	3.37	0.971	0.23	5.99	0.987
8%	0.07	1.51	0.994	0.06	3.24	0.973	0.24	5.97	0.987
12%	0.07	1.49	0.994	0.06	3.12	0.975	0.24	5.92	0.988
15%	0.07	1.48	0.995	0.06	3.00	0.976	0.24	5.88	0.988

Figure 4 shows the results of the linear model for the clean and damaged compressor at two different levels of compressor degradation and flow coefficient (8 and 12%, respectively), when small PLA perturbations were enforced according to Figure 3. For each case, a sudden increase in fuel flow rate caused a rise in turbine exhaust gas temperature (Figure 4a) and turbine speed (Figure 4b), and, consequently, a higher power at the turbine shaft (Figure 4c). The higher shaft speed increased the compressor pressure ratio, as shown in Figure 4d. The inverse behavior for the same output variables was obtained when the fuel flow rate was reduced.

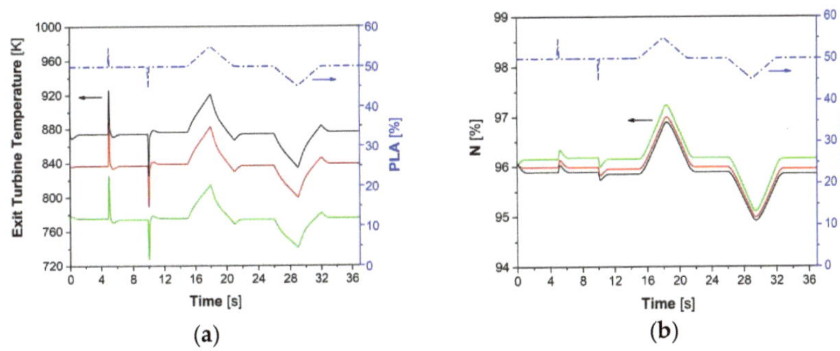

Figure 4. *Cont.*

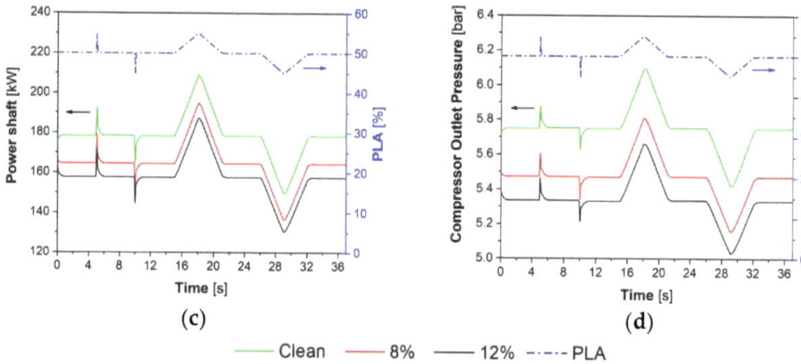

Figure 4. Comparison between the GSP and linear model output variables. (**a**) Exhaust gas temperature; (**b**) shaft speed; (**c**) shaft power; (**d**) compressor outlet pressure. The black arrow indicates the primary axis variables; the blue arrow refers to the secondary axis (PLA).

The linearization matrices of Equation (12) were computed for each of the analyzed cases, and the coefficients of matrices *A*, *B*, *C* and *D* were plotted in Figures 5–8, respectively. It can be observed that the trend of each coefficient, with increasing the compressor degradation level, is approximately linear for most of the cases. This is a very interesting and favorable occurrence from a controller point of view, as it allows one to implement, for each matrix element, a linear law with the health status of the component. These laws, together with the R^2, are summarized in Table 6.

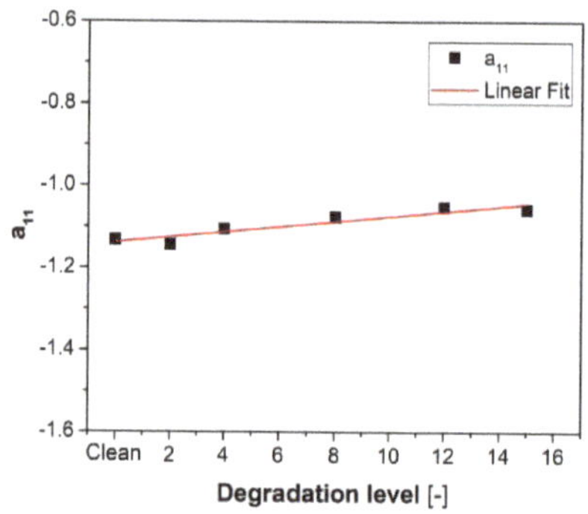

Figure 5. Coefficient a_{11} computed from linearization for different levels of compressor degradation (dots) and the linear fit curve.

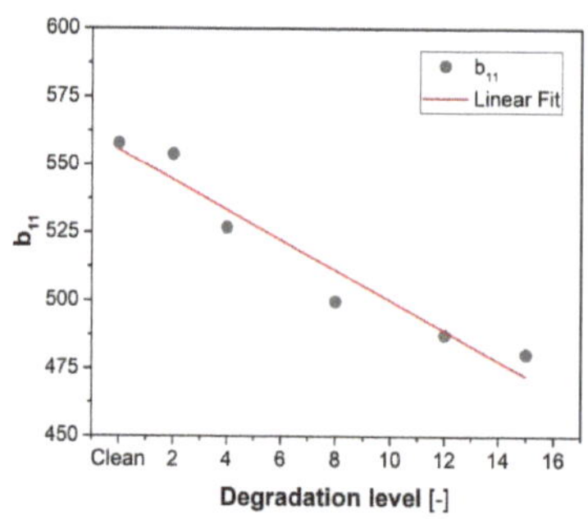

Figure 6. Coefficient b_{11} computed from linearization for different levels of compressor degradation (dots) and the linear fit curve.

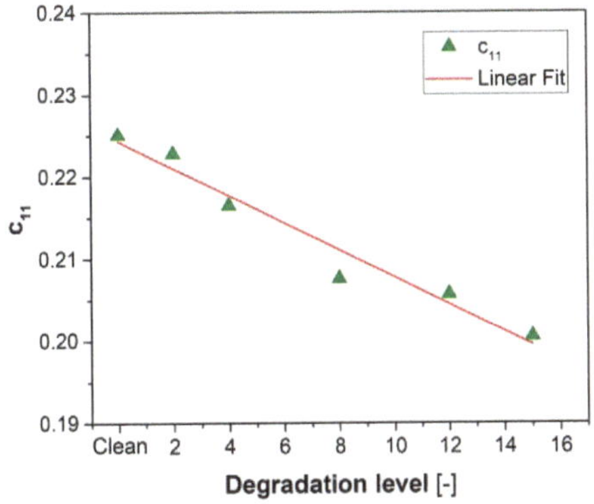

Figure 7. Coefficient c_{11} of Matrix C computed from linearization for different levels of compressor degradation (dots) and the linear fit curve.

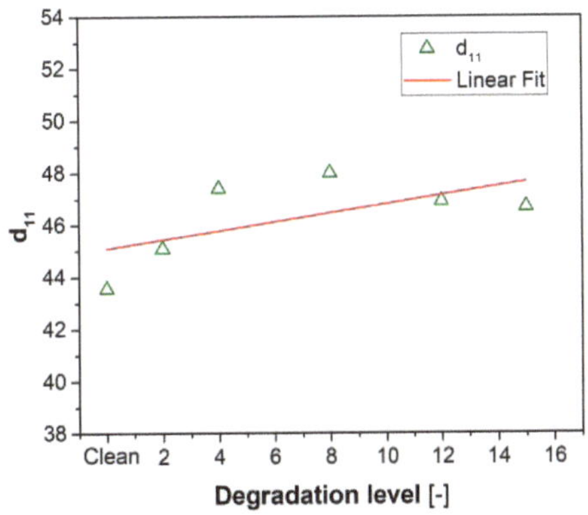

Figure 8. Coefficient d_{11} of Matrix D computed from linearization for different levels of compressor degradation (dots) and the linear fit curve.

Table 6. Matrices linear laws with degradation degree, *h*.

Matrix Coefficient	Linear Law	R^2
a_{11}	$a_{11} = 0.003h - 1.129$	0.862
b_{11}	$b_{11} = -5.912h + 562.5$	0.902
c_{11}	$c_{11} = -0.002h + 0.225$	0.981
c_{21}	$c_{21} = -9.52h - 0.278$	0.998
c_{31}	$c_{31} = -0.179h + 27.28$	0.929
c_{41}	$c_{41} = -0.111h + 16.972$	0.929
d_{11}	$d_{11} = 0.134h + 44.96$	0.853
d_{21}	$d_{21} = 198.5h + 17{,}714$	0.994
d_{31}	$d_{31} = 9.526h + 8047.3$	0.778
d_{41}	$d_{41} = 6.749h + 4999.7$	0.739

As Equation (11) shows, each term of the linearization matrices represents a physical property of the aero-engine system. As for Equation (11), for instance, a_{11} represents the time constant of the engine through the correlation $\tau = -\frac{1}{a_{11}}$ [12]. Hence, a_{11} was expected to be negative; moreover, with increasing compressor degradation, a_{11} increased (Figure 6). This indicates that, as expected, a damaged compressor contributes to retarding the time of response of the shaft. As for b_{11}, this coefficient correlates the shaft acceleration to a fuel flow rate variation. Therefore, when an increase in fuel flow rate is applied, a shaft acceleration is expected. When the compressor is damaged, the higher the degradation level, the slower the shaft accelerates. The linear trend is observed in Figure 6.

As for the output equation (Equation (11)), the terms of matrices C and D are the partial derivatives of each output variable, and represent the slopes of a small change in the output variable with respect to a small change in the state variable, N_{HP}, and with respect to a small change in the input variable, W_f, respectively. These coefficients varied with the level of degradation of the compressor, as plotted in Figures 7 and 8. For instance, with reference to c_{11}, which represents the rate of change in compressor pressure PT_3 to a small variation in shaft speed, it is clear that by increasing the shaft speed, the compressor outlet pressure increases accordingly; therefore, $c_{11} > 0$. If the compressor is damaged, the same increase in shaft speed corresponds to a lower pressure rise. This was confirmed by the decreasing trend in c_{11}, as displayed in Figure 7. Similar considerations can be made for the other coefficients.

Finally, the achieved results allowed us to carry out an analysis on the effects of engine degradation on the output variables. A steady-state operating condition, in which fuel flow rate and engine speed were fixed (fixed PLA), was considered for different values of compressor degradation. Figure 9 (top) shows that the more the compressor was damaged, the lower the pressure and the higher the air temperature at the compressor exit. In particular, an efficiency drop of 8% caused a pressure reduction of about 4% and a temperature rise of 2.8%; if the compressor efficiency dropped by 12%, then the pressure reduced by about 7% and the temperature increases by about 4.5%. When the compressor is damaged, its performance map undergoes considerable changes. Therefore, at a fixed speed, lower air flow rates were processed by the compressor with respect to the clean case. Air flow rates reductions of about 8% and 12% with respect to the clean case were recorded for compressor efficiency reductions of about 8% and 12%, respectively. The lower air flow rates, at fixed fuel, caused a rise in the turbine inlet and exhaust temperature (Figure 10). Therefore, for a compressor degradation of about 8%, the inlet and outlet turbine temperatures rose by about 7%, while for a compressor degradation of about 12%, the temperatures rose by about 10%. Overall, engine performance deteriorated and lower power was obtained, as reported in Table 7. These results are in agreement with the literature data [46,50].

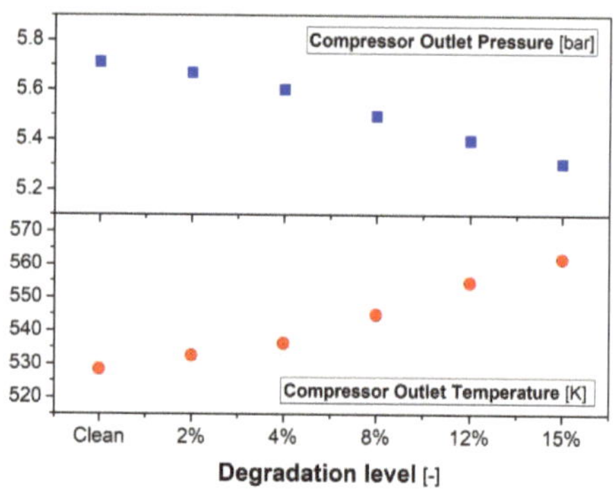

Figure 9. Compressor pressure (**top**) and outlet compressor temperature (**bottom**) for damaged compressor.

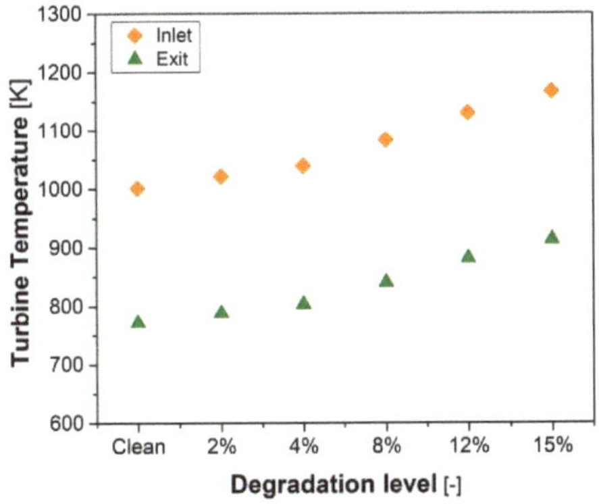

Figure 10. Turbine inlet and exit temperatures for damaged compressor.

Table 7. Power reduction with increasing compressor degradation for operating point (PLA) of about 50%, altitude h = 50 m and Mach = 0.09, at 6000 rpm.

Degradation	Output Power [kW]	% Power Variation
clean	178	
2%	175	−1.7
4%	171	−3.9
8%	164	−7.9
12%	157	−11.8
15%	152	−14.6

3.2. Linearized Turboshaft Model with System Identification (SI)

The model linearization was carried out through the Matlab System Identification toolbox. In this case, the input and output data vectors were given to the software using a GUI, and the state-space model was selected from among the available ones (transfer functions, polynomial, non-linear and other dynamic models). Two algorithms for SI are available: SSEST, which allows for the estimation of a continuous-time state-space model, and N4SID, which allows for the definition of a discrete-time state space model. In the latter case, the discrete-time identified state-space model has the following form:

$$\begin{aligned} x(t+T_s) &= Ax(t) + Bu(t) + Ke(t) \\ y(t) &= Cx(t) + Du(t) + e(t) \end{aligned} \tag{14}$$

where A, B, C, D and K are state-space matrices, $u(t)$ is the input, $y(t)$ is the output, $e(t)$ is the disturbance and $x(t)$ is the vector of n_x states, which represent the order of the system. In the present case, the N4SID algorithm with $n_x = 4$ was selected. It is worth pointing out that $n_x = 4$ was selected on the basis of a trial-and-error approach in order to achieve the best agreement between the linear model output and the experiments. Moreover, this procedure does not require the definition of the state variables, only their order (four in this case). Therefore, the physical meaning of the matrices elements, which were obtained by adopting the small perturbation approach, were lost in this case.

The comparison between the SI output variables and the GSP results is included in Table 8. It can be observed that by adopting this approach, the mean error was also below 1% for all of the output variables; moreover, a maximum error lower than 3%was recorded for the shaft power. The values of this approach were even better than those for the previous approach; therefore, similarly to the previous approach, these results allow one to conclude that the linear model obtained through the SI provides optimal predictive capabilities.

Table 8. Linear model validation.

OUTPUTVAR. / DEGRAD. LEVEL	PT_3 [bar]			TT_{45} [K]			Power [kW]		
	Mean Err. %	Max Err. %	R^2	Mean Err. %	Max Err. %	R^2	Mean Err. %	Max Err. %	R^2
Clean	0.08	1.08	0.995	0.04	0.19	0.999	0.10	0.30	0.999
2%	0.08	0.83	0.997	0.04	0.27	0.999	0.10	0.32	0.999
4%	0.09	0.67	0.997	0.05	0.26	0.998	0.11	0.33	0.999
8%	0.13	0.91	0.991	0.06	0.49	0.997	0.22	2.08	0.997
12%	0.10	0.95	0.997	0.07	0.61	0.996	0.22	2.29	0.997

Matrices A, B, C and D, for the analyzed case, had the following dimensions: 4×4, 4×1, 4×4 and 4×1, respectively. As in the previous case, each term of the matrices was analyzed with varying the compressor degradation level, in order to extrapolate a degradation law. Unfortunately, the coefficients varied in an unpredictable manner. For each term, the R^2 was computed, and this clearly indicated the poor fitting of the data with a linear trend. This allows one to conclude that this approach is not suitable for implementation in controller design when prediction of components degradation is included.

4. Conclusions

In this study, two approaches for obtaining reliable linear models of a turboshaft aircraft engine were analyzed. The main purpose of developing such linear models relied on the possibility of developing advanced control systems for direct fuel control instead of adopting measurable parameters. Moreover, the possibility of including components degradation in the linear model, in order to reduce control correction error, was investigated. The adopted techniques were Small Perturbation and System Identification. The models were validated by a comparison with simulation results, under the same operating conditions, obtained through a high-fidelity simulation tool (GSP).

Both approaches yielded quite reliable results in predicting the output variables for all of the degradation conditions. In fact, from a comparison between the linear model and data, the Small Perturbation approach yielded a recorded mean error below 1% for all of the output variables, and the maximum error, recorded for shaft power, was about 6%. For System Identification, even better results were obtained. In fact, the mean error was below 1% for all of the output variables, and the maximum error, recorded in this case for shaft power, was lower than 3%.

A simple correlation between engine performance and components degradation was also demonstrated; in particular, the achieved results allow one to conclude that the Small Perturbation approach is the best candidate for controller development when a prediction of components degradation is included.

Author Contributions: Conceptualization, T.C., A.F. and S.B.; methodology, A.F. and T.C.; software, D.P. and L.S.; validation, L.S. and D.P.; resources, D.P. and L.S.; data curation, T.C.; writing—original draft preparation, T.C.; writing—review and editing, T.C. and A.F.; supervision, S.B. and A.F.; project administration, A.F.; funding acquisition, A.F. All authors have read and agreed to the published version of the manuscript.

Funding: This research was funded by the Italian Ministry of University and Research, Project PON "SMEA", Cod. PON03PE_00067_5 CUP: B74C12000310005.

Data Availability Statement: The data used in this paper are publicly available from the sources cited in the text.

Conflicts of Interest: The authors declare no conflict of interest.

References

1. Élimination des Émissions Nettes de Carbone d'ici 2050 IATA COMMUNIQUÉ No: 66. Available online: https://www.iata.org/contentassets/dcd25da635cd4c3697b5d0d8ae32e159/2021-10-04-03-fr.pdf (accessed on 1 December 2022).
2. Flamm, J.; James, K.; Bonet, J. Overview Of ERA Integrated Technology Demonstration (ITD) 51A Ultra-High Bypass (UHB) Integration for Hybrid Wing Body (HWB). In Proceedings of the 54th AIAA Aerospace Sciences Meeting, San Diego, CA, USA, 4–8 January 2016.
3. Donateo, T.; De Pascalis, C.L.; Strafella, L.; Ficarella, A. Off-line and On-line Optimization of the Energy Management Strategy in a Hybrid Electric Helicopter for Urban Air-Mobility. *Aerosp. Sci. Technol.* **2021**, *113*, 106677. [CrossRef]
4. Dinc, A. NOx Emissions of Turbofan Powered Unmanned Aerial Vehicle for Complete Flight Cycle. *Chin. J. Aeronaut.* **2020**, *33*, 1683–1691. [CrossRef]
5. Cabrera, E.; de Sousa, J.M.M. Use of Sustainable Fuels in Aviation—A Review. *Energies* **2022**, *15*, 2440. [CrossRef]
6. Montazeri-Gh, M.; Rasti, A.; Jafari, A.; Ehteshami, M. Ehteshami, Design and Implementation of MPC for Turbofan Engine Control System. *Aerosp. Sci. Technol.* **2019**, *92*, 99–113. [CrossRef]
7. Wei, Z.; Jafari, S.; Zhang, S.; Nikolaidis, T. Hybrid Wiener model: An On-board Approach Using Post-Flight Data for Gas Turbine Aero-Engines Modelling. *Appl. Therm. Eng.* **2021**, *184*, 116350. [CrossRef]
8. Sun, X.; Jafari, S.; Fashandi, S.A.M.; Nikolaidis, T. Compressor Degradation Management Strategies for Gas Turbine Aero-Engine Controller Design. *Energies* **2021**, *14*, 5711. [CrossRef]
9. Imani, A.; Montazeri-Gh, M. Improvement of Min–Max Limit Protection in Aircraft Engine Control: An LMI approach. *Aerosp. Sci. Technol.* **2017**, *68*, 214–222. [CrossRef]
10. Mohammadi, E.; Montazeri-Gh, M. Active Fault Tolerant Control with Self-Enrichment Capability for Gas Turbine Engines. *Aerosp. Sci. Technol.* **2016**, *56*, 70–89. [CrossRef]
11. Chen, Y.; Guo, Y.; Li, R. State Feedback Control for Partially Distributed Turboshaft Engine with Time Delay and Packet Dropouts. In Proceedings of the 37th Chinese Control Conference, Wuhan, China, 25–27 July 2018; pp. 25–27.
12. Mattingly, J.; Jaw, L. *Aircraft Engine Controls: Design, System Analysis, and Health Monitoring*, 1st ed.; AIAA: Reston, VA, USA, 2009.
13. Connolly, J.W.; Csanky, J.; Chicatelliz, A. Advanced Control Considerations for Turbofan Engine Design. In Proceedings of the 52nd AIAA/SAE/ASEE Joint Propulsion Conference, Salt Lake City, UT, USA, 25–27 July 2016; pp. 1–18.
14. Lu, F.; Qian, J.N.; Huang, J.; Qiu, X. In-flight Adaptive Modeling using Polynomial LPV Approach for Turbofan Engine Dynamic Behavior. *Aerosp. Sci. Technol.* **2017**, *64*, 223–236. [CrossRef]
15. Castiglione, T.; Morrone, P.; Perrone, D.; Bova, S. Application of a model-based controller for improving internal combustion engines fuel economy. *Energies* **2020**, *13*, 1148. [CrossRef]
16. DeCastro, J. Rate-based Model Predictive Control of Turbofan Engine Clearance. *J. Propuls. Power* **2007**, *23*, 804–813. [CrossRef]
17. Zheng, Q.; Xu, Z.; Zhang, H.; Zhu, Z. A Turboshaft Engine NMPC Scheme for Helicopter Autorotation Recovery Maneuver. *Aerosp. Sci. Technol.* **2018**, *76*, 421–432. [CrossRef]
18. Peng, K.; Fan, D.; Yang, F.; Fu, Q.; Li, Y. Active Generalized Predictive Control of Turbine Tip Clearance for Aero-Engines. *Chin. J. Aeronaut.* **2013**, *26*, 1147–1155. [CrossRef]
19. Pang, S.; Li, Q.; Ni, B. Improved Nonlinear MPC for Aircraft Gas Turbine Engine Based on Semi-Alternative Optimization Strategy. *Aerosp. Sci. Technol.* **2021**, *118*, 106983. [CrossRef]
20. Yin, K.; Zhou, W.X.; Qiao, K.; Wang, H.J. Research on Methods of Improving Real-time Performance for Aero-engine Component-level Model. *J. Propuls. Technol.* **2017**, *38*, 199–206.
21. Junior, H.G.; Bringhenti, C.; Barbosa, J.R.; Tomita, J.T. Real-time Gas Turbine Model for Performance Simulations. *J. Aerosp. Technol. Manag.* **2017**, *9*, 346–356. [CrossRef]
22. Lu, J.; Guo, Y.Q.; Zhang, S.G. Research on the Iteration Methods in Aero-Engine Non-Linear Model Real-Time Computation. *J. Aerosp. Power* **2010**, *25*, 681–686.
23. Khorasani, K.; Kiakojoori, S. Dynamic Neural Networks for Gas Turbine Engine Degradation Prediction, Health Monitoring and Prognosis. *Neural Comput. Appl.* **2016**, *27*, 2157–2192. [CrossRef]
24. De Giorgi, M.G.; Quarta, M. Hybrid MultiGene Genetic Programming—Artificial Neural Networks Approach for Dynamic Performance Prediction of an Aeroengine. *Aerosp. Sci. Technol.* **2020**, *103*, 105902. [CrossRef]
25. Asgari, H.; Chen, X.; Morini, M.; Pinelli, M.; Sainudiin, R.; Spina, P.; Venturini, M. NARX Models for Simulation of the Start-Up Operation of a Single-Shaft Gas Turbine. *Appl. Therm. Eng.* **2016**, *93*, 368–376. [CrossRef]
26. Zheng, Q.; Pang, S.; Zhang, H.; Hu, Z. A study on Aero-Engine Direct Thrust Control with Nonlinear Model Predictive Control Based on Deep Neural Network. *Int. J. Aeronaut. Space Sci.* **2019**, *20*, 933–939. [CrossRef]
27. Du, X.; Sun, X.-M.; Wang, Z.-M.; Dai, A.-N. A Scheduling Scheme of Linear Model Predictive Controllers for Turbofan Engines. *IEEE Access* **2017**, *5*, 24533–24541. [CrossRef]
28. Du, X.; Guo, Y.Q.; Chen, X.L. Multivariable Constrained Predictive Control and its Application to a Commercial Turbofan Engine. *Adv. Mater. Res.* **2014**, *99*, 281–287. [CrossRef]
29. Saluru, D.; Yedavalli, R.; Belapurkar, R. Active Fault Tolerant Model Predictive Control of a Turbofan Engine Using C-MAPPS40k. In Proceedings of the Asme 5th Annual Dynamic Systems and Control Division Conference and JSME 11th Motion and Vibration Conference, Fort Lauderdale, FL, USA, 17–19 October 2012; pp. 349–358. [CrossRef]

30. Sugiyama, N. Derivation of System Matrices From Nonlinear Dynamic Simulation of Jet Engines. *J. Guid. Control. Dyn.* **1994**, *17*, 1320–1326. [CrossRef]
31. Chung, G.; Prasad, J.; Dhingra, M.; Meisner, R. Real Time Analytical Linearization of Turbofan Engine Model. *ASME J. Eng. Gas Turbines Power* **2014**, *136*, 011201. [CrossRef]
32. Kulikov, G.; Thompson, H. *Dynamic Modelling of Gas Turbines: Identification, Simulation, Condition Monitoring and Optimal Control*; Springer: London, UK, 2004.
33. Armstrong, J.; Simon, D. *Constructing an Efficient Self-Tuning Aircraft Engine Model for Control and Health Management Applications*; Report No.: NASA/TM- 2012-217806; Glenn Research Center, National Aeronautics and Space Administration: Cleveland, OH, USA, 2012.
34. Montazeri-Gh, M.; Rasti, A. Analyzing Different Numerical Linearization Methods for the Dynamic Model of a Turbofan Engine. *Mech. Ind.* **2019**, *20*, 303. [CrossRef]
35. Luppold, R.; Roman, J.; Gallops, G.; Kerr, L. Estimating In-Flight Engine Performance Variations Using Kalman Filter Concepts. In Proceedings of the AIAA/ASME/SAE/ASEE 25th Joint Propulsion Conference, Monterey, CA, USA, 10–12 July 1989; p. 2584.
36. Simon, D.; Garg, S. Optimal Tuner Selection For Kalman Filter-Based Aircraft Engine Performance Estimation. *J. Eng. Gas Turbines Power* **2009**, *132*, 031601. [CrossRef]
37. Csank, J.; Connolly, J. Model-Based Engine Control Architecture with An Extended Kalman Filter. In Proceedings of the AIAA Guidance, Navigation, and Control Conference, San Diego, CA, USA, 4–8 January 2016.
38. Nayfeh, A. *Perturbation Methods*; Wiley-VCH Verlag GmbH & Co. KGaA: Weinheim, Germany, 2004.
39. Holmes, M. *Introduction to Perturbation Methods*, 2nd ed.; Springer Science+Business Media: New York, NY, USA, 2013.
40. Tavakolpour-Saleh, A.; Nasib, S.; Sepasyan, A.; Hashemi, S. Parametric and Nonparametric System Identification of an Experimental Turbojet Engine. *Aerosp. Sci. Technol.* **2015**, *43*, 21–29. [CrossRef]
41. Ljung, L. *System Identification—Theory for the User*, 2nd ed.; Printice Hall: Hoboken, NJ, USA, 1999.
42. Keesman, K.J. *System Identification—An Introduction*; Springer: London, UK, 2011.
43. GSP 11 User Manual National Aerospace Laboratory NLR 2016. Available online: https://www.gspteam.com/Files/manuals/UM/GSP_UM_11.pdf (accessed on 1 July 2021).
44. Janikovic, J. Gas Turbine Transient Performance Modeling for Engine Flight Path Cycle Analysis. Ph.D. Thesis, Cranfield University, Bedford, UK, 2010.
45. Sogut, M.Z.; Yalcin, E.; Karakoc, T.H. Assessment of degradation effects for an aircraft engine considering exergy analysis. *Energy* **2017**, *140*, 141–1426. [CrossRef]
46. Kurz, R.; Brun, K. Degradation in Gas Turbine Systems. *ASME J. Eng. Gas Turbines Power* **2001**, *123*, 70–77. [CrossRef]
47. Zhao, N.-B.; Yang, J.-L.; Li, S.-Y.; Sun, Y.-W. A GM (1, 1) Markov Chain-Based Aeroengine Performance Degradation Forecast Approach Using Exhaust Gas Temperature. *Math. Probl. Eng.* **2014**, *2014*, 832851. [CrossRef]
48. DeCastro, J.; Litt, J.; Frederick, D. A Modular Aero-Propulsion System Simulation Of A Large Commercial Aircraft Engine. In Proceedings of the 44th AIAA/ASME/SAE/ASEE Joint Propulsion Conference and Exhibit, Hartford, CT, USA, 21–23 July 2008.
49. *Matlab Users Guide*; University of New Mexico: Albuquerque, NM, USA, 1982.
50. Igie, U.; Goiricelaya, M.; Nalianda, D.; Minervino, O. Aero engine compressor fouling effects for short- and long-haul missions. *Proc. Inst. Mech. Eng. Part G J. Aerosp. Eng.* **2016**, *230*, 1312–1324. [CrossRef]

energies

MDPI

Article

Dimensioning Air Reactor and Fuel Reactor of a Pressurized CLC Plant to Be Coupled to a Gas Turbine: Part 2, the Fuel Reactor

Wang Lu [1], Pietro Bartocci [2,*], Alberto Abad [2], Aldo Bischi [3], Haiping Yang [1], Arturo Cabello [2], Margarita de Las Obras Loscertales [2], Mauro Zampilli [4] and Francesco Fantozzi [4]

1 State Key Laboratory of Coal Combustion, Huazhong University of Science and Technology, Wuhan 430074, China; luwang721@hust.edu.cn (W.L.); yhping2002@163.com (H.Y.)

2 Instituto de Carboquímica (C.S.I.C.), C. Miguel Luesma Castán 4, 50018 Zaragoza, Spain; abad@icb.csic.es (A.A.); acabello@icb.csic.es (A.C.); mobras@icb.csic.es (M.d.L.O.L.)

3 Department of Energy, Systems, Territory and Construction Engineering, University of Pisa, Largo Lucio Lazzarino 1, 56122 Pisa, Italy; aldo.bischi@unipi.it

4 Department of Industrial Engineering, University of Perugia, Via G. Duranti 67, 06125 Perugia, Italy; mauro.zampilli@unipg.it (M.Z.); francesco.fantozzi@unipg.it (F.F.)

* Correspondence: pbartocci@icb.csic.es

Abstract: Bioenergy with Carbon Capture and Storage (BECCS) technologies are fundamental to reach negative CO_2 emissions by removing it from the atmosphere and storing it underground. A promising solution to implement BECCS is pressurized Chemical Looping Combustion (CLC), which involves coupling a pressurized CLC reactor system to a turboexpander. The typical configuration chosen is to have an air reactor and a fuel reactor based on coupled circulating fluidized beds. The fluidization regime in both reactors is preferred to be fast fluidization to enhance gas particle contact and solids circulation among reactors. To design the two reactors, Aspen Plus software was used, given that the new version has a module for fluidized bed modeling. At first, the oxygen carrier was designed ex novo, but given that it is a composite compound mainly made by nickel oxide freeze-granulated on alumina (Ni40Al-FG), the molecular structure has been inserted in Aspen Plus. Then, based on the power of the gas turbine, the power output per kg of evolving fluid (in this case, depleted air) is calculated using Aspen Plus. Once the nitrogen content in the depleted air is known, the total air at the inlet of the air reactor is calculated. The fuel reactor is modeled by inserting the reduction reactions for nickel-based oxygen carriers. The paper presents a useful methodology for developing pressurized Chemical Looping Combustors to be coupled to gas turbines for power generation. The provided data will be cross-validated with 0D-models and experimental results.

Keywords: carbon negative technologies; gas turbines; pressurized chemical looping combustor; biofuels; BECCS

Citation: Lu, W.; Bartocci, P.; Abad, A.; Bischi, A.; Yang, H.; Cabello, A.; de Las Obras Loscertales, M.; Zampilli, M.; Fantozzi, F. Dimensioning Air Reactor and Fuel Reactor of a Pressurized CLC Plant to Be Coupled to a Gas Turbine: Part 2, the Fuel Reactor. *Energies* **2023**, *16*, 3850. https://doi.org/10.3390/en16093850

Academic Editors: Roberta De Robbio and Maria Cristina Cameretti

Received: 3 August 2022
Revised: 7 November 2022
Accepted: 18 April 2023
Published: 30 April 2023
Corrected: 23 July 2024

1. Introduction

According to Kang et al. [1], more than 120 nations have made carbon neutrality pledges at current times [2,3]. In this context, Carbon Capture and Storage (CCS) is considered an irreplaceable technology to achieve this target [4]. Together with CCS, the development of Negative Emissions Technologies (NET) is also advocated [5]. These are key technologies that are able to absorb CO_2 directly from the atmosphere and store it underground, while CCS is used "only" to prevent CO_2 from reaching the atmosphere and increasing the carbon dioxide concentration. Examples of NET are afforestation and reforestation, the increase of carbon content in soils, storing carbon in the soils in the form of biochar, Bioenergy with Carbon Capture and Storage (for BECCS see also [6]), Direct Air Carbon Capture and Storage (DACCS), and enhanced weathering and ocean fertilization [7]. However, what is the Technology Readiness Level of the many NET? In an interesting report from the Royal Society and Royal Academy of Engineering, the TRL of NETs is displayed in Table 1 [8].

Table 1. TRL of most important NETs [8].

NET Technology	TRL
Afforestation, reforestation, and forest management [9–11]	8–9
Wetland, peatland, and coastal habitat restoration [10]	5–6
Soil carbon sequestration [9,12]	8–9
Biochar [9,12,13]	3–6
Bioenergy with carbon capture and storage [11,14]	Bioenergy: 7–9 CCS: 4–7
Ocean fertilization [15,16]	1–5
Building with biomass [17]	8–9
Enhanced terrestrial weathering [11]	1–5
Mineral carbonation [18]	3–8
Ocean alkalinity [19,20]	2–4
Direct air capture [12,21,22]	4–7
Low-carbon concrete [18,23,24]	6–7

It can be seen in Table 1 that the TRL of Bioenergy with Carbon Capture and Storage (BECCS) technologies depend on the TRL of bioenergy production technologies. These are needed to produce biofuel (which can be solid, gaseous, and liquid), and the TRL of Carbon Capture and Storage technologies (which are needed to capture and store carbon dioxide). While bioenergy technologies have a high technology readiness level, carbon capture technologies have a lower TRL level, which is comparable to that of direct air capture. On the other hand, there is no agreement in the literature on the TRL of BECCS because this is a category of technologies that have different TRLs [25]. For example, from the data reported by Möllersten [25], we can infer that 2G bioethanol with carbon capture can be considered TRL 5–6, while biomethane with carbon capture can be considered TRL 7–8; 1G bioethanol with CCS has a TRL ranging from 7 to 9; biomass power with CCS has a TRL of 7–8. For what has been said above, the question "What is the TRL of BECCS?" implies another question, which is: "What is the TRL of CCS?" An interesting classification of CCS technologies is proposed in Table 2 [26].

Table 2. Classification of Carbon Capture and Storage based on their TRL [26].

TRL	Technology
high technology readiness level—TRL (7–9)	1. Chemical Absorption 2. Physical Absorption 3. Membrane Permeation 4. Pre-combustion 5. Cryogenic distillation
low technology readiness level—TRL (≤7)	1. Hybrid coupling among cryogenic distillation, membrane permeation, and chemical or physical absorption 2. Enhanced chemical or physical absorption 3. Gas-liquid membrane contractors 4. Adsorption 5. Oxy-combustion 6. Chemical Looping Combustion 7. Mineralization

This paper is particularly focused on the Chemical Looping Combustion (CLC) process. In the work of Di Giuliano et al. [27], it is stated that the TRL of CLC is probably about 6, this is in agreement with what reported also in the report of the Global CCS Institute "Technology Readiness and costs of CCS" [28]. Chemical Looping Combustion is a technology in which combustion oxygen is provided by a metal oxide (named oxygen carrier), which is reduced while the fuel is oxidized. This happens in a reactor named FUEL

REACTOR (see Figure 1). Then, the oxygen carrier is regenerated in the AIR REACTOR (see Figure 1) oxidizing with air and the reduced metal. Oxidation and the reduction of the metal happen separately in two different reactor grants and, from the air reactor, a flow of depleted air is generated (mainly containing less oxygen than the incoming air due to the fact that some oxygen is consumed by oxidation). Additionally, a flow of carbon dioxide and water vapor is obtained from the fuel reactor. By condensing water vapor, a pure flow of carbon dioxide is obtained, which can be compressed and stored.

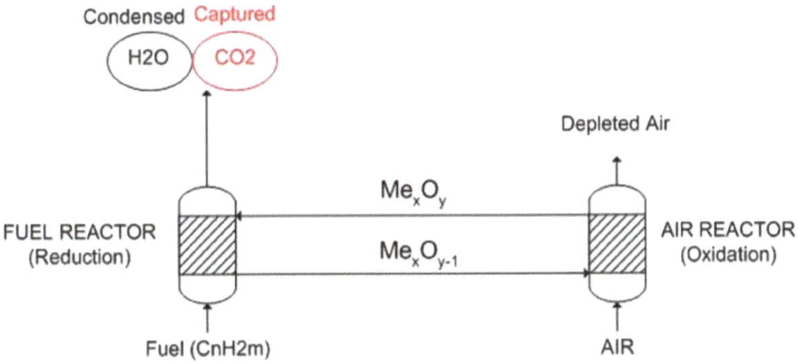

Figure 1. The Chemical Looping Combustion process [29].

The advantages of Chemical Looping Combustion are represented by the fact that when technological feasibility is proved and when plants will be scaled up the cost of carbon capture with this technology will be probably the lowest on the market and also the energy penalty linked with it. To develop Bioenergy with Carbon Capture and Storage (BECCS) technology based on Chemical Looping Combustion, two main routes can be followed: one is to produce steam that can be expanded into a steam turbine and the other is to couple the chemical looping combustor with a gas turbine. The latter configuration is supposed to achieve higher efficiency of the final power generation plant. On the other hand, when coupling a chemical looping combustor with a gas turbine, pressurizing both the chemical reactor and the air reactor is needed. This represents a technological barrier, which is addressed in the Marie Curie project: GTCLC-NEG. In this project, the following configuration of plant is proposed (see Figure 2).

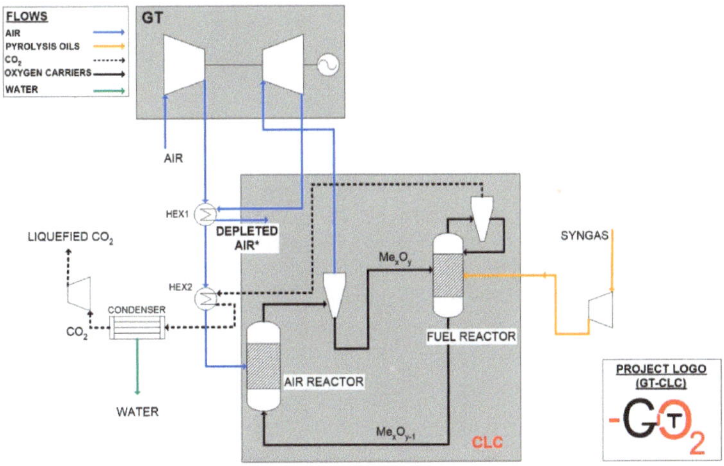

Figure 2. The GTCLC-NEG concept [29].

We can see in Figure 2 that air is at first compressed to about 12 bars and then enters the air reactor, which is a fluidized bed working at fast fluidization regime. Then, depleted air passes through a cyclone and expands in the gas turbine. The process is similar to what happens also in externally fired gas turbines. The expanded air has still a relevant temperature, so the waste heat is recovered in the first heat exchanger (HEX1). The cooled air is released to the environment. The fuel reactor can be fed with different biofuels (such as syngas or biogas or even solid biofuels, such as wood chips). These have to also be pressurized in a compressor at 12 bars. When the biofuels react with the oxygen carriers, they are converted to carbon dioxide and water vapor. After passing through a cyclone, the exhaust gases produced at the fuel reactor are also heated up to a temperature at about 1200 °C. Usually, the oxidation reaction is exothermic, while the reduction reaction can be either endothermic or exothermic. The waste heat in this case is recovered by the second heat exchanger (HEX2) which is also used to separate the water vapor from the CO_2 by condensing the water vapor. We obtain in this way a pressurized flow of carbon dioxide that can be further compressed to liquify it and then transported to the storage location.

To optimize the efficiency of the plant, an ASPEN Plus model has been realized in this paper we present the results of the design phase of the fuel reactor, while in [29], the results of the design phase of the air reactor is presented. We leave out of the analysis the cyclone, which is an important component and needs to be analyzed.

2. Materials and Methods

2.1. Methodology Followed to Design the Fuel Reactor

The methodology followed for the design of the fuel reactor is similar to that used for the air reactor, presented in [29] and adapted from [30,31]; see Figure 3. We see in Figure 3 that the design process is divided into two parts:

- The part that uses fuel characteristics, oxygen carrier characteristics, and oxidation and reduction reactions kinetics to calculate mass balances, reactor inventory, and solids circulation rate. This part first takes into account the air reactor because it is the one which is directly connected to the gas turbine and then the fuel reactor mass flows are determined based on stoichiometric calculations referred to the fuel needed to reduce the oxidized oxygen carrier. The dimensions of the reactor are set considering the optimal velocity of fluidization and the entrainment of particles;
- In the second part, the dimensions of the reactor and the optimal velocities are double checked using the Grace diagram to calculate the fluidization regime and checking that this is close to the fast fluidization regime.

The fuel reactor has one more criticality respect to the air rector: the chemical reactions happening significantly change the properties of the gases inside the reactor. In fact, in the case of the air reactor depleted air has quite similar properties respect the air fed into the reactor at the inlet. On the other hand, the density, viscosity, and velocities of the fuel are completely different from those of the reduction products, such as: carbon dioxide and water vapor. For this reason, the introduction of chemical reactions in the fuel reactor is of key importance and so it is the use of reactions kinetics derived at pressurized conditions. Simulation of reactions in fluidized bed is not new in Aspen Plus. Research have focused mainly on biomass gasification [32–35] and coal combustion in circulating fluidized beds [36]. The advantage of chemical looping combustion respect to biomass gasification in the specific case examined in this paper is that we use already syngas composed by mainly hydrogen and carbon monoxide and so the reactions which happen in the reactor result to be simpler. The reactions and the reaction kinetic constants are introduced directly in the "Fluidbed" model which makes the simulation also easier to perform.

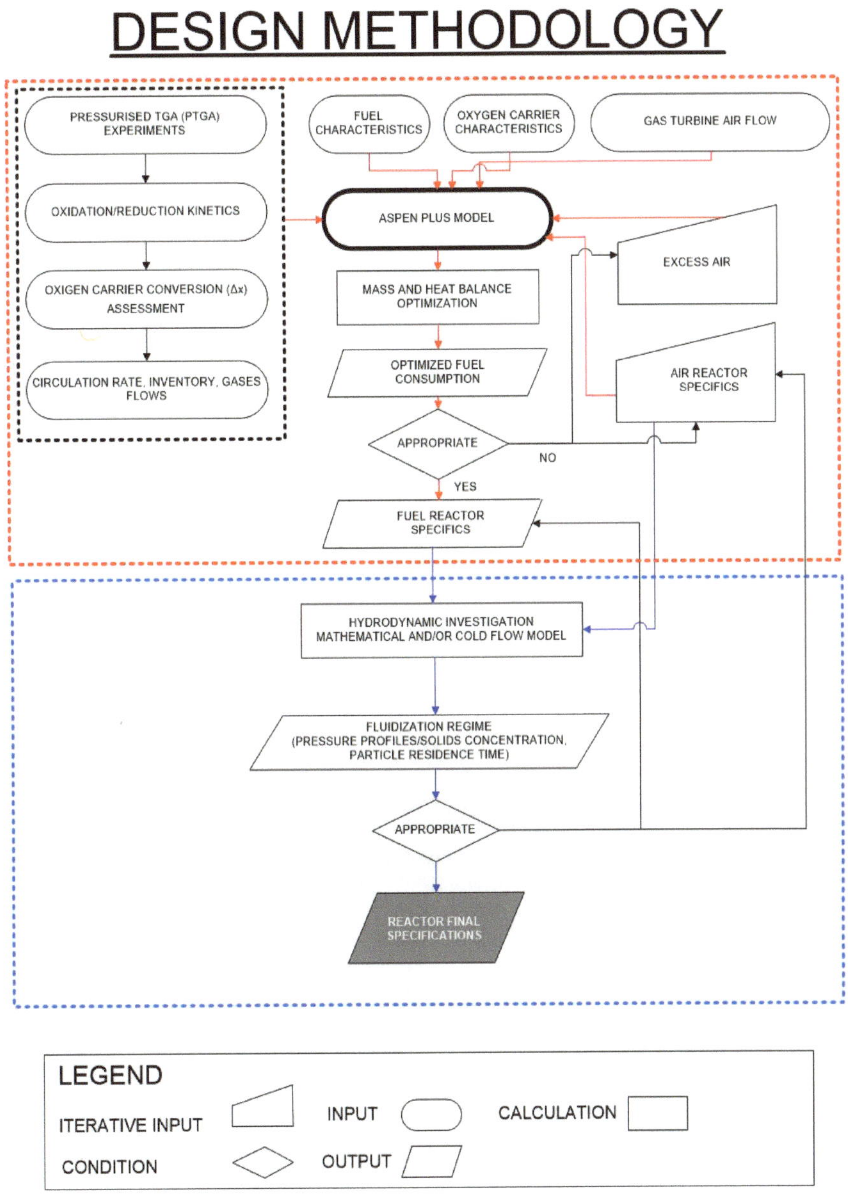

Figure 3. Methodology used to design the fuel reactor [29].

2.2. *Oxygen Carrier Main Characteristics*

The main characteristics of the oxygen carrier used are shown in Table 3. The oxygen carrier is derived from a freeze granulation (FG) preparation process, performed at Chalmers University, according to what was reported in [37], and it contains 40% of nickel oxide in weight. It is identified accordingly with the abbreviation Ni40Al-FG.

Table 3. Ni40Al-FG Oxygen carrier characteristics [38].

Parameter	Value	Unit of Measure
Active NiO content	40	wt%
Oxygen transport capacity, R_0	0.084	-
Particle size	0.2	μm
Porosity	0.36	%
Specific surface area (BET)	0.8	m^2/g
Solid density	5380	kg/m^3

On the Aspen Plus software components section, when selecting an oxygen carrier to be inserted as a component the ID should be introduced in form of the chemical formula of the compound (e.g., Al_2O_3-NiO); if this identity is not recognized as a compound contained in the database linked with Aspen Plus software, the compound has to be user defined, specifying the following parameters:

- Component ID and Alias are identified as the compound chemical formula;
- Compound state (in the case of the oxygen carrier we consider solid state);
- Molecular weight;
- Normal boiling point;
- Solid enthalpy of formation;
- Solid Gibbs energy of formation;
- Molar volume data (here also the density of the solid can be inserted);
- Vapor pressure data;
- Extended Antoine vapor pressure coefficients;
- Solid heat capacity data;
- Solid heat capacity polynomial coefficient.

The final data, after they have been inserted in the software, are proposed in Table 4. Dealing with the solid heat capacity coefficient, the solid molar volume, these are estimated based on the polynomial of alumina (Al_2O_3) and Nickel (Ni and NiO), which are assumed to be inserted separately, given that it is assumed that there is no chemical interaction between the two. Another step is represented by the necessity to model the Particle Size Distribution (PSD) inside the air reactor a PSD mesh is used, and it can be seen in Table 4, as calculated with Apen Plus V11. To realize the mesh a distribution function is built. This is based on the GGS (Gates–Gaudin–Schuhmann) approach; see [39]. The dispersion parameter is set to 1.5 and the maximum diameter is set to 0.4 mm. The minimum value of the distribution is set to 0.1 mm.

Table 4. Kinetic analysis of the oxygen carrier reduction [38].

Interval	Lower Limit	Upper Limit (μm)	Weight Fraction (μm)	Cumulative Weight Fraction
1	100	130	0.185279	0.185279
2	130	160	0.0677037	0.252982
3	160	190	0.0743889	0.327371
4	190	220	0.0805198	0.407891
5	220	250	0.086215	0.494106
6	250	280	0.0915561	0.585662
7	280	310	0.0966021	0.682264
8	310	340	0.101397	0.783661
9	340	370	0.105975	0.889637
10	370	400	0.110363	1

The image of the PSD distribution is shown in Figure 4.

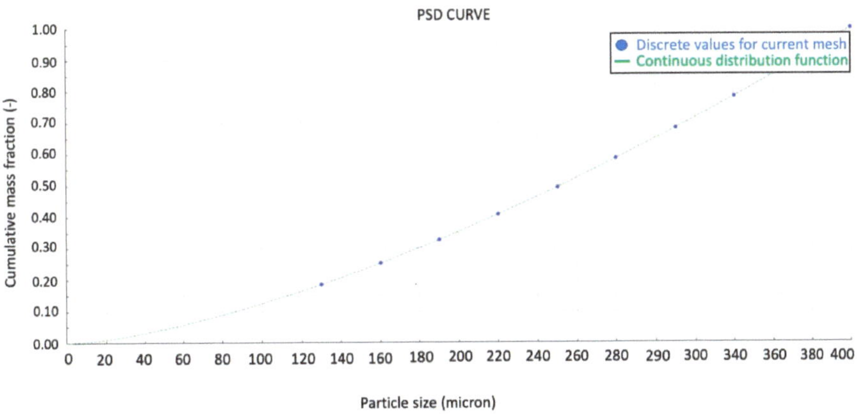

Figure 4. PSD distribution of Ni particles.

2.3. Inventory and Circulation Rate Calculations

Dealing with the circulation rate, this is defined as the mass flow of oxygen carrier which is completely oxidized [40]. This is influenced by:

- Type of oxygen carrier;
- Conversion variation obtained in the oxygen carrier;
- Type of fuel;
- Fuel reactor characteristics.

The process followed is already explained in [29], and the final value of 111 kg/s is obtained.

On the other hand, the inventory of the fuel reactor is different from that of the air reactor, even though they are both calculated in the same way as conducted in [29]. The entire process is explained in Figure 5.

As it can be seen in Figure 5, the inventory is calculated based on three main parameters:

- Characteristic circulation rate;
- Time for complete solid conversion in the reduction reaction;
- Characteristic reactivity of the FUEL REACTOR.

The time for complete solid conversion in the reduction reaction is influenced by the type of metal that is used as an oxygen carrier. This can have kinetics, which follows two main models: the plate-like geometry or the spherical grain geometry. To calculate the time for completed solid conversion in the reduction reaction, it is required that the parameter: average concentration of reacting gas in the reactor, which is calculating integrating the gas conversion rates. The parameter ε_g is the coefficient of expansion of the gas mixture and it is equal to 2 in the case of methane, 0 for hydrogen and carbon monoxide, and -0.21 in the case of oxidation. The final inventory calculated in this way is about 8000 kg of oxygen carrier. This is less than the inventory required for the air reactor and indicated in [29]. If we consider the value of inventory reported for the fuel reactor in [38] this is about 34 kg/MW with syngas composed of 45 vol% CO, 30 vol% H_2, 10 vol% CO_2, and 15 vol% H_2O and operating at a pressure of 20 bar and at a temperature of 1223 K, also assuming that water gas shift occurs. This value is referred to the power of the fuel fed into the reactor so to convert it in total kilograms we need to know the efficiency of the plant, assuming a 30% efficiency we obtain a total inventory for the fuel reactor of 1360 kg. This is reasonable when compared to our value because in [29], we have already noted that the calculated value is a theoretical one which, in practice, has to be multiplied for a safety factor that ranges between 2 and 10 times.

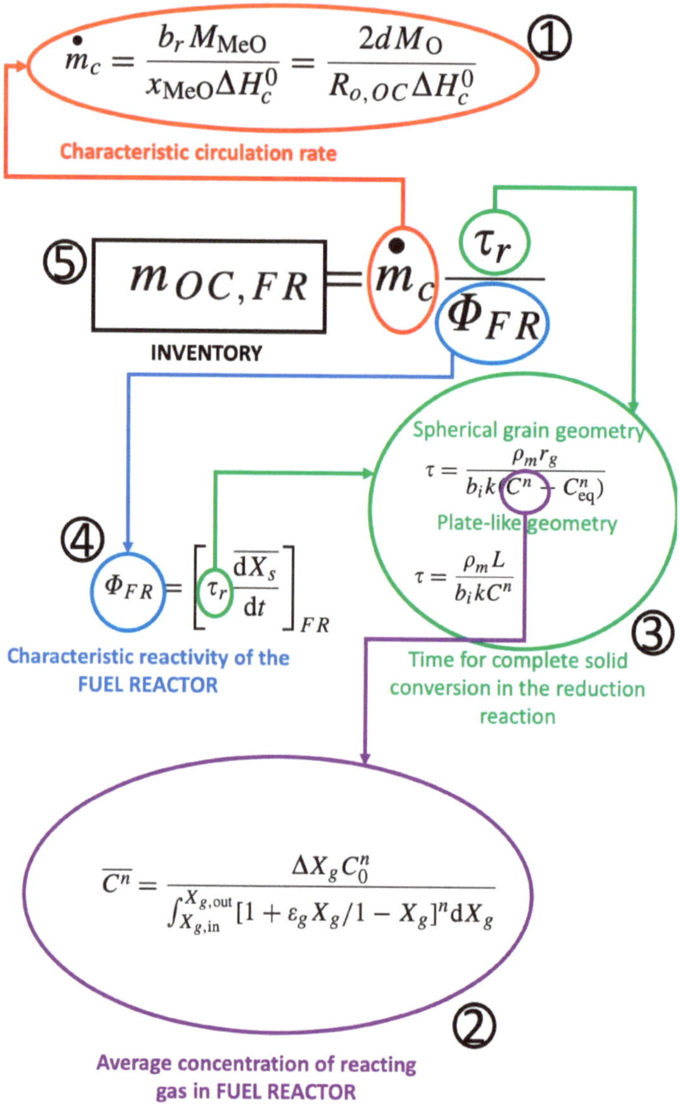

Figure 5. Method used to calculate the inventory.

2.4. Oxygen Carrier Reduction and Oxidation Reactions

From the point of view of the reactions, Equations (1) and (2) indicate reduction of carbon monoxide and hydrogen respectively (assuming that syngas is used as a fuel). Reaction 3 indicates oxidation.

$$NiO + CO \rightarrow Ni + CO_2 \tag{1}$$

$$NiO + H_2 \rightarrow Ni + H_2O \tag{2}$$

$$2Ni + O_2 \rightarrow 2\ NiO_2 \tag{3}$$

When nickel, as in this case, is used, water gas shift reaction has to also be taken into account. The kinetic of the reduction reactions is proposed in [38] and it is reported in Table 5. Where Equation (4) is the following:

$$k_{0,p} = k_0(P/P_0)^q \quad (4)$$

where $k_{0,p}$ is the preexponential factor of the chemical reaction rate constant at pressurized conditions ($mol^{1-n}\ m^{3n-2}\ s^{-1}$).

Table 5. PSD inside the fuel reactor calculated with ASPEN Plus V11.

	Symbol	H_2	CO
Molar density of metal oxide in the solid (mol MeO m^{-3} solid)	ρ_m	89,290	89,290
Stoichiometric factor in the reduction reaction of metal oxide (moles of MeO per mole of fuel gas)	b	1	1
Preexponential factor of the chemical reaction rate constant ($mol^{1-n}\ m^{3n-2}\ s^{-1}$)	k_0	9.3×10^{-3}	5.2×10^{-3}
Activation Energy (J mol^{-1})	E	26	25
Reaction order	n	0.5	0.8
Exponent in Equation (4)	q	−0.47	−0.93
dXr/dt at 0.1 MPa	-	0.040	0.028
dXr/dt at 2.0 MPa	-	0.044	0.019

The two reduction reactions using hydrogen and carbon monoxide as fuels have been inserted in the Aspen Plus model, first defining the stoichiometry and then the kinetics; see Table 5.

3. Results

3.1. Final Design of the Fuel Reactor

The final design of the fuel reactor has been modeled with Aspen Plus, V11. As already said, this version of the software already has a model to implement a fluidized bed which is available in the section "solids" with the reactor "Fluidbed". The input parameters used to model the fuel reactor fluidized beds are reported in Table 6. Dealing with the fuel reactor the dimensions are the following: diameter: 3.0 m; height: 8 m. As can be seen in Table 7, the inventory and the circulation rate are the values that have been calculated with the help of the procedure explained in [29] and in Figure 5.

Table 6. Fluidized bed model input parameters.

Parameter	Value	Unit of Measure
Voidage at minimum fluidization	0.5	-
Geldart classification	B	-
Minimum fluidization velocity calculation method	Ergun [41]	-
Transport disengagement Height Model	George and Grace [42]	-
Maximum dCv/dh	1×10^{-5}	-
Elutriation model	Tasirin and Geldart [43]	-
Decay constant	3	-
TG parameter A1	23.7	-
TG parameter A2	14.5	-
TG parameter B1	2.5	-
TG parameter B2	2.5	-
TG parameter C1	−5.4	-
TG parameter C2	−5.4	-
Constant diameter	-	-
Cross section	circular	-
Solids discharge location	95% of total height	-
Gas distribution	injectors	-
Distributor pressure drop	0.04	bar

3.2. Profiles of Velocity, Solids Volume Fraction

In this paragraph, we present the final profiles of superficial velocity (m/s); interstitial velocity (m/s); solids volume fraction (−); and pressure (bar). The velocities inside the fuel reactor are proposed in Figure 6.

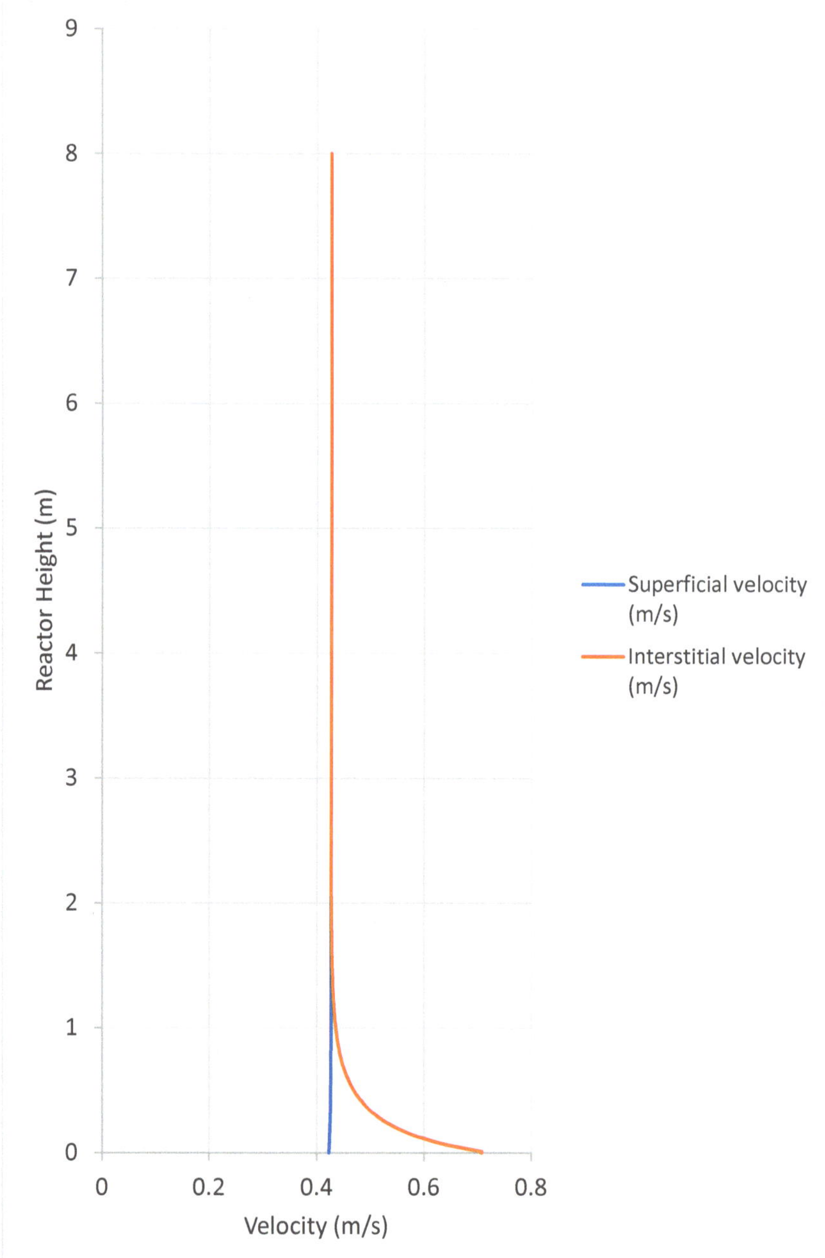

Figure 6. Velocity profiles inside the fuel reactor.

As can be seen in Figure 6, the bed of the fuel reactor is concentrated in the first 1 m of the height of the reactor. This can be seen from the interstitial velocity, which is initially higher than the superficial velocity and decreases rapidly when the freeboard begins. The final velocity value of the fuel reactor is about 0.4 m/s. The solids volume fraction in the fuel reactor is presented in Figure 7.

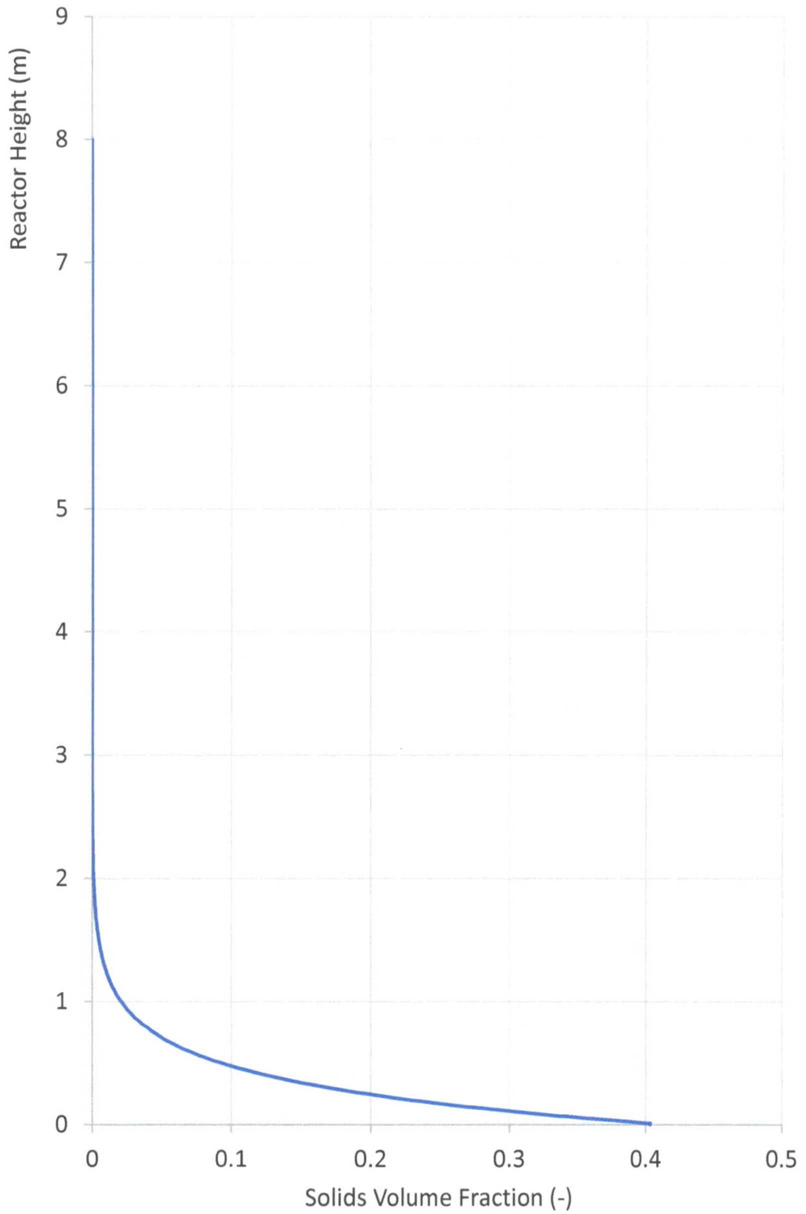

Figure 7. Solids volume fraction in the fuel reactor.

In the fuel reactor for the above-said reasons, there is less entrainment than in the air reactor.

Figure 8 confirms the data mainly reported in Table 7. The fuel reactor has high pressure in the bed zone and the pressure decreases steeply when we enter the freeboard zone.

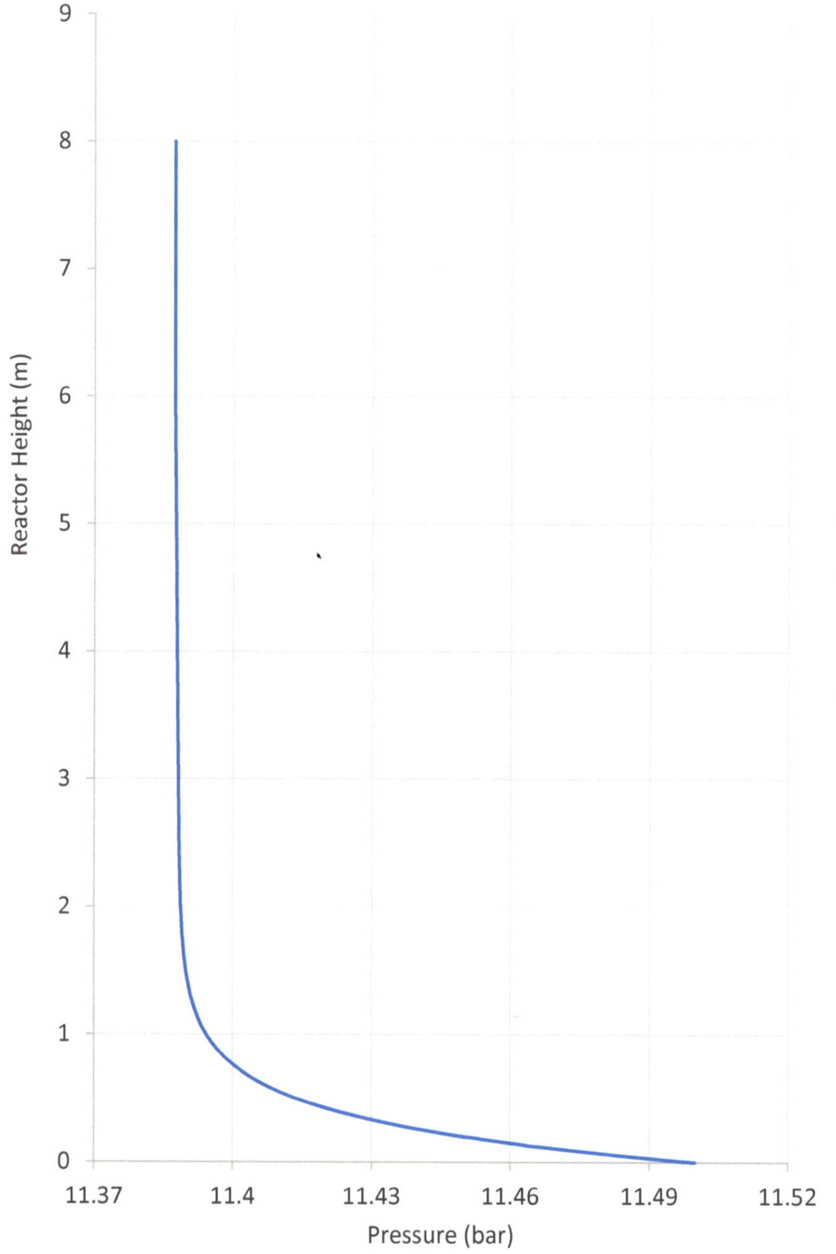

Figure 8. Pressure trend in the fuel reactor.

The decrease in the initial pressure at the inlet of the two reactors obviously represents a pressure loss, which can also have a penalty on the whole plat energy efficiency.

3.3. Profiles of Gases Concentrations

Figure 9 shows the details of the gaseous streams inside the whole fuel reactor.

It is interesting to note that the mass fractions of the gases appear to have a nonlinear trend at the height of 1 m, which is more or less when the bed transitions to the freeboard area. Depending on the position of the bubbles and of the solids, the model tries to simulate some oscillation on gas concentrations.

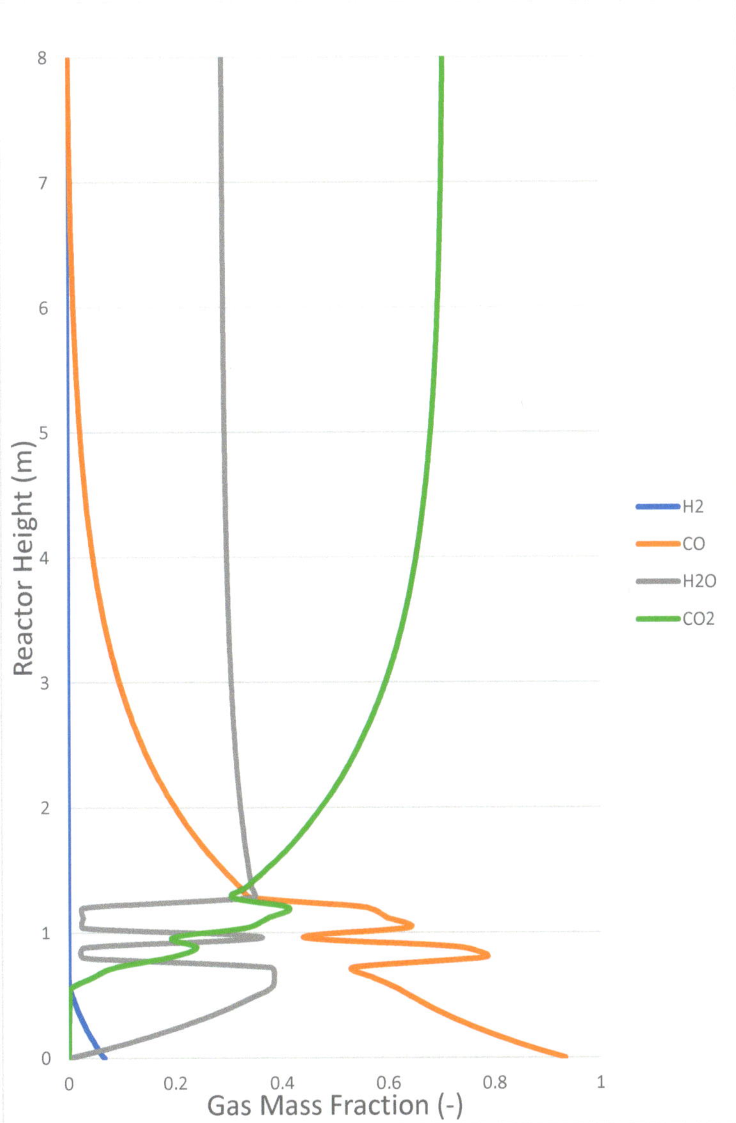

Figure 9. Gases concentrated inside the fuel reactor.

Table 7. Reactor characteristics are derived from the ASPEN Plus V11 model (Bedford, Massachusetts, U.S).

Parameter	Fuel Reactor	Unit of Measure
Total reactor height	8	m
Reactor diameter	3.0	m
Inventory	8000	kg
Circulation rate	111	kg/s
Operating pressure	12	bar
Height of bottom zone	0.01	m
Height of freeboard	5.99	m
Transport Disengaging Height calculated by correlation	5.33	m
Transport Disengaging Height based on solids volume fraction profile	3.89	m
Number of particles in bed	3.3×10^{10}	-
Surface area	26,040	sqm
Minimum fluidization velocity	0.04	m/s

3.4. Grace Diagram in the Fuel Reactor

The final calculation of the Grace diagram parameters for the fuel reactor is proposed in Figure 10.

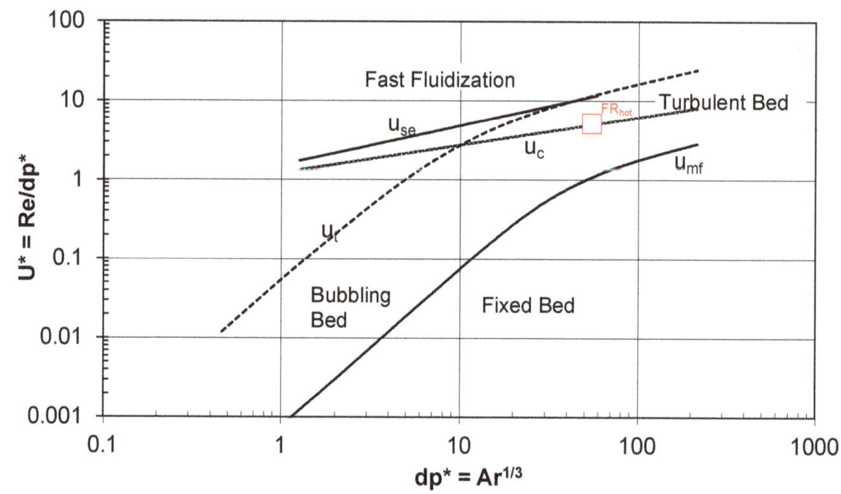

Figure 10. Grace diagram in the fuel reactor.

The two parameters described in the graph are given by the following equations, according to what is presented in [29]:

$$u^* = Re/dp^* \tag{5}$$

$$dp^* = Ar^{1/3} \tag{6}$$

$$Ar = {d_p}^3 \rho_g\,(\rho_p - \rho_g)g/\upsilon p \tag{7}$$

4. Conclusions

In this paper, the second part of the methodology, which can be used to design a Chemical Looping Combustor to be coupled to a gas turbine, is presented. In the first part, the air mass flow was calculated based on the final gas turbine power capacity; then, based on this data, the final air reactor geometry was optimized, also taking into account the oxygen carrier kinetics parameters (which influenced the circulation rate and the inventory of the air reactor). After having calculated the air needed in the air reactor and having assumed an excess air first tentative value of 3, the stoichiometric mass flow of fuel was calculated. Based on this, the geometry of the fuel reactor was determined. Finally, the Grace diagram is used to determine the fluidization regime. The innovative aspect of this paper is the introduction of the detailed chemistry of the reactions based on experimental data derived through Pressurized Thermogravimetric Balance analysis. The reactions have been effectively modelled in Aspen plus having good information on the gas concentrations along the reactor length.

Author Contributions: Aspen calculations W.L., H.Y. balances correction, Conceptualization, A.A. and F.F.; methodology, M.Z. and A.B.; software, A.C., M.d.L.O.L. and P.B. paper writing and conceptualization. All authors have read and agreed to the published version of the manuscript.

Funding: This work has been partially funded by the GTCLC-NEG project that has received funding from the European Union's Horizon 2020 research and innovation program under the Marie Sklodowska-Curie grant agreement No. 101018756.

Data Availability Statement: Data are publicly available in the project repository in Zenodo: https://zenodo.org/deposit?page=1&size=20 (accessed on 3 April 2023).

Acknowledgments: This work has been funded by the GTCLC-NEG project that has received funding from the European Union's Horizon 2020 research and innovation program under the Marie Sklodowska-Curie grant agreement No. 101018756. Special acknowledgments are given to William A. Rogers of the National Energy Technology Laboratory (US) for helping with the concluding remarks on the arrangements needed to carefully model the effect of the pressure on the PCLC plant.

Conflicts of Interest: The authors declare no conflict of interest.

References

1. Kang, J.-N.; Wei, Y.-M.; Liu, L.-C.; Wang, J.-W. Observing technology reserves of carbon capture and storage via patent data: Paving the way for carbon neutral. *Technol. Forecast. Soc. Chang.* **2021**, *171*, 120933. [CrossRef]
2. Mi, Z.; Meng, J.; Guan, D.; Shan, Y.; Song, M.; Wei, Y.-M.; Liu, Z.; Hubacek, K. Chinese CO_2 emission flows have reversed since the global financial crisis. *Nat. Commun.* **2017**, *8*, 1712. [CrossRef] [PubMed]
3. Black, R.; Cullen, K.; Fay, B.; Hale, T.; Lang, J.; Mahmood, S.; Smith, S. Taking Stock: A Global Assessment of Net Zero Targets. 2021. Available online: https://www.google.com.hk/url?sa=t&rct=j&q=&esrc=s&source=web&cd=&ved=2ahUKEwis1dW5zdD-AhUlmFYBHdzRAe4QFnoECA0QAQ&url=https%3A%2F%2Fca1-eci.edcdn.com%2Freports%2FECIU-Oxford_Taking_Stock.pdf&usg=AOvVaw1ZhRcYFX86SGMCS1CayncH (accessed on 3 April 2023).
4. Vergragt, P.J.; Markusson, N.; Karlsson, H. Carbon capture and storage, bio-energy with carbon capture and storage, and the escape from the fossil-fuel lock-in. *Glob. Environ. Change* **2011**, *21*, 282–292. [CrossRef]
5. Gasser, T.; Guivarch, C.; Tachiiri, K.; Jones, C.D.; Ciais, P. Negative emissions physically needed to keep global warming below 2 °C. *Nat. Commun.* **2015**, *6*, 7958. [CrossRef] [PubMed]
6. Gough, C.; Upham, P. Biomass energy with carbon capture and storage (BECCS or Bio-CCS). *Greenh. Gases Sci. Technol.* **2011**, *1*, 324–334. [CrossRef]
7. Rueda, O.; Mogollón, J.M.; Tukker, A.; Scherer, L. Negative-emissions technology portfolios to meet the 1.5 °C target. *Glob. Environ. Change* **2021**, *67*, 102238. [CrossRef]
8. *Greenhouse Gas Removal*; Royal Society and Royal Academy of Engineering: London, UK, 2019; ISBN 978-1-78252-349-9.
9. Smith, P. Soil carbon sequestration and biochar as negative emission technologies. *Glob. Change Biol.* **2016**, *22*, 1315–1324. [CrossRef] [PubMed]
10. Griscom, B.; Adams, J.; Ellis, P.; Houghton, R.; Lomax, G.; Miteva, D.; Schlesinger, W.; Shoch, D.; Smith, P.; Woodbury, P. Natural climate solutions. *Earth Atmos. Planet. Sci.* **2017**, *114*, 11645–11650. [CrossRef]
11. Smith, P.; Davis, S.J.; Creutzig, F.; Fuss, S.; Minx, J.; Gabrielle, B.; Kato, E.; Jackson, R.B.; Cowie, A.; Kriegler, E. Biophysical and economic limits to negative CO_2 emissions. *Nat. Clim. Change* **2016**, *6*, 42–50. [CrossRef]
12. Fuss, S.; Lamb, W.F.; Callaghan, M.W.; Hilaire, J.; Creutzig, F.; Amann, T.; Beringer, T.; de Oliveira Garcia, W.; Hartmann, J.; Khanna, T. Negative emissions—Part 2: Costs, potentials and side effects. *Environ. Res. Lett.* **2018**, *13*, 063002. [CrossRef]
13. Woolf, D.; Amonette, J.; Street-Perrott, F.; Lehmann, J.; Joseph, S. Sustainable Biochar to Mitigate Global Climate Change. *Nat. Commun.* **2010**, *1*, 56. [CrossRef] [PubMed]
14. Bhave, A.; Taylor, R.H.; Fennell, P.; Livingston, W.R.; Shah, N.; Mac Dowell, N.; Dennis, J.; Kraft, M.; Pourkashanian, M.; Insa, M. Screening and techno-economic assessment of biomass-based power generation with CCS technologies to meet 2050 CO_2 targets. *Appl. Energy* **2017**, *190*, 481–489. [CrossRef]
15. *Climate Intervention: Carbon Dioxide Removal and Reliable Sequestration*; Committee on Geo engineering Climate: Technical Evaluation and Discussion of Impacts; The National Academies Press: Washington, DC, USA, 2015.

16. Harrison, D.P. A method for estimating the cost to sequester carbon dioxide by delivering iron to the ocean. *Int. J. Glob. Warm.* **2013**, *5*, 231–254. [CrossRef]
17. McLaren, D. A comparative global assessment of potential negative emissions technologies. *Process Saf. Environ. Prot.* **2012**, *90*, 489–500. [CrossRef]
18. Sanna, A.; Uibu, M.; Caramanna, G.; Kuusik, R.; Maroto-Valer, M. A review of mineral carbonation technologies to sequester CO_2. *Chem. Soc. Rev.* **2014**, *43*, 8049–8080. [CrossRef]
19. González, M.F.; Ilyina, T. Impacts of artificial ocean alkalinization on the carbon cycle and climate in Earth system simulations. *Geophys. Res. Lett.* **2016**, *43*, 6493–6502. [CrossRef]
20. Renforth, P.; Jenkins, B.; Kruger, T. Engineering challenges of ocean liming. *Energy* **2013**, *60*, 442–452. [CrossRef]
21. Chan, G.G.; Koch, C.M.; Connors, L.H. Blood proteomic profiling in inherited (ATTRm) and acquired (ATTRwt) forms of transthyretin-associated cardiac amyloidosis. *J. Proteome Res.* **2017**, *16*, 1659–1668. [CrossRef]
22. Keith, D.W.; Holmes, G.; Angelo, D.S.; Heidel, K. A process for capturing CO_2 from the atmosphere. *Joule* **2018**, *2*, 1573–1594. [CrossRef]
23. Ghouleh, Z.; Guthrie, R.I.; Shao, Y. Production of carbonate aggregates using steel slag and carbon dioxide for carbon-negative concrete. *J. CO2 Util.* **2017**, *18*, 125–138. [CrossRef]
24. Huijgen, W.J.; Comans, R.N.; Witkamp, G.-J. Cost evaluation of CO_2 sequestration by aqueous mineral carbonation. *Energy Convers. Manag.* **2007**, *48*, 1923–1935. [CrossRef]
25. Möllersten, K. Assessment of classes of CDR methods: Technology Readiness, Costs, Impacts and Practical Limitations of Biochar as Soil Additive and BECCS. *Energy* **2022**, *2004*, 2965.
26. Araújo, O.d.Q.F.; de Medeiros, J.L. Carbon capture and storage technologies: Present scenario and drivers of innovation. *Curr. Opin. Chem. Eng.* **2017**, *17*, 22–34. [CrossRef]
27. Di Giuliano, A.; Capone, S.; Anatone, M.; Gallucci, K. Chemical Looping Combustion and Gasification: A Review and a Focus on European Research Projects. *Ind. Eng. Chem. Res.* **2022**, *61*, 14403–14432. [CrossRef]
28. Kearns, D.; Liu, H.; Consoli, C. *Technology Readiness and Costs of CCS*; Global CCS Institute: Docklands, Australia, 2021.
29. Bartocci, P.; Abad, A.; Bischi, A.; Wang, L.; Cabello, A.; de Las Obras Loscertales, M.; Zampilli, M.; Yang, H.; Fantozzi, F. Dimensioning Air Reactor and Fuel Reactor of a Pressurized Chemical Looping Combustor to Be Coupled to a Gas Turbine: Part 1, the Air Reactor. *Energies* **2023**, *16*, 2102. [CrossRef]
30. Bischi, A.; Langørgen, Ø.; Saanum, I.; Bakken, J.; Seljeskog, M.; Bysveen, M.; Morin, J.-X.; Bolland, O. Design study of a 150 kWth double loop circulating fluidized bed reactor system for chemical looping combustion with focus on industrial applicability and pressurization. *Int. J. Greenh. Gas Control* **2011**, *5*, 467–474. [CrossRef]
31. Bartocci, P.; Abad, A.; Mattisson, T.; Cabello, A.; de las Obras Loscertales, M.; Negredo, T.M.; Zampilli, M.; Taiana, A.; Serra, A.; Arauzo, I. Bioenergy with Carbon Capture and Storage (BECCS) developed by coupling a Pressurised Chemical Looping combustor with a turbo expander: How to optimize plant efficiency. *Renew. Sustain. Energy Rev.* **2022**, *169*, 112851. [CrossRef]
32. Nikoo, M.B.; Mahinpey, N. Simulation of biomass gasification in fluidized bed reactor using ASPEN PLUS. *Biomass Bioenergy* **2008**, *32*, 1245–1254. [CrossRef]
33. Abdelouahed, L.; Authier, O.; Mauviel, G.; Corriou, J.-P.; Verdier, G.; Dufour, A. Detailed modeling of biomass gasification in dual fluidized bed reactors under Aspen Plus. *Energy Fuels* **2012**, *26*, 3840–3855. [CrossRef]
34. Kaushal, P.; Tyagi, R. Advanced simulation of biomass gasification in a fluidized bed reactor using ASPEN PLUS. *Renew. Energy* **2017**, *101*, 629–636. [CrossRef]
35. Puig-Gamero, M.; Pio, D.; Tarelho, L.; Sánchez, P.; Sanchez-Silva, L. Simulation of biomass gasification in bubbling fluidized bed reactor using aspen plus®. *Energy Convers. Manag.* **2021**, *235*, 113981. [CrossRef]
36. Sotudeh-Gharebaagh, R.; Legros, R.; Chaouki, J.; Paris, J. Simulation of circulating fluidized bed reactors using ASPEN PLUS. *Fuel* **1998**, *77*, 327–337. [CrossRef]
37. Cho, P.; Mattisson, T.; Lyngfelt, A. Comparison of iron-, nickel-, copper- and manganese-based oxygen carriers for chemical-looping combustion. *Fuel* **2004**, *83*, 1215–1225. [CrossRef]
38. Abad, A.; García-Labiano, F.; de Diego, L.F.; Gayán, P.; Adánez, J. Reduction Kinetics of Cu-, Ni-, and Fe-Based Oxygen Carriers Using Syngas (CO + H2) for Chemical-Looping Combustion. *Energy Fuels* **2007**, *21*, 1843–1853. [CrossRef]
39. Schuhmann, R. *Technical Publication 1189*; American Institute of Mining and Metallurgical Engineers: New York, NY, USA, 1940.
40. Abad, A.; Adánez, J.; García-Labiano, F.; de Diego, L.F.; Gayán, P.; Celaya, J. Mapping of the range of operational conditions for Cu-, Fe-, and Ni-based oxygen carriers in chemical-looping combustion. *Chem. Eng. Sci.* **2007**, *62*, 533–549. [CrossRef]
41. Ergun, S. Fluid flow through packed columns. *Chem. Eng. Prog.* **1952**, *48*, 89–94.
42. George, S. Entrainment of particles from aggregative fluidized beds. *AIChE Symp. Ser.* **1978**, *74*, 67–74.
43. Tasirin, S.; Geldart, D. Entrainment of FCC from fluidized beds—A new correlation for the elutriation rate constants Ki∞. *Powder Technol.* **1998**, *95*, 240–247. [CrossRef]

energies

MDPI

Article

Direct Numerical Simulation of a Reacting Turbulent Hydrogen/Ammonia/Nitrogen Jet in an Air Crossflow at 5 Bar

Eugenio Giacomazzi [1,*], Donato Cecere [1], Matteo Cimini [2] and Simone Carpenella [3]

1 Laboratory of Processes and Systems Engineering for Energy Decarbonisation, ENEA, 00124 Rome, Italy; donato.cecere@enea.it

2 Department of Mechanical and Aerospace Engineering, Sapienza University of Rome, 00184 Rome, Italy; matteo.cimini@uniroma1.it

3 Department of Civil, Computer Science and Aeronautical Technologies Engineering, Roma Tre University, 00146 Rome, Italy; simone.carpenella@uniroma3.it

* Correspondence: eugenio.giacomazzi@enea.it

Abstract: The article aims to analyze the fluid dynamics and combustion characteristics of a non-premixed flame burning a fuel mixture derived from ammonia partial decomposition injected in an air crossflow. Nominal pressure (5 bar) and inlet air temperature (750 K) conditions are typical of micro-gas turbines. The effects of strain on the maximum flame temperature and NO generation in laminar non-premixed counter-flow flames are initially explored. Then, the whole three-dimensional fluid dynamic problem is investigated by setting up a numerical experiment: it consists of a Direct Numerical Simulation, based on accurate transport, chemical, and numerical models. The flow topology of the specific reacting jet in crossflow configuration is described in terms of its main turbulent structures, like shear layers, ring, and horse-shoe vortices, as well as of its leeward recirculation region anchoring the flame. The reacting region is characterized by providing instantaneous spatial distributions of temperature, heat release, and some transported chemical species, including NO, and calculating the Flame Index to identify non-premixed and premixed combustion local conditions. The latter is quantified by looking at the distribution of the volume fraction associated with a certain Flame Index versus the Flame Index and at the distribution of the average values of both the Heat Release Rate and NO versus the Flame Index and the mixture fraction.

Keywords: hydrogen; ammonia; DNS; jet in crossflow; combustion

Citation: Giacomazzi, E.; Cecere, D.; Cimini, M.; Carpenella, S. Direct Numerical Simulation of a Reacting Turbulent Hydrogen/Ammonia/Nitrogen Jet in an Air Crossflow at 5 Bar. *Energies* **2023**, *16*, 7704. https://doi.org/10.3390/en16237704

Academic Editors: Maria Cristina Cameretti and Roberta De Robbio

Received: 20 October 2023
Revised: 13 November 2023
Accepted: 20 November 2023
Published: 22 November 2023

1. Introduction

In recent years, the pursuit of alternatives to fossil fuels has greatly expanded. Among these, ammonia has emerged as a noteworthy substance for storing hydrogen [1,2], which can be easily broken down and used as a sustainable fuel for both transportation fuel cells and power generation.

Technically speaking, NH_3 offers considerable advantages when compared to hydrogen in terms of cost and convenience as a fuel because of its higher density, simple production, simplified storage, handling, and distribution; furthermore, it benefits from an already well-established infrastructure, thus displaying potential for commercial viability. Notably, the hydrogen density of NH_3 is four times higher than that of the most advanced methods involving metal hydrides [3]. Although ammonia is toxic, it poses a considerably reduced fire risk compared to hydrogen and gasoline [4,5]. Moreover, the cost per unit of stored energy for ammonia is more cost-effective than that of hydrogen [2,6]. Additionally, ammonia has demonstrated its merit as a viable alternative for hydrocarbon fuels in gas turbine engines, showcasing the potential for achieving high efficiencies [7].

Nonetheless, there are notable combustion difficulties linked to the use of NH_3. These challenges stem from its low flame speed, temperature, and reactivity, as well as its potential for substantial NO_x production [8–11]. To tackle these challenges, supplementary fuels such as H_2 have shown improvements in enhancing and stabilizing operational performance [12]. However, conventional combustor designs require fundamental reassessment and development to reduce NO_x emissions before NH_3 can be seriously regarded for extended use in cleaner and more efficient technologies, like gas turbines [13,14].

Several studies have indicated that a blend of approximately 30% H_2 or CH_4 with 70% NH_3 can enhance combustion stability. Such a blending strategy can be adopted to overcome the problem of NH_3 combustion poor reactivity and to manage mixtures with combustion properties close to methane, useful features for gas turbine applications. Raising the ammonia fraction in the fuel blend promotes lean blowout but diminishes the tendency to flashback. This latter impact is particularly noticeable when the volume fraction of ammonia in the fuel blend surpasses 0.7. Under these conditions, boosting the equivalence ratio at a constant bulk velocity prevents flashback, leading to rich blowout instead, resulting in a significantly broader spectrum of stable equivalence ratios [15].

In the realm of gas turbine operations, Valera-Medina et al. [13] have presented studies involving a generic swirl burner fueled by various ammonia/methane blends: although reduced NO_x and CO emissions were found at high equivalence ratios (>1.10), it was also stressed that conventional injection methods are unsuitable for ammonia-based blends. Hayakawa et al. [16] have explored reaction enhancement and stability of ammonia flames by adding hydrogen, demonstrating the potential for efficient NH_3 combustion using ammonia blends. Research conducted by Pugh et al. [17], Zhang et al. [18], and Franco et al. [19] has examined the product gas characteristics of ammonia/hydrogen flames within industrially relevant swirl burner flame configurations.

Sorrentino et al. [20] demonstrated the feasibility of MILD (Moderate or Intense Low-oxygen Dilution) combustion for pure ammonia, showing good stability and emissions performance when the reactor temperature exceeded 1300 K. Following studies [21] have shown that introducing water into MILD ammonia combustion holds promise in enhancing both NO_x emissions and stability, especially under fuel-lean conditions.

Experiments conducted on an industrial micro-gas turbine system have suggested a two-stage rich-lean combustion setup for pure ammonia [22], ammonia/methane blends [23], ammonia/hydrogen/nitrogen blends [24], and liquid ammonia [25] combustion. Numerical studies suggest that optimizing the equivalence ratio in the first (rich) combustion stage is essential for achieving low emissions at the outlet [26]. Furthermore, chemical kinetics investigations demonstrated that using a staged combustion (consisting of a rich stage, followed by a relaxation stage before the final lean stage) coupled with high-pressure conditions (>20 bar), it is fundamentally possible to achieve NO_x emissions as low as in natural gas turbines [27].

Due to the convenience of distributing and storing ammonia compared to hydrogen, it is plausible to consider that in many instances, when relying on co-firing ammonia with hydrogen, the necessary hydrogen can be locally generated by breaking down a portion of the available ammonia. Several researchers have investigated the structure of partially cracked ammonia-air flames [28], leading to the development of thermal cracking processes. Additionally, other investigators [29,30] have recorded successful power generation through catalytic processes using the cracked ammonia molecule. However, cracking 2 moles of ammonia produces 3 moles of hydrogen and 1 mole of nitrogen. The nitrogen can either be separated or, to avoid separation costs, introduced into the burner alongside hydrogen and ammonia. The nitrogen introduced into the burner will serve as a diluent, causing chemical effects and most likely impacting emissions and flame stabilization. Khateeb's research [15,31] demonstrated that the introduction of nitrogen into the ammonia-hydrogen blend decreases NO emissions, promotes lean blowout, and retards flashback. Furthermore, the computational investigation in [32] unveiled that for nearly equivalent adiabatic flame temperature, thickness, and propagation velocities, premixed laminar $NH_3/H_2/N_2$-air flames demonstrate significantly greater resistance to strain-induced extinction compared to laminar CH_4-air flames [33].

Since NO_x formation times are of the order of milliseconds, reducing the flame length as it is typically done in micro-mixing combustion systems could be a successful strategy for reducing it. Such technology is currently under investigation for hydrogen combustion [34–36]. In this work, the authors would like to check if it can be extended to ammonia/hydrogen blends by investigating a non-premixed jet flame of $NH_3/H_2/N_2$ injected in an air crossflow at 5 bar, simulated in the Direct Numerical Simulation (DNS) context. Despite the simplicity of the geometric configuration, its fluid dynamics are complex and make it possible to highlight the effects of turbulent mixing and strain on flame structure and stabilization and on NO formation.

The paper is organized as follows: Section 2 quantifies NO formation in laminar diffusion flames with the fuel having the same composition of the DNS jet mixture at different strain rates; Section 3 describes the DNS configuration; Section 4 presents the results, highlighting the flow and flame structures, characterizing the combustion mode by means of the Flame Index and mixture fraction, and quantifying the pollutants production; conclusions follow in Section 5.

2. Laminar Diffusion Flames

In a high-velocity diffusion flame, fuel and oxidant are subjected, in the region where they mix, to unsteady strain due to turbulent vortices. This section investigates the effect of strain-rate variation on laminar non-premixed counter-flow flames with the fuel side having the same composition mixture as the jet in the DNS numerical experiment. In particular, a numerical study with OpenSMOKE++ software [37] (https://www.opensmokepp.polimi.it/, accessed on 19 November 2023) is conducted to measure the maximum flame temperature and NO concentration as a function of the strain rate up to the quenching value. The species transport equations incorporated thermal diffusion (Soret effect), and individual species mixture-averaged diffusion coefficients were taken into account [38]. The counter-flow flame in its axisymmetric configuration is shown in the panel of Figure 1. The global strain rate χ is obtained as follows [39]:

$$\chi = \frac{2|u_O|}{L}\left[1 + \frac{|u_F|}{|u_O|}\sqrt{\frac{\rho_F}{\rho_O}}\right], \quad (1)$$

where ρ is the density, u the velocity of fuel (F) and oxidizer (O) fluxes separated by the distance L.

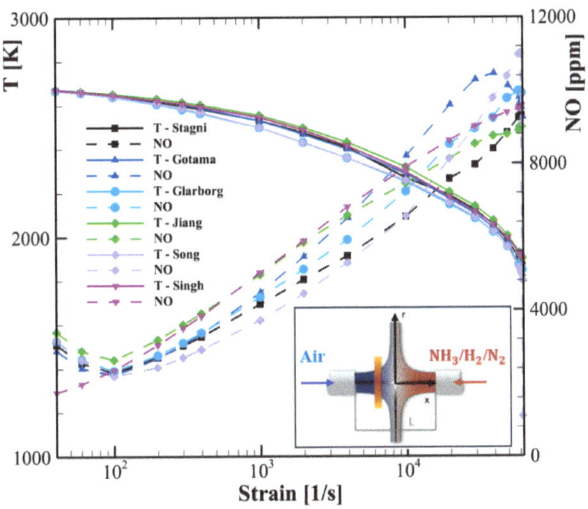

Figure 1. Solution of laminar non-premixed counter-flow flames at 5 bar as a function of the strain rate [1/s] for several mechanisms; (solid line) temperature peak; (dashed line) NO peak emissions.

The adaptive grid refinement technique was employed during the simulations. The relative and absolute errors of the nonlinear equations iterative process during the simulations were fixed to 10^{-6} and 10^{-12}, respectively. In ammonia-hydrogen and air flames, the flame is positioned towards the oxidizer side of the stagnation plane. Extinction strain rates were determined by initially simulating a stable low-strain flame and gradually increasing the strain rate in a series of simulations until reaching the point of extinction. The kinetic scheme used matches the DNS simulation.

Figure 1 shows temperature and NO concentration trend as a function of the strain rate for different chemical mechanisms. The trends are the same for all mechanisms; the Gotama one shows a slight deviation at high strains; it was adopted in the present DNS since it is specifically improved for laminar burning velocities under a fuel-rich flame, i.e., up to an equivalence ratio of 1.8, and pressure in the range 0.1–0.5 MPa [40]. Despite the decreasing temperature with increasing strain rate up to the quenching, NO concentration increases. In fact, the NO Thermal formation path becomes less important as the temperature decreases, while the weight of the fuel pathway increases, as shown in Figure 2, which depicts the normalized NO formation rate for each reaction distinctive of different pathways [41]. The panel of Figure 2 depicts the relative importance [42] of each path in the NO formation for two different strains. So even though temperature decreases as strain increases, and thermal path loses importance in the NO formation, 70% of it is formed via the fuel path reaching ~10,000 ppm before quenching.

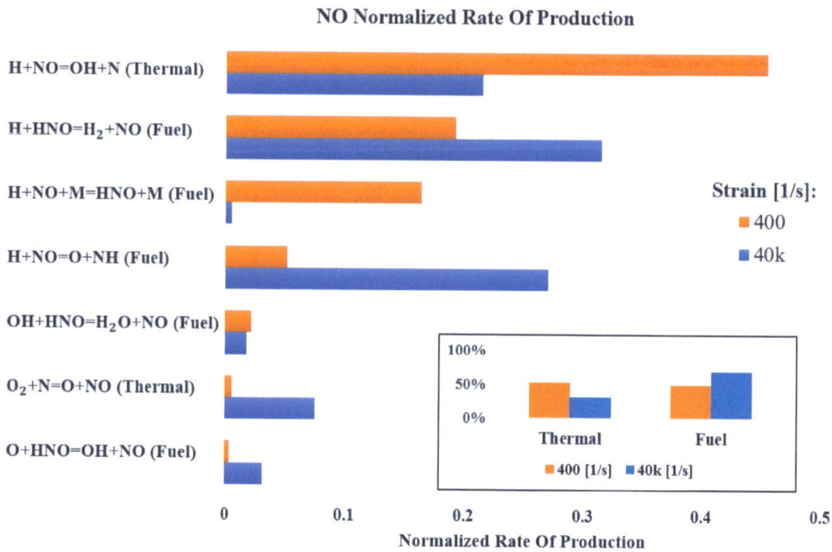

Figure 2. NO rate of production for several reactions involved in Thermal and Fuel formation paths and relative weight of NO formation pathway (analysis based on the Gotama chemical mechanism).

3. The Numerical Experiment and Its Computational Setup

The three-dimensional compressible Navier–Stokes equations are solved to study the combustion behavior of a reactive transverse jet of $NH_3/H_2/N_2$ fuel and a crossflow of heated non-vitiated air at the pressure of 5 bar.

In principle, for applications, it would be convenient to have a fuel blend with nominal flame properties identical to methane while gaining the positive effects of hydrogen [43]. Such a mixture is adopted for the present case: 40% NH_3, 45% H_2, and 15% N_2 by volume. This mixture is discharged into a crossflow of air through a round nozzle with d = 1 mm, positioned flush with a smooth solid surface. The jet center aligns with the spanwise symmetric plane, and it is located along the streamwise direction at x = 0. For the sake of clarity, the origin of the system coincides with the jet center. Nominal conditions of the inlet flows are summarized in Table 1. The crossflow air temperature is $T_{cf} = 850\,K$ (a usual value at the end of the compression stage in stationary gas turbines), while the fuel jet is at $T_j = 750\,K$ (a value coherent with the thermal cracking of ammonia). The crossflow freestream velocity $u_{cf} = 50\,m/s$ and the jet velocity $u_j = 150\,m/s$ result in a velocity ratio $r = 3$ and a momentum flux ratio $J = 1.97$, and the jet Reynolds number is $Re_j = \rho u_j d/\mu \approx 5042$ (see Table 2).

Table 1. Flow inlets conditions.

	Jet Flow	Crossflow
Species	$NH_3/H_2/N_2$	O_2/N_2
Composition, (% by vol.)	40/45/15	21/79
Pressure, (bar)	5	5
Velocity, (m/s)	150	50
Temperature, (K)	750	850
Density, (kg/m^3)	0.9558	2.0411
Viscosity, (kg/m/s)	2.8437×10^{-5}	3.8510×10^{-5}

Table 2. Some characteristics of the numerical experiment.

Jet exit diameter, d (m)	0.001
Domain extent, $L_x \times L_y \times L_z$	$25d \times 20d \times 14d$
Grid size, $N_x \times N_y \times N_z$	$1008 \times 1008 \times 672$
Velocity ratio, $r = u_j/u_{cf}$	3
Momentum flux ratio, $J = \rho_j u_j^2/\rho_{cf} u_{cf}^2$	1.97
Jet Reynolds number, Re_j	5042

The boundary conditions are set as follows. The incoming air crossflow is introduced into the computational domain through an inflow boundary condition (where velocity, temperature, and composition are fixed, while pressure and density evolve with the flow inside the computational domain). The velocity fluctuations prescribed at the inlet of the domain are of the order of 5% of the mean inflow air velocity. No turbulent velocity fluctuations are imposed on the jet [44]. This assumption arises from the jet's relatively low Reynolds number, which has a minor influence on the jet features with respect to the significant impact exerted by the turbulence from the incoming crossflow and their interaction. At the lower wall boundary ($y = 0$), a no-slip adiabatic solid surface condition is imposed, while the upper wall ($y = 0.02\,m$) is treated as an Eulerian (slip) adiabatic solid surface. The flame leaves the domain through a non-reflecting outlet boundary condition. Periodicity is enforced in the spanwise direction. A schematic of the applied boundary conditions is exposed in Figure 3.

The physical dimensions of the DNS domain are $L_x \times L_y \times L_z = 25d \times 20d \times 14d$, respectively, in the streamwise, wall-normal, and spanwise directions. The choice of grid spacing aims to adequately resolve both the flame and turbulence structures. A grid with $\Delta x = \Delta y = \Delta z \approx 10\,\mu m$ is employed in the region defined by $x/d \in [-1, 9]$, $y/d \in [1, 8]$, and $z/d \in [-1.5, 1.5]$, encompassing most of the flame regions. Algebraic stretching is then applied in all three spatial directions, reaching a maximum grid size of 130 µm. The first cell height in the wall-normal direction is set to 3 µm, ensuring $y^+ = 1$ at the lower wall. The Kolmogorov scale, denoted as $\eta = L_t/Re_{T,cf}^{0.75}$, exhibits a minimum value of 54 µm. Here, the used length scale is $L_t = 0.02$ m and $Re_{T,cf}$ is the crossflow turbulent Reynolds number, where the imposed velocity fluctuation was adopted, i.e., the 5% of the crossflow inlet velocity. The dissipative length scale calculated through the jet parameters seems to be not of practical usage in this study since, as discussed further below, no turbulent velocity fluctuations are applied on the jet inlet boundary condition, as motivated by the study of Kolla [44]. The same reasoning can be made about the flame zone, where the increase in viscosity with temperature raises the value of η. The final mesh has a total number of grid points that is $N_x \times N_y \times N_z = 1008 \times 1008 \times 672$, resulting in approximately 683 M cells.

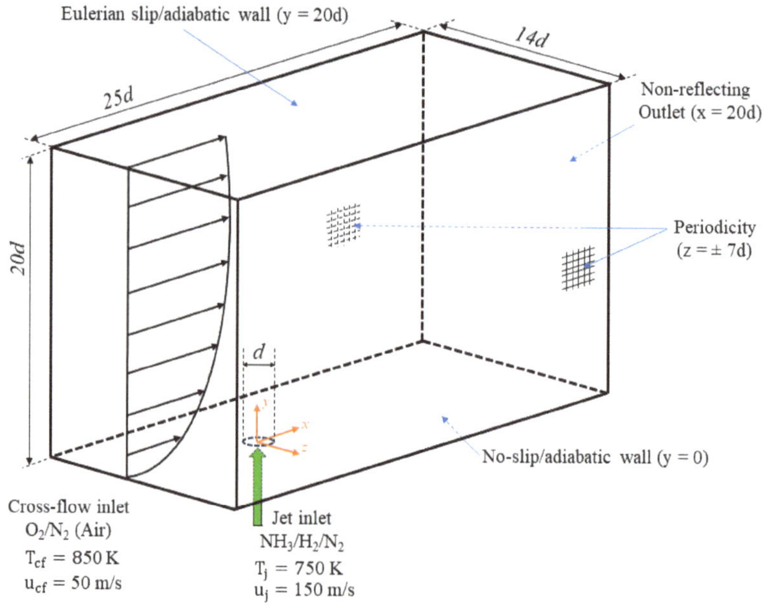

Figure 3. Schematic of the computational domain and boundary conditions for the DNS simulation.

Molecular transport in the multicomponent mixture is modeled using the Hirschfelder and Curtiss approximate formula [45], taking into account the Soret thermo-diffusive effect and the pressure gradient diffusion. At run-time, binary diffusion coefficients and the n-th species thermo-diffusion coefficients are computed using kinetic theory expressions. Dynamic viscosity and thermal conductivity for individual chemical species are pre-calculated using software libraries [46,47]. Mixture-average properties are estimated using Wilke's formula with Bird's correction for viscosity [48,49], and Mathur's expression for thermal conductivity [48,50]. Preferential diffusion is modeled based on the Hirschfelder and Curtiss law [45]. It is worth pointing out that this study does not take into consideration radiative heat transfer effects.

The chemical reactions are modeled using the Gotama [51] chemical kinetics mechanism specifically designed for ammonia, blended with hydrogen, and combustion in air. This mechanism encompasses 26 distinct species and 119 elementary reaction steps.

The numerical simulations are conducted using the in-house parallel code HeaRT and ENEA's supercomputing facility CRESCO [52]. The code solves the compressible Navier–Stokes equations, discretized using staggered finite-difference schemes. Each chemical species has its individual transport equation. Diffusive fluxes are calculated using a second-order accurate centered scheme; convective terms are modeled through the $AUSM^{+}$-up method [53], coupled with a third/fifth-order accurate $WENO$ interpolation to minimize spurious oscillations. Extended non-reflecting boundary conditions [54–56] are employed at open boundaries to consider the impact of variable transport properties [57] and local heat release [58]. At the airflow inlet, a synthetic turbulence generator is utilized [59].

4. Results

4.1. Flow-Field Organization and Features

In this section, a brief overview of the fluid dynamics of the present jet in crossflow configuration (JICF) is provided, based on the analysis of instantaneous quantities from DNS. Figure 4 shows some iso-surfaces of the NH_3 concentration and temperature, demonstrating the complex topology of the configuration. Looking at these iso-surfaces, the large vortical structures of the windward shear layer developing from the fuel jet are clearly visible: they produce mixing with the crossflow air stream by increasing the contact surfaces of the two streams. More downstream, such coherent structures develop to smaller turbulent scales, enhancing mixing at finer levels. The projection of the Heat Release Rate (HRR) of the central x−y plane is also depicted on the periodic plane and that of the y−z plane at x = 7d on the outlet plane. High HRR and temperature are observed within the leeward branch of the flow; this aspect will be considered in more detail later in this section. Moreover, in the periodic plane of the same picture, the stoichiometric mixture fraction iso-line is also shown (white line). The mixture fraction is calculated following the study of Tang [60] on a non-premixed $NH_3/H_2/N_2$ jet flame as follows:

$$\xi = \frac{(Y_H - Y_{H,2})/2W_H - (Y_O - Y_{O,2})/W_O}{(Y_{H,1} - Y_{H,2})/2W_H - (Y_{O,1} - Y_{O,2})/W_O}, \tag{2}$$

where $Y_{O(H)}$ and $W_{O(H)}$ are the local elemental mass fractions and atomic masses of the species O and H. In the same expression, the subscripts 1 and 2 indicate the fuel jet and the air crossflow, respectively. Although the stoichiometric conditions are reached on both the windward and leeward sides of the fuel jet, no flame develops on the first region due to too short convective times and high strain rates, while the flame can develop around the stoichiometric iso-line in the latter region due to the longer residence times.

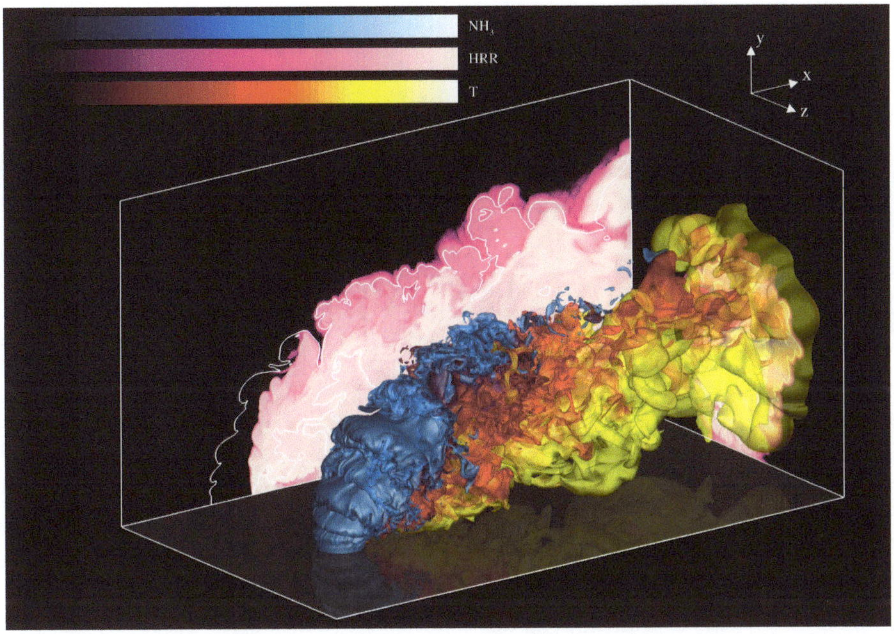

Figure 4. Volume rendering of temperature (black body colormap) and NH_3 mass fraction (blue colormap), and Heat Release Rate slices at z = 0 and x = 7d (violet colormap) at t = 0.00646 s from the start of the simulation. The white solid line on the two slices represents the stoichiometric mixture fraction ($\xi = 0.14$).

Figure 5 shows a snapshot of the vorticity magnitude (ω) projected on one of the periodic plane: the spatial distribution sheds light on the complex interaction between the crossflow and the transverse jet. The flow field is characterized by the formation and expansion of the shear layer, which then rolls up into a clear pattern of Kelvin-Helmholtz hydrodynamic instabilities. These vortical structures are believed to play a crucial role in governing key JICF processes like mixing and penetration [61]. As the jet evolves, these rolling vortices break into smaller and smaller structures as a result of the energy cascade phenomenon and are finally convected downstream, out of the domain, while being elongated in the streamwise direction. Vortex pairing is also visible, occurring on both the leeward and the windward side of the jet. These instantaneous turbulent structures are also represented three-dimensionally in the same figure, identified through the Q-criterion and colored by temperature. The formation of ring-like vortices (RLV) in the shear layer is clearly visible, as well as the horse-shoe vortex (HSV), which is in front of the windward side exit of the jet as a result of the interaction with the crossflow. This vortex further develops around the jet and dissipates far downstream. The overall described configuration is typical of this JICF problem with nominally $r > 1$. In fact, when the velocity ratio is considerably lower than the unity, another flow configuration with the presence of hairpin vortices is more likely to occur [62].

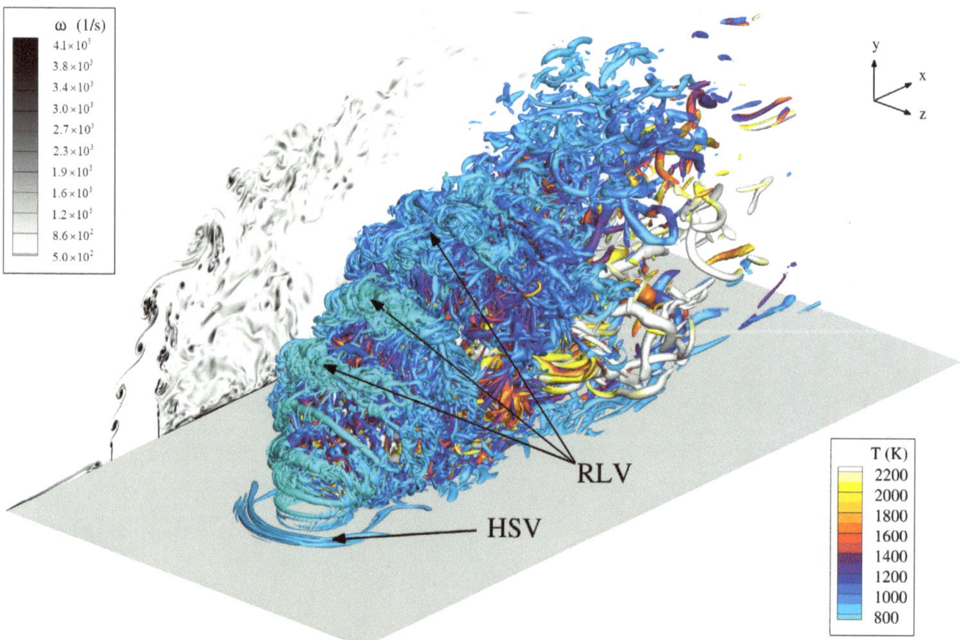

Figure 5. Visualization of instantaneous turbulent structures identified through an iso-surface of the Q-criterion colored by the local value of temperature (magma colormap). On the bottom periodic plane, there is the related vorticity magnitude projected from the mid-plane at $z = 0$ (grayscale colormap). The snapshot refers to $t = 0.00646$ s from the start of the simulation.

Figure 6 displays the instantaneous spatial distribution of the velocity components in the streamwise direction (u_x) and the wall-normal direction (u_y) within the central $x-y$ plane. The distribution of the streamwise velocity is closely intertwined with the behavior of the crossflow and the transverse jet. Specifically, as the crossflow encounters the jet, it triggers the formation of shear layer vortices and leads to a notable reduction in the magnitude of u_x immediately ahead of the jet, resulting in a recirculation zone that develops into an HSV. The figure also highlights an increase in u_x values along the jet trajectory. Additionally, negative values of u_x are found in the immediate vicinity of the jet on its leeward side, indicative of the entrainment of the crossflow and the presence of a recirculation zone. This recirculation zone effectively captures the high-temperature products generated by combustion on the leeward side, therefore contributing to the stabilization of the flame. The strongly different behavior of the windward and leeward regions of the fuel jet is also clear from the intensity of the vortex shedding in the related shear layers: the one on the leeward side of the jet is less pronounced compared to the windward side, attributable to an upsurge in flow viscosity around the flame. Conversely, the wall-normal velocity distribution indicates that the jet, characterized by significant values of u_y, is notably affected by the crossflow. This emphasizes the role of the crossflow in augmenting the mixing between the jet and the surrounding flow. From Figure 6b, the fuel jet penetration can also be deduced: its potential core reaches nearly ∼5 mm in the y-direction (nearly one quarter of the channel height), while its plume goes slightly above one third of the channel height.

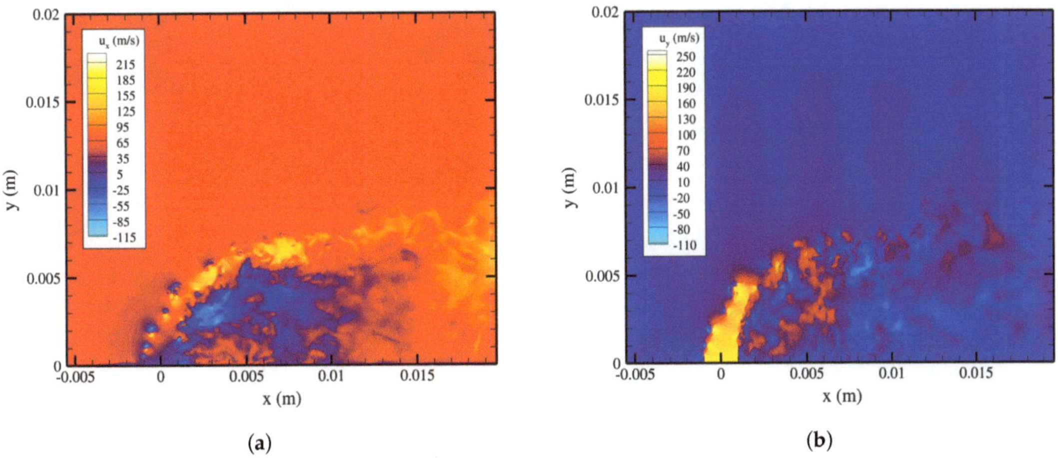

Figure 6. Contours of instantaneous streamwise (**a**) and wall−normal (**b**) velocities in the central longitudinal x−y plane.

Figure 7 shows the temperature field in six x−z planes at different distances from the bottom wall, demonstrating the wake behind the fuel jet. At shorter distances, the flame is limited to a small area and does not appear continuous, showing some dispersed hot spots asymmetrically released downstream. However, at greater distances from the wall, the reacting area significantly expands due to the accelerated rate of crossflow fluid entrainment [63,64] and gas expansion caused by the chemical reactions. Moreover, the temperature field clearly shows that the flame is well anchored to the leeward side of the jet.

A different view of the temperature field is provided in Figure 8, referring to the central x−y plane. The windward flame is absent and not situated within the shear layer, where the crossflow interacts with the jet. This is confirmed by the HRR (see Figure 4), which reaches its maximum values on the leeward side. The leeward branch of the flame is anchored within the shear layer near the jet exit and a broader reaction zone can be observed, undergoing to gas expansion with strong density gradients and with temperatures above 2400 K. In fact, the flame is situated in an area with lower airflow velocity and, consequently, lower strain rates. This region is characterized by longer residence times, allowing radicals and temperature to reach their highest levels. These results agree with those reported in the study of Grout [65], where the flame was anchored on the leeward side of the jet, coinciding with a region of low velocity and near stoichiometric mixture conditions. There was no distinct flame branch observed on the windward side, contrary to the study of Lyra [66], where a H_2 JICF with hotter vitiated air was considered.

To close the overview of the topology of this reacting jet in crossflow, Figure 9a,b shows the instantaneous spatial evolution of the jet mixture fuel components, H_2 and NH_3, respectively. As the jet evolves and interacts with the crossflow, the fuel is consumed due to the combustion process, thus producing radicals and more stable combustion products. Figure 9c,d report the H and NO instantaneous field, respectively. It is worth noticing the high level of NO emissions, exceeding 20,000 ppm, and the strong correlation with high-temperature regions.

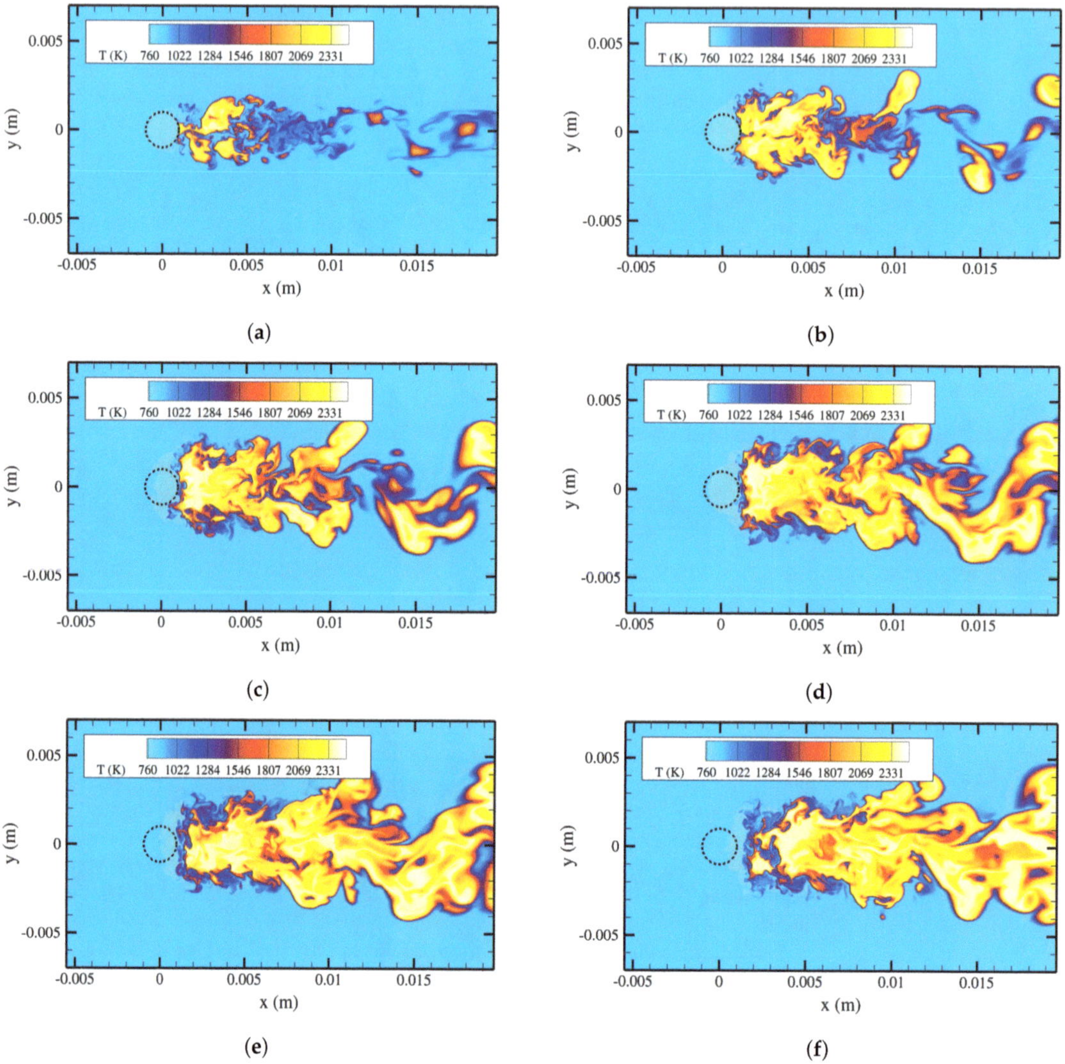

Figure 7. Contours of instantaneous temperature in the x−z planes at different heights: (**a**) y = 0.44d, (**b**) y = 0.98d, (**c**) y = 1.54d, (**d**) y = 2.12d, (**e**) y = 2.71d and (**f**) y = 3.33d. The black dashed line represents the round shape of the jet exit.

4.2. Analysis of the Combustion Modes

The turbulent structures deriving from the complex interaction between the fuel jet and the air jet in crossflow produce an intense mixing of reactants. Hence, combustion is likely to happen in both non-premixed and premixed modes, depending on the flow region.

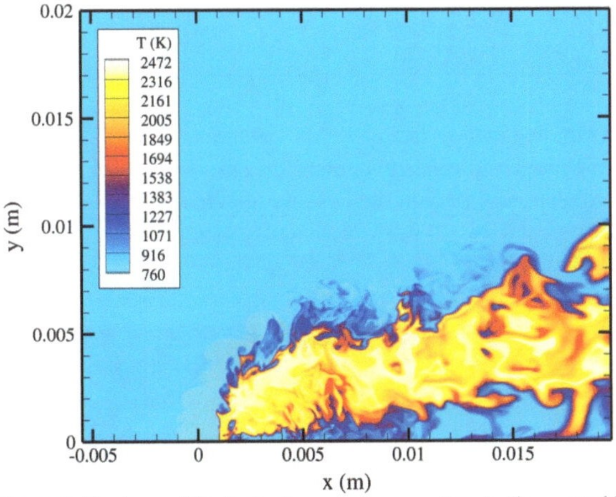

Figure 8. Contour of the instantaneous temperature in the central longitudinal x−y plane.

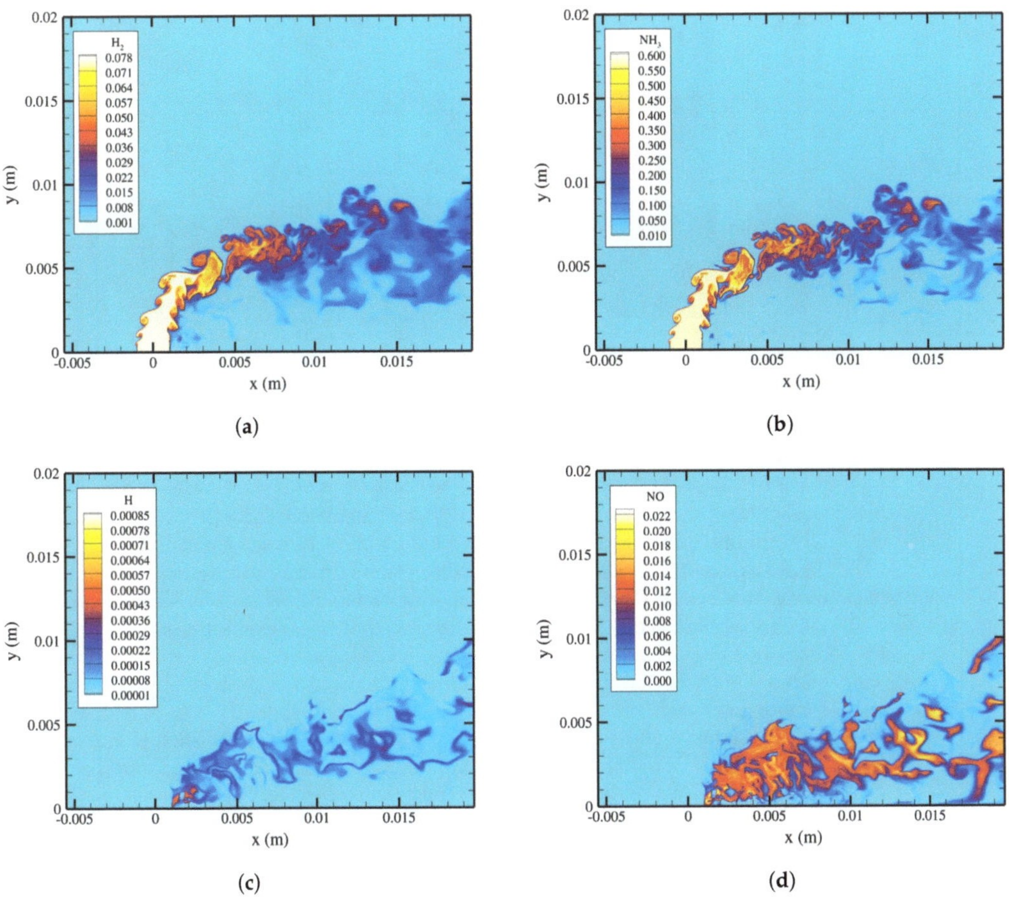

Figure 9. Contours of the instantaneous species mass fraction concentrations in the central longitudinal x−y plane: (**a**) H_2, (**b**) NH_3, (**c**) H_2O and (**d**) NO.

The Flame Index helps in identifying such combustion modes. In particular, the Takeno index is calculated in its revised fuel multispecies formulation as $FI = \nabla Y_F \cdot \nabla Y_{O_2} / (|\nabla Y_F||\nabla Y_{O_2}|)$, ∇Y_F being the gradient of the fuel species ($Y_F = Y_{NH_3} + Y_{H_2}$) [67]. Figure 10 reports an instantaneous Flame Index distribution in the central longitudinal x−y plane. It can be observed that $FI \in [-1, 1]$: this means that, although the overall flame is non-premixed, in fact, reacting structures are associated with both non-premixed ($FI < 0$) and premixed ($FI > 0$) combustion modes. At the same time, it must be stressed that no flame is exhibited on the windward side of the fuel jet, as shown in Figure 8. Hence, from Figure 10 it is hard to say which of the combustion modes is the most probable and relevant for flame anchoring.

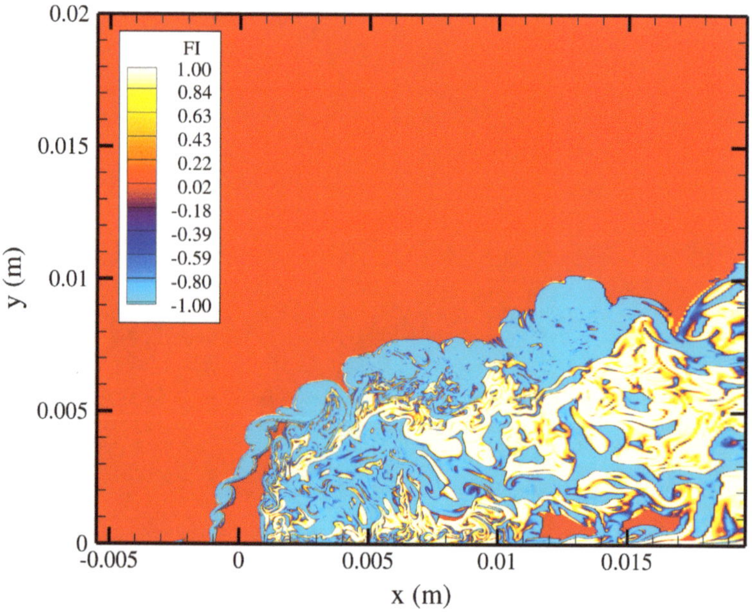

Figure 10. Contour of the instantaneous Flame Index in the central longitudinal x−y plane.

At first, to understand which of the two combustion modes is the most relevant in terms of occupied volume, the volume fraction distribution in the computational domain as a function of the Flame Index $FI \in [-1, 1]$ was calculated, i.e., $V(FI) = \sum_{n=1}^{N_i} V_n(FI)/V_{Tot}$, N_i being the number of volumes at a fixed value of FI. Such a distribution is shown in Figure 11a, and it reveals that most of the volume can be attributed to the non-premixed mode. However, this analysis is based only on the mixing of reactants and does not consider the effects of local convective times and strains that can negatively affect the local ignition of the potentially reactive mixture. Hence, as already observed for the spatial distribution of FI in Figure 10, this analysis is not sufficient to draw conclusions on the most effective combustion mode.

From the previous analysis, to understand which of the combustion modes is the most relevant in this nominally non-premixed configuration, the Flame Index must be correlated with the local heat release. Figure 11b reports the mean Heat Release Rate distribution as a function of the Flame Index and mixture fraction ξ. It is observed that the maximum heat release is found in regions near the stoichiometric mixture fraction ($\xi = 0.14$), but, unexpectedly, for positive values of the Flame Index associated with premixed combustion. These regions are located several diameters downstream of the fuel jet where reactants had time to mix. The heat released from the diffusion flame is located immediately downstream of the fuel jet, and is mainly responsible for the flame anchoring. Here, vortices of the jet diameter size contribute to bringing part of the crossflow air immediately downstream of the injection region and to intensively stretch the newly formed diffusion flames, thus reducing the peak value of the heat release in comparison to the premixed regions.

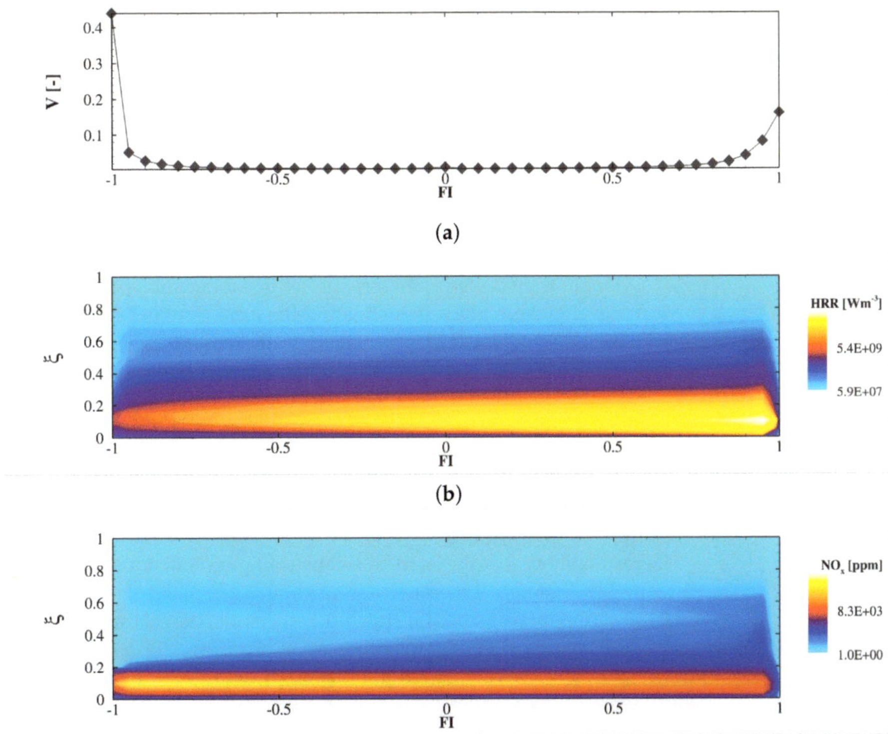

(**a**)

(**b**)

(**c**)

Figure 11. (**a**) Volume fraction distribution as a function of the Flame Index FI; (**b**) distribution of the mean Heat Release Rate as a function of the Flame Index and Mixture Fraction ξ; (**c**) distribution of the mean NO concentration [dppm] as a function of the Flame Index and Mixture Fraction.

What was found from the combustion modes analysis has interesting implications on NO emissions. Figure 11c shows the mean NO pollutant concentration in dry ppm (dppm) as a function of the Flame Index and mixture fraction ξ. As expected, it is mostly formed in the near stoichiometric regions, but the maximum values attain to the non-premixed combustion mode (FI < 0). Since the present jet in crossflow configuration exhibits such diffusion flames in zones characterized by high strain, this seems in agreement with findings from laminar diffusion flames where higher NO concentrations correspond to higher strains. The NO concentrations are well above the permitted limits; this is due to the presence of hydrogen in the fuel, which enhances thermal NO by increasing the maximum temperatures and fuel-NO by increasing the concentration of O and H radicals. This suggests working with H_2 concentrations in fuel mixtures lower than the present one. Further reduction of NO can be achieved by increasing the nominal pressure [27]. This will be the aim of future work.

5. Conclusions

The article investigated the turbulent non-premixed combustion of a fuel mixture derived from NH_3 partial decomposition in an air crossflow at 5 bar by means of DNS.

Preliminary chemical kinetics simulations of laminar flames in a non-premixed counterflow reactor demonstrated that the maximum temperature decreases as the applied strain increases, while the maximum NO increases: in fact, under higher strains, the fuel-NO production is enhanced, while the thermal-NO becomes less important.

The three-dimensional DNS instantaneous flow field was analyzed, demonstrating the complex structure of this ideally simple jet in crossflow configuration. The flame develops and anchors on the leeward side of the fuel jet. The flow topology was described, demonstrating the different intensities of the windward and leeward shear layers, the latter having shed vortices with shorter lifetimes due to the damping effect of the reacting region. The recirculation zone behind the jet anchors the flame. The main turbulent structures, like ring and horse-shoe vortices, were identified. The analysis provided temperature, heat release, and main transported chemical species instantaneous spatial distributions.

More insights were given investigating the local combustion modes experienced in the flame, i.e., non-premixed and premixed. Although the configuration is globally non-premixed, the Flame Index distribution demonstrated both non-premixed ($FI < 0$) and premixed ($FI > 0$) combustion modes, locally. The non-premixed mode is the most relevant from a volumetric point of view, but the premixed one is the most relevant in terms of heat release. The local premixed combustion volumes belong to a low-strain region on the leeward side of the jet. Coherently, the NO concentration is higher in the local non-premixed combustion volumes belonging to a region of high strain, in agreement with the findings from laminar frames.

The results show that the jet in crossflow configuration, nominally non-premixed, can be potentially adopted to generate a non-negligible premixed combustion region. How to control the actual mixing and the stoichiometry of the desired premixed reacting region becomes the next critical step for the exploitation of the investigated strategy in applications. Reducing the hydrogen content in the fuel mixture can be of help: in fact, increasing the ignition delay time provides more time for turbulent mixing to act before combustion takes place in a non-premixed mode. Lower hydrogen content also helps in lowering NO emission; further reduction can be achieved by operating with higher nominal pressures. Moreover, aiming at a $NH_3/H_2/N_2$ fuel blend with less hydrogen reduces the cracking requirements of ammonia.

Author Contributions: E.G., D.C., M.C. and S.C.: original draft preparation and writing; D.C. and M.C.: fluid dynamic simulation; S.C.: laminar chemical kinetics simulations. All authors have read and agreed to the published version of the manuscript.

Funding: This study was performed within the Italian project "RICERCA E SVILUPPO DI TECNOLOGIE PER LA FILIERA DELL'IDROGENO POR-H2" ("Research and Development of Technologies for Hydrogen Chain, POR-H2, WP2, LA2.2.5"), funded by the Italian Ministry of Environment and Energy Security, under the National Recovery and Resilience Plan (PNRR, Mission 2, Component 2, Investment 3.5, project code I83C22001170006).

Data Availability Statement: Interested researchers can contact the authors.

Acknowledgments: The computational resources and associated technical support utilized for this study were made available by the CRESCO/ENEAGRID High-Performance Computing infrastructure and its staff [52]. This computing infrastructure is funded by ENEA, the Italian National Agency for New Technologies, Energy, and Sustainable Economic Development, along with contributions from Italian and European research programs (visit http://www.cresco.enea.it/english, accessed on 20 October 2023).

Conflicts of Interest: The authors declare no conflict of interest.

References

1. Karabeyoglu, A.; Evans, B.; Stevens, J.; Cantwell, B. Development of ammonia based fuels for environmentally friendly power generation. In Proceedings of the 10th International Energy Conversion Engineering Conference AIAA, Atlanta, GE, USA, 29 July–1 August 2012; Paper AIAA2012-4055.
2. Zamfirescu, C.; Dincer, I. Ammonia as a green fuel and hydrogen source for vehicular applications. *Fuel Process. Technol.* **2009**, *90*, 729–737. [CrossRef]
3. Schlapbach, L.; Züttel, A. Hydrogen-Storage Materials for Mobile Applications. *Nature* **2001**,*414*, 353–358. [CrossRef]
4. Karabeyoglu, A.; Stevens, J.; Geyzel, D.; Cantwell, B.; Micheletti, D. High performance hybrid upper stage motor. In Proceedings of the 47th AIAA/ASME/SAE/ASEE Joint Propulsion Conference & Exhibit, San Diego, CA, USA, 31 July–3 August 2011; pp. 1–24.
5. Brown, T. Ammonia Fuel Risk Levels A summary of the two existing studies of Ammonia Fuel Safety. In Proceedings of the NH_3 Fuel Conference 2015, Chicago, IL, USA, 20–23 September 2015.
6. Vitse, F.; Cooper, M.; Botte, G.G. On the use of ammonia electrolysis for hydrogen production. *J. Power Sources* **2005**, *142*, 18–26. [CrossRef]
7. Pratt, D.T. *Performance of Ammonia-Fired Gas-Turbine Combustors*; Report TS-67-5; California Univ. Berkeley THERMAL SYSTEMS Div.: Berkeley, CA, USA, 1967; p. 35. Available online: https://apps.dtic.mil/sti/citations/AD0657585 (accessed on 19 November 2023).
8. Pfahl, U.J.; Ross, M.C.; Shepherd, J.E.; Pasamehmetoglu, K.O.; Unal, C. Flammability limits, ignition energy, and flame speeds in H_2–CH_4–NH_3–N_2O–O_2–N_2 mixtures. *Combust. Flame* **2000**, *123*, 140–158. [CrossRef]
9. Mathieu, O.; Petersen, E.L. Experimental and modeling study on the high-temperature oxidation of Ammonia and related NOx chemistry. *Combust. Flame* **2015**, *162*, 554–570. [CrossRef]
10. Hayakawa, A.; Arakawa, Y.; Mimoto, R.; Kunkuma, K.D.; Somarathne, A.; Kudo, T.; Kobayashi, H. Experimental investigation of stabilization and emission characteristics of ammonia/air premixed flames in a swirl combustor. *Int. J. Hydrogen Energy* **2017**, *42*, 14010–14018. [CrossRef]
11. Hayakawa, A.; Goto, T.; Mimoto, R.; Arakawa, Y.; Kudo, T.; Kobayashi, H. Laminar burning velocity and Markstein length of ammonia/air premixed flames at various pressures. *Int. J. Hydrogen Energy* **2015**, *159*, 98–106. [CrossRef]
12. Mei, B.; Zhang, J.; Shi, X.; Xi, Z.; Li, Y., Enhancement of ammonia combustion with partial fuel cracking strategy: Laminar flame propagation and kinetic modeling investigation of $NH_3/H_2/N_2$/air mixtures up to 10 atm. *Combust. Flame* **2021**, *231*, 111472. [CrossRef]
13. Valera-Medina, A.; Pugh, D.; Marsh, R.; Bulat, G.; Bowen, P. Preliminary study on lean premixed combustion of ammonia-hydrogen for swirling gas turbine combustors. *Int. J. Hydrogen Energy* **2017**, *42*, 24495–24503. [CrossRef]
14. Verkamp, F.J.; Hardin, M.C.; Williams, J.R. Ammonia combustion properties and performance in gas-turbine burners. *Proc. Comb. Inst.* **1967**, *11*, 985–992. [CrossRef]
15. Khateeb, A.A.; Guiberti, T.F.; Zhu, X.; Younes, M.; Jamal, A.; Roberts, W.L. Stability limits and NO emissions of technically premixed ammonia-hydrogen-nitrogen-air swirl flames. *Int. J. Hydrogen Energy* **2020**, *45*, 22008–22018. [CrossRef]
16. Hayakawa, A.; Ichikawa, A.; Arakawa, Y. Enhancement of Reaction and Stability of Ammonia Flames using Hydrogen Addition and Swirling Flows. In Proceedings of the 2015 NH_3 Fuel Conference, Chicago, IL, USA, 20–23 September 2015.
17. Pugh, D.; Bowen, P.; Valera-Medina, A.; Giles, A.; Runyon, J.; Marsh, R. Influence of steam addition and elevated ambient conditions on NO_x reduction in a staged premixed swirling NH_3/H_2 flame. *Proc. Combust. Inst.* **2019**, *37*, 5401–5409. [CrossRef]
18. Zhang, M.; An, Z.; Wang, L.; Wei, X.; Jianayihan, B.; Wang, J.; Huang, Z.; Tan, H. The regulation effect of methane and hydrogen on the emission characteristics of ammonia/air combustion in a model combustor. *Int. J. Hydrogen Energy* **2021**, *46*, 21013–21025. [CrossRef]
19. Franco, M.C.; Rocha, R.C.; Costa, M.; Yehia, M. Characteristics of NH_3/H_2/air flames in a combustor fired by a swirl and bluff-body stabilized burner. *Proc. Combust. Inst.* **2020**, *38*, 5129–5138. [CrossRef]
20. Sorrentino, G.; Sabia, P.; Bozza, P.; Ragucci, R.; de Joannon, M. Low-NOx conversion of pure ammonia in a cyclonic burner under locally diluted and preheated conditions. *Appl. Energy* **2019**, *254*, 113676. [CrossRef]
21. Ariemma, G.B.; Sabia, P.; Sorrentino, G.; Bozza, P.; de Joannon, M.; Ragucci, R. Influence of water addition on MILD ammonia combustion performances and emissions. *Proc. Combust. Inst.* **2020**, *38*, 5147–5154. [CrossRef]
22. Okafor, E.C.; Somarathne, K.D.K.A.; Hayakawa, A.; Kudo, T.; Kurata, O.; Iki, N.; Kobayashi, H. Towards the development of an efficient low-NO_x ammonia combustor for a micro gas turbine. *Proc. Combust. Inst.* **2019**, *37*, 4597–4606. [CrossRef]
23. Okafor, E.C.; Somarathne, K.D.K.A.; Ratthanan, R.; Hayakawa, A.; Kudo, T.; Kurata, O.; Iki, N.; Tsujimura, T.; Furutani, H.; Kobayashi, H. Control of NO_x and other emissions in micro gas turbine combustors fuelled with mixtures of methane and ammonia. *Combust. Flame* **2020**, *211*, 406–416. [CrossRef]
24. Heggset, T.; Meyer, O.; Tay-Wo-Chong, H.L.; Ciani, A.; Gruber, A. Numerical Assessment of a Rich-Quench-Lean Staging Strategy for Clean and Efficient Combustion of Partially Decomposed Ammonia in the Constant Pressure Sequential Combustion System. *J. Eng. Gas Turbines Power* **2023**. [CrossRef]
25. Okafor, E.C.; Kurata, O.; Yamashita, H.; Inoue, T.; Tsujimura, T.; Iki, N.; Hayakawa, A.; Ito, S.; Uchida, M.; Kobayashi, H. Liquid ammonia spray combustion in two-stage micro gas turbine combustors at 0.25 MPa; Relevance of combustion enhancement to flame stability and NO_x control. *Appl. Energy Combust. Sci.* **2021**, *7*, 100038. [CrossRef]

26. Guteša, B.M.; Mashruk, S.; Zitouni, S.; Valera-Medina, A. Humidified ammonia/hydrogen RQL combustion in a trigeneration gas turbine cycle. *Energy Convers. Manag.* **2021**, *227*, 113625. [CrossRef]
27. Gubbi, S.; Cole, R.; Emerson, B.; Noble, D.; Steele, R.; Sun, W.; Lieuwen, T. Air Quality Implications of Using Ammonia as a Renewable Fuel: How Low Can NOx Emissions Go? *ACS Energy Lett.* **2023**, *8*, 4421–4426. [CrossRef] [PubMed]
28. Normand, A. Method for Generating a Reducing Atmosphere for Heat-Treating Installations. U.S. Patent US4069071A, 17 January 2000.
29. Sørensen, R.Z.; Nielsen, L.J.E.; Jensen, S.; Hansen, O.; Johannessen, T.; Quaade, U.; Christensen, C.H. Catalytic ammonia decomposition: Miniaturized production of COx-free hydrogen for fuel cells. *Catal. Commun.* **2005**, *6*, 229–232. [CrossRef]
30. Ganley, J.C.; Seebauer, E.G.; Masel, R.I. Development of a microreactor for the production of hydrogen from ammonia. *J. Power Sources* **2004**, *137*, 53–61. [CrossRef]
31. Khateeb, A.A.; Guiberti, T.F.; Wang, G.; Boyette, W.R.; Younes, M.; Jamal, A. Stability limits and NO emissions of premixed swirl ammonia-air flames enriched with hydrogen or methane at elevated pressures. *Int. J. Hydrogen Energy* **2021**, *46*, 11969–11981. [CrossRef]
32. Chen, J.H.; Choudhary, A.; de Supinski, B.; DeVries, M.; Hawkes, E.R.; Klasky, S.; Liao, W.K.; Ma, K.L.; Mellor-Crummey, J.; Podhorski, N.; et al. Terascale direct numerical simulations of turbulent combustion using s3d. *Comput. Sci. Discov.* **2009**, *2*, 015001. [CrossRef]
33. Jiang, Y.; Gruber, A.; Seshadri, K.; Williams, F.A. An updated short chemical-kinetic nitrogen mechanism for carbon-free combustion applications. *Int. J. Energy Res.* **2020**, *44*, 795–810. [CrossRef]
34. Noble, D.; Wu, D.; Emerson, B.; Sheppard, S.; Lieuwen, T.; Angello, L. Assessment of Current Capabilities and Near-Term Availability of Hydrogen-Fired Gas Turbines Considering a Low-Carbon Future. *J. Eng. Gas Turbines Power* **2021**, *143*, 041002. [CrossRef] [CrossRef]
35. Funke, H.H.W.; Beckmann, N.; Abanteriba, S. An overview on dry low NO_x micromix combustor development for hydrogen-rich gas turbine applications. *Int. J. Hydrogen Energy* **2019**, *44*, 6978–6990. [CrossRef] [CrossRef]
36. Devriese, C.; Pennings, W.; de Reuver, H.; Penninx, G.; de Ruiter, G.; Bastiaans, R.; De Paepe, W. The Design and Optimisation of a Hydrogen Combustor for a 100 Kw Micro Gas Turbine, Paper ID Number: 23-IGTC21, Gas turbines in a carbon-neutral society. In Proceedings of the 10th International Gas Turbine Conference, ETN, Brussels, Belgium, 11–15 October 2021.
37. Cuoci, A.; Frassoldati, A.; Faravelli, T.; Ranzi, E. OpenSMOKE++: An object-oriented framework for the numerical modeling of reactive systems with detailed kinetic mechanisms. *Comput. Phys. Commun.* **2015**, *192*, 237–264. [CrossRef] [CrossRef]
38. Cuoci, A.; Frassoldati, A.; Faravelli, T.; Ranzi, E. Numerical Modeling of Laminar Flames with Detailed Kinetics Based on the Operator-Splitting Method. *Energy Fuels* **2013**, *27*, 7730–7753. [CrossRef]
39. Puri, I.; Seshadri, K. Extintion of diffusion flames burning diluted methane and diluted propane in diluted air. *Combust. Flame* **1986**, *65*, 137–150. [CrossRef]
40. Alnasif, A.; Mashruk, S.; Shi, H.; Alnajideen, M.; Wang, P.; Pugh, D.; valera-Medina, A. Evolution of ammonia reaction mechanisms and modeling parameters: A review. *Appl. Energy Combust. Sci.* **2023**, *15*, 100175.
41. Cecere, D.; Giacomazzi, E.; Di Nardo, A.; Calchetti, G. Gas Turbine Combustion Technologies for Hydrogen Blends. *Energies* **2023**, *16*, 6829. [CrossRef]
42. Xu, S.; Jin, S.; Tong, Y.; Shi, B.; Tu, Y.; Liu, H. Quantitative evaluation of NO formation and destructionroutes during methane MIL combustion using an improved calculation method. *Fuel* **2022**, *324*, 124593. [CrossRef]
43. Wiseman, S.; Rieth, M.; Gruber, A.; Dawson, J.R.; Chem, J.H. A comparison of the blow-out behavior of turbulent premixed ammonia/hydrogen/nitrogen-air and methane–air flames. *Proc. Combust. Inst.* **2021**, *38*, 2869–2876. [CrossRef]
44. Kolla, H.; Grout, R.W.; Gruber, A.; Chen, J.H. Mechanisms of flame stabilization and blowout in a reacting turbulent hydrogen jet in cross-flow. *Combust. Flame* **2012**, *159*, 2755–2766. [CrossRef]
45. Hirschfelder, J.O.; Curtiss, C.F.; Bird, R.B.; Spotz, E.L. *The Molecular Theory of Gases and Liquids*; John Wiley: New York, NY, USA, 1954.
46. Ern, A.; Giovangigli, V. *Multicomponent Transport Algorithms, Lecture Notes in Physics*; Springer: Heidelberg, Germany, 1994; Volume M24.
47. Ern, A.; Giovangigli, V. Fast and accurate multicomponent transport property evaluation. *J. Comput. Phys.* **1995**, *120*, 105–116. [CrossRef]
48. Bird, R.B.; Stewart, W.E.; Lightfoot, E.N. *Transport Phenomena*, 2nd ed.; Wiley International Edition; John Wiley and Sons: New York, NY, USA, 2002.
49. Wilke, C.R. A Viscosity Equation for Gas Mixtures. *J. Chem. Phys.* **1950**, *18*, 517–519. [CrossRef]
50. Mathur, S.; Tondon, P.K.; Saxena, S.C. Thermal conductivity of binary, ternary and quaternary mixtures of rare gases. *Mol. Phys.* **1967**, *12*, 569. [CrossRef]
51. Gotama, G.J.; Hayakawaa, A.; Okaford, E.C.; Kanoshimaa, R.; Hayashia, M.; Kudoa, T.; Kobayashia, H. Measurement of the laminar burning velocity and kinetics study of the importance of the hydrogen recovery mechanism of ammonia/hydrogen/air premixed flames. *Combust. Flame* **2022**, *236*, 111753. [CrossRef]
52. Ponti, G. The role of medium size facilities in the HPC ecosystem: The case of the new CRESCO4 cluster integrated in the ENEAGRID infrastructure. In Proceedings of the 2014 International Conference on High Performance Computing and Simulation, Bologna, Italy, 21–25 July 2014; pp. 1030–1033.

53. Liou, M.S. A Sequel to AUSM, Part II: AUSM$^+$-up for All Speeds. *J. Comput. Phys.* **2006**, *214*, 137–170. [CrossRef]
54. Rudy, D.H.; Strikwerda, J.C. Boundary conditions for subsonic compressible Navier-Stokes calculations. *Comput. Fluids* **1981**, *9*, 327–338. [CrossRef]
55. Thompson, K.W. Time-dependent boundary conditions for hyperbolic systems. *J. Comput. Phys.* **1990**, *89*, 439–461. [CrossRef]
56. Poinsot, T.J.; Lele, S.K. Boundary Conditions for Direct Simulations of Compressible Viscous Flows. *J. Comput. Phys.* **1992**, *101*, 104–129. [CrossRef]
57. Baum, M.; Poinsot, T.; Thévenin, D. Accurate Boundary Conditions for Multicomponent Reactive Flows. *J. Comput. Phys.* **1995**, *116*, 247–261. [CrossRef]
58. Sutherland, J.C.; Kennedy, C.A. Improved Boundary Conditions for Viscous, Reacting, Compressible Flows. *J. Comput. Phys.* **2003**, *191*, 502–524. [CrossRef]
59. Klein, M.; Sadiki, A.; Janicka, J. A digital filter based generation of inflow data for spatially developing direct numerical or large eddy simulations. *J. Comput. Phys.* **2003**, *186*, 652–665. [CrossRef]
60. Tang, H.; Yang, C. ; Wang, G.; Krishna, Y.; Guiberti, T.F.; Roberts, W.L.; Magnotti, G. Scalar structure in turbulent non-premixed NH3/H2/N2 jet flames at elevated pressure using Raman spectroscopy. *Combust. Flame* **2022**, *244*, 112292. [CrossRef]
61. Karagozian, A.R. Transverse jets and their control. *Prog. Energy Combust. Sci.* **2010**, *36*, 531–553. [CrossRef]
62. Mahesh, K. The interaction of jets with crossflow. *Annu. Rev. Fluid Mech.* **2013**, *45*, 379–407. [CrossRef]
63. Hasselbrink, E.F.; Mungal, M.G. Transverse jets and jet flames. Part 2. Velocity and OH field imaging. *J. Fluid Mech.* **2001**, *443*, 27–68. [CrossRef]
64. Muniz, L.; Mungal, M.G. Effects of heat release and buoyancy on flow structure and entrainment in turbulent nonpremixed flames. *Combust. Flame* **2001**, *126*, 1402–1420. [CrossRef]
65. Grout, R. W.; Gruber, A.; Kolla, H.; Bremer, P.T.; Bennett, J.C.; Gyulassy, A.; Chen, J.H. A direct numerical simulation study of turbulence and flame structure in transverse jets analysed in jet-trajectory based coordinates. *J. Fluid Mech.* **2012**, *706*, 351–383. [CrossRef]
66. Lyra, S.; Wilde, B.; Kolla, H.; Seitzman, J.; Lieuwen, T.C.; Chen, J.H. *Structure and Stabilization of Hydrogen-Rich Transverse Jets in a Vitiated Turbulent Flow*; Sandia National Lab. (SNL-CA): Livermore, CA, USA, 2014.
67. Zirwes, T.; Zhang, F.; Habisreuther, P.; Hansinger, M.; Bockhorn, H.; Pfitzner, M.; Trimis, D. Identification of Flame Regimes in Partially Premixed Combustion from a Quasi-DNS Dataset. *Flow Turbul. Combust.* **2021**, *106*, 373–404. [CrossRef]

energies

Review

Micro Gas Turbine Role in Distributed Generation with Renewable Energy Sources

Roberta De Robbio

Department of Industrial Engineering, University of Naples Federico II, Via Claudio 21, 80125 Naples, Italy; roberta.derobbio@unina.it

Abstract: To become sustainable, the production of electricity has been oriented towards the adoption of local and renewable sources. Distributed electric and thermal energy generation is more suitable to avoid any possible waste, and the Micro Gas Turbine (MGT) can play a key role in this scenario. Due to the intrinsic properties and the high flexibility of operation of this energy conversion system, the exploitation of alternative fuels and the integration of the MGT itself with other energy conversion systems (solar field, ORC, fuel cells) represent one of the most effective strategies to achieve higher conversion efficiencies and to reduce emissions from power systems. The present work aims to review the results obtained by the researchers in the last years. The different technologies are analyzed in detail, both separately and under a more complete view, considering two or more solutions embedded in micro-grid configurations.

Keywords: micro gas turbine; syngas; hydrogen; solar field; ORC; SOFC; distributed energy system; hybrid energy systems; micro-grid

Citation: De Robbio, R. Micro Gas Turbine Role in Distributed Generation with Renewable Energy Sources. *Energies* **2023**, *16*, 704. https://doi.org/10.3390/en16020704

Academic Editor: Andrea De Pascale

Received: 7 December 2022
Revised: 27 December 2022
Accepted: 4 January 2023
Published: 7 January 2023

1. Introduction

The awareness of the environmental and climatic damage caused by decades of excessive energy consumption with the use of fossil fuels has been achieved. Following the well-known international treaties [1], there are currently many countries involved in formulating strategies to control fuel consumption and pollution emissions, to limit the anthropogenic impact [2].

To obtain a sustainable production of electricity, research has been oriented towards the adoption of renewable local resources such as biomass, solid residues from various industrial processes, or towards the use of hydrogen. To this end, the Micro Gas Turbine (MGT) can significantly contribute to a more flexible employment and efficient conversion of energetic sources [3–5]. Even if simple cycle MGTs feature lower efficiency levels with respect to reciprocating internal combustion engines (ICE), modifications can be made to the plant architecture to overcome this issue [6], such as regeneration, post-combustion, vapour injection and air humidification [7], and the gap between these two technologies can be reduced by the implementation of advanced materials [8] or even by using different working media (air, nitrogen, hydrogen, helium, xenon and carbon dioxide) [9].

With the main purpose is to reduce emissions and to better control the reaction zone, improvements can still be made especially on the combustor design through additive manufacturing [4] and the external combustion-air adjustment [10]. In addition, a proper modeling of the complex dynamic characteristics of micro gas turbine components allows for the identification of the lower fuel consumption areas, including those during the start-up process [3,11]. Moreover, it is well-known that gas turbines are supplied with a substantial excess of air, and in general produce high emissions of nitrogen oxides (NO_x). To overcome this issue, innovative combustor concepts were implemented to limit the formation of pollutants such as the Dry Low NO_x (DLN) or the Rich-Burn, Quick-Mix, Lean Burn (RQL). Unfortunately, MGTs are not a simple scale down of standard gas turbines and, due to the required reduced dimensions and reduced costs, such complex combustors cannot always be adopted, or at least they must be redesigned for turbines with a different architecture like MGT. However, an applicable solution to MGTs is also represented by the exhaust gas recirculation (EGR). Indeed, it is experimentally observed that EGR has a positive influence on the heat release rate, reducing the non-uniformities in temperature distribution [12]. EGR used in humified air turbines (HAT) increases CO_2 partial pressure, thereby leading to a more efficient post-combustion capture [13]. Alternatively, CO_2 concentrations in the exhausts could be enhanced by employing a selective exhaust recirculation By using a CO_2-selective system, it is possible to separate CO_2 from the flue gas and recirculate it back to the inlet of the compressor. In this way, a saving in economics and energy for the downstream CO_2 capture plant can be attained as well [14].

Nevertheless, the addition of carbon capture and storage (CCS) systems complicates the plant that on the contrary is required to be compact and not-excessively expensive. Therefore, considering that an acceptable level of the thermal efficiency can be achieved only with regeneration, and that the maximum value cannot be reached at maximum power output [15], the economic success of micro gas turbine applications strongly depends on the optimal usage of the system with a proper compromise on greenhouse gas emissions. Indeed, the main efforts are addressed to a full exploitation of the fuel, obtained by embedding heat recovery in the overall setup, that is, with a combined heat and power plant (CHP) [16]. In this regard, vapour injection and HAT cycle was evidenced to further increase the flexibility of an MGT plant, to enhance the heat capacity of the mixture entering the combustor, to lower fuel consumption and a more efficient recovery of heat in the recuperator [17,18].

The current trend is to create small-scale plants with the aim of exploiting alternative sources and recovering waste energy, which means moving towards a decentralized production with micro-grids capable of self-producing electric and thermal energy, and at the same time, are connected directly to the distribution network, with reduced transmission losses (grid-connect) or, in some cases, are totally consumed by the self-producer (standalone) [19]. The electrical grids are, therefore, required to function as bidirectional systems, managing to accommodate the energy production coming from multiple generation nodes. Through distributed generation, electricity is then produced directly on the site. It is, therefore, a decentralized production with a power rating of 25–500 kW, suitable for small- and medium-sized companies [20] and residential utilities [21].

Despite lower efficiency and higher prices compared to ICE, MGT offers major potential advantages in small-scale power generation. For instance, a compact size, a low weight per unit power, a small number of moving parts, lower noise, multi-fuel capabilities and opportunities for lower emissions due to the high-quality recoverable heat at the turbine exhaust [16]. This last characteristic makes the MGT implementation feasible not only for the combined production of heat and power [22] but with the addition of an absorption chiller, cooling generation (CCHP) is possible as well [23].

In the last several years, many manufacturers have proposed some packaged micro-gas turbines with the benefit of lower maintenance costs and high availability. The main application for micro-gas turbines is in packaged small-scale CHP which can be installed as single or multiple units. Typical performance features of the main MGTs used are summarized in Table 1.

Table 1. Technical characteristics of leading microturbine manufacturers [8,24].

Manufacturer	Model	Power Output	Efficiency	Pressure Ratio	TIT/EGT	Nominal Speed	CO_2
		kW_{el}	%		°C	rpm	g/kWh_{el}
ABB [25]	MT 100	100	30	4.5	950/300	70,000	660
Bladon Jets	MTG12TL	12	-	-	-	134,000	-
Bowman	-	80	27	4.3	680/300	68,000	733.33
	C30	30	26	3.2	871/275	96,000	761.54
Capstone [26]	C65	65	28	3.2	871/309	85,000	707.14
	C200 HP	200	33	3.2	870/280	45,000	600
	TA 45	45	30	4.0	871/280	116,000	660
Elliott Energy System	TA 80	80	30	4.0	871/280	68,000	660
	TA 200	200	30	4.0	871/280	43,000	660
ET Group [27]	-	45	28	-	-	60,000	707.14
Honeywell (AlliedSignal) [28]	Parallon 75	75	28.5	3.7	900/250	85,000	694.74
Ingersoll-Rand	-	70	33	3.3	870/200	60,000	594.59
MTT BV [29]	-	3	16	2.8	-	240,000	1237.5
Nissan	-	2.6	8–10	2.8	850/250	100,000	2200
Power Works™	-	70	30 (HHV)	3	704/-	-	660
Turbec (Ansaldo Energia) [30]	AE-T100	100	30	4.5	950/270	70,000	660

In the last column for each MGT the emissions of CO_2 in gram per unit of electric energy produced are calculated considering pure CH_4 as fuel with a LHV of 50 MJ/kg in a stoichiometric reaction:

$$CH_4 + 2O_2 \rightarrow CO_2 + 2H_2O + 800\ \text{MJ/kmole} \tag{1}$$

$$\text{SFC}\left[\frac{g_{CH_4}}{\text{kWh}_{el}}\right] = \frac{3600}{\eta_G \cdot LHV} \tag{2}$$

$$EICO_2\left[\frac{g_{CO_2}}{\text{kWh}_{el}}\right] = \frac{44}{16}\cdot\text{SFC} = 2.75\cdot\text{SFC} \tag{3}$$

Table 1 highlights that, if only the electric production is accounted for, the emissions of CO_2 are unacceptable in today's climate where environmental sustainability requires that greenhouse emissions are reduced. Conversely, a significant exhaust heat recovery to produce additional power, heat or cooling increases the total efficiency (or the fuel utilization factor), thus allowing a substantial savings in fuel and a reduction in CO_2.

Moreover, a further reduction in emissions can be obtained from the already mentioned high flexibility operation of MGTs [31–33]. Indeed, apart from the most used natural gas (NG), they can be supplied with carbon-neutral fuels such as synthetic gases obtained from the processing of biomass feedstocks. However, due to the different chemical compositions and the lower energetic contents, a deep analysis of the system behaviour when using these fuels is mandatory [34–36]. Furthermore, the increasing interest towards hydrogen derives not only from the absence of carbon but also from the beneficial effects on performance and flame stabilization that are induced by a moderate presence of this species in the syngas [37], whereas the well-known problem of flashback at higher hydrogen concentrations can be easily prevented by using water dilution [38].

Finally, micro gas turbines are well suited to be coupled with power and heat generation systems supplied with renewable medium-temperature energy sources. Indeed, the integration of MGT with biomass gasification, solar fields, Organic Rankine Cycle (ORC) units and Fuel Cell (FC) allows for the building of self-sustaining plants, the so-called "micro-grid", that is able to support the distributed generation of heat and power, even in isolated areas with a consistent fuel saving. As a matter of fact, it should be emphasized that the evolution of small-size gas turbines, together with the use of renewable sources, are in favor of distributed energy in micro-grid configurations, which is capable of satisfying the needs of autonomous electricity supply and heat in an environmentally friendly manner.

To this end, several studies are focused on improving and optimizing MGT-based systems and verifying the plant's economic feasibility [16,39,40].

In this review, an investigation on different plant layouts integrating the MGT with several renewable energy sources that stand out with regard to optimal energetic performance and low ambient impact is presented. An attempt is made to classify innovative elements by typology on an MGT plant that could be integrated into a micro-grid. However, a clear division between these types of hybrid systems is not always possible since there is often an overlap of such elements, complicating the installation but increasing the efficiency and reducing the environmental impact.

The work is organized as follows. In Section 2, the utilization of alternative fuel in a micro gas turbine is examined in terms of combustion stability, performance, and emission; then different hybrid systems in which the MGT is integrated are described. Section 3 focuses on solar fields, Section 4 on ORC while fuel cells are illustrated in Section 5. Finally, some hints regarding the application of MGTs in transport sector are provided in Section 6.

2. Alternative Fuels in Micro Gas Turbines

MGT represents a viable solution to the progressive increase in natural gas costs and the necessity of reducing greenhouse gas emissions. Indeed, it can be supplied with alternative liquids (kerosene and diesel) and gaseous fuels (syngas and hydrogen). Moreover, the use of fuels obtained from biomass, although usually characterized by a higher price than natural gas, guarantees the same production of dispatchable electricity while maintaining net zero carbon emissions.

2.1. *Syngas*

Directive 2001/77/EC defines as biomass "the biodegradable fraction of products, waste and residues from agriculture (including vegetal and animal substances), forestry and related industries, as well as the biodegradable fraction of industrial and municipal waste" [41]. It is considered a renewable energy source that is respectful of the environment because it is "carbon-neutral". According to the Intergovernmental Panel on Climate Change (IPCC), carbon neutrality is obtained when anthropogenic emissions of greenhouse gases are compensated by an equal number of emissions reduced, avoided, or sequestered within a certain time frame [42]. Basically, the CO_2 released by the energy processes of biomass would be the same as that biologically released in nature during its "life cycle".

The different pathways for producing energy from biomass often lead to an intermediate product such as synthesis gas (or syngas) and biofuels, which are typically rich in hydrogen and carbon monoxide. However, depending on the original feedstock and process (pyrolysis, anaerobic, gasification, partial oxidation, just to name a few), chemical and physical properties of the final fuel can be very diverse. Consequently, combustion stability, temperature, efficiency, and emissions are strongly affected by the composition as well. Although these conversion processes are optimized [43], usually these kinds of fuels are characterized by a reduced lower heating value (LHV) with respect to natural gas. The latter aspect must be accounted for especially when syngas feeds a MGT designed to work with NG. Indeed, to maintain a constant energy supply, the reduced LHV imposes an enhancement of fuel mass flow and thus leading the turbine and the compressor to part load-like conditions. This means that not only the combustion itself, but the whole system's efficiency is affected by varying fuel characteristics.

Delattin et al. [34] discussed the experimental and imaging results found for a combustor burning NG, syngas (from a steel mill) or a mixture of both. It was found that the beneficial effects from the presence of H_2 and CO are surpassed by the negative influence of water, and differences were noted in terms of flame shape, flame temperature and anchored position. Near walls temperature remained within the allowable limits, avoiding problems with material resistance for hydrogen-rich mixtures as well. The location of the flame tended to move towards the pilot cup causing flashback phenomena. However, no problems concerning blowoff, dynamic instability or auto-ignition were detected. Also, at equal thermal loads, the combustion of the syngas produced remarkably low NO_x and CO emissions.

To overcome the limitations of syngas supply in [44] the FLOX-combustion system was tested in the Turbec T100 micro turbine. Such a combustion system is based on the flameless oxidation (FLOX) concept. Similar to the original combustion system of the T100, the fuel was distributed through two lines: pilot and main stages. However, due to the higher fuel mass flow rate required, bigger valves were installed in the test rig and an additional adjustable nozzle ensured a constant pilot fuel mass flow, used for ignition and flame stabilization. Instead, the main stage presented ten annular air nozzles with coaxial fuel jets. The high momentum generated by the injected jets created a wide central recirculation zone towards the exit of the jets, allowing a strong mixing of air and exhaust gases before the reaction zone. In turn, the jets of the main stage entrained the hot gases from the pilot stage, that, in this way, assisted the combustion inside the main region. This combustion system ensured a stable operation between 50 and 100 kW_{el} while enabling a reliable start-up procedure of the micro turbine supplied with a 5.0 MJ/kg LHV fuel. Moreover, the pressure losses across the combustor remained below 4% but, most importantly, the emissions of CO and NO_x were less than 30 ppm and 6 ppm, respectively. Another flameless combustion concept that is attracting the interest of researchers is the Moderate and Intense Low-Oxygen Dilution (MILD) combustion since it proved to enhance efficiency and decrease emissions. To establish the MILD regime, strong exhaust gas recirculation is required to abate oxygen reactivity. Additionally, in this case, this is obtained via a proper geometry of the burner and the design of the gas inlet nozzles. Fortunato et al. [45] discussed the performances of the combustion chamber (CC) when burning methane and biogas. In both cases the prototype of the burner demonstrated flexible behaviour and performed well in terms of emissions, achieving ignition and stable combustion for various air-to-fuel ratios and different air inlet temperatures.

Ref. [46] investigated the effects of mixtures of natural gas and syngas derived from the gasification of rice hulls in a micro turbine of 30 kW. The MGT presented an efficiency of 26% at the maximum load when supplied with only NG; with a blend of 25% made of syngas the efficiency decreased to 24.2%; with a further enhancement of syngas volume share (50%) the efficiency fell to 22.5%. The same trend was observed in the maximum output power that dropped to 22.94 kW. Furthermore, at the same power level, the chemiluminescence of OH radicals of NG/syngas blends showed a weaker flame with respect to NG flame; this also implies that the overall heat release was weaker resulting in higher presence of unburned gas in the exhaust (CO). In [47], the effects of the fuel injector position and the pilot split on combustion performance were experimentally examined. Indeed, jet depth and direction have a significant role in the reduction of NO_x emissions since they affect the mixing distance and uniformity of fuel and air.

Parallel to experimental research, combustion of synthetic fuels has been widely studied with numerical tools, adapting already used models or implementing new mechanisms. Cameretti and Tuccillo [33] examined the behaviour of a MGT combustor when fed with different syngases: the first one obtained from a gasification process with oxygen, the second one from an anaerobic digestion, and the last from solid waste pyrolysis. Compositions and thermochemical properties of these biofuels differ significantly from those of NG, as listed in Table 2.

Table 2. Natural gas and biogas chemical properties.

Fuel Compos. [%, molar]	NG	BIOM O	BIOM AD	SW
CH_4	92.00	18.00	65.00	7.00
C_2H_6	3.70	2.00	- -	7.00
C_3H_8	1.00	2.00	- -	7.00
C_4H_{10}	0.25	2.00	- -	- -
N_2	2.90	8.00	- -	- -
H_2	- -	25.00	- -	18.00
CO	- -	33.00	- -	61.00
CO_2	0.15	10.00	35.00	- -
H_2O	- -	- -	- -	- -
Mol. Mass [g/mol]	17.34	21.92	25.83	23.76
LHV [kJ/kg]	47,182	19,198	20,183	21,697
f_{st}	0.0620	0.1680	0.145	0.1530
T_{af} [K]	2220	2231	2126	2300
f	0.0087	0.0208	0.0198	0.0184
φ	0.137	0.123	0.136	0.1203

Preliminarily, a thermodynamic analysis was carried out on a 110 kW micro gas turbine where a recuperator by-pass option and an EGR circuit were introduced (Figure 1). The results obtained were employed in a second phase as boundary conditions for the CFD simulations on the tubular lean-premixed burner.

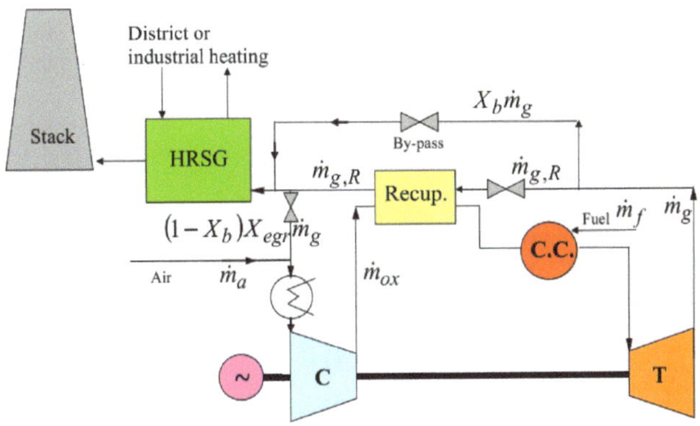

Figure 1. Plant layout of the recuperated micro-gas turbine with the EGR option [33].

With such a complex system characterized by important flexibility of the MGT in terms of electrical and thermal outputs, off-design conditions were constantly induced by the variation of the by-pass ratio (X_b), EGR ratio (X_{egr}) and mass flow rate. In particular, the reduction of the calorific value of biogas could lead to a dramatic enhancement in the fuel/air ratio, turbine choking and compressor stall, exerting a negative effect on the micro gas turbine thermal efficiency.

The same authors found that the exhaust gas recirculation allowed the achievement of combustion regimes, similar to MILD and flameless ones, with smoother temperature profiles and reduced nitrogen oxide formation. However, such benefits were obtained at the expense of the efficiency. To overcome this problem, a modification of the pilot location was attempted by testing the effectiveness of the internal EGR. CFD results demonstrated that by exploiting the main vortex inside the combustor, the temperature peaks migrated in a zone characterized by high concentrations of inert species, without affecting the reaction progress.

The same biofuels were used in paper [35] but with reference to an annular reverse flow combustor of a 30 kW MGT [26]. The combustor geometry was realized with a

CAD (Figure 2a) while the mesh (Figure 2b) and CFD simulations were realized with ANSYS software.

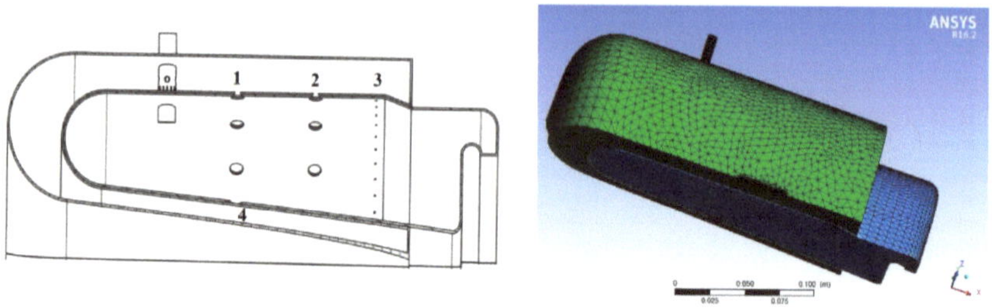

(**a**) (**b**)

Figure 2. Combustor geometry (**a**) and mesh (**b**).

The injector crosses the external liner and through holes and slots present on the walls, it allows a first partial fuel premixing with air inside the injector itself. In addition, in the original geometry the first row of holes (1 in Figure 2a) is not present. Indeed, in the first phase, the authors examined the thermo-fluid dynamic field inside the burner by testing different solutions, such as staging air introduction or air swirl intensity.

The results demonstrate that by varying air inlet in the primary zone, the residence time in the combustion core changed significantly, affecting NO and CO formation. In particular, by closing the air passage through the first row of holes, the combustion process developed under nearly RQL conditions. Indeed, in Figure 3a, it is evident the great vortex in the secondary region (different from Figure 3b) that identified the quick-mix of partially burned fuel that was in rich condition from the primary region and the air coming from the rows of holes 2 and 4. Finally, a very lean-mixture was formed thanks to the presence of the dilution in holes 3.

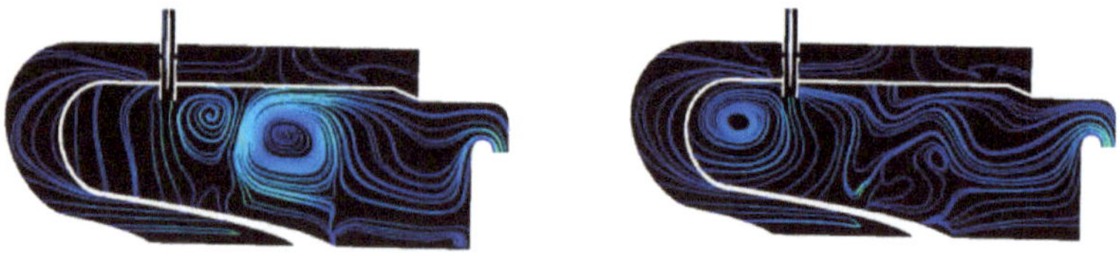

(**a**) (**b**)

Figure 3. Streamlines with the first row of holes is closed (**a**) and open (**b**) [35].

After choosing the air staging asset that guaranteed a better control on emissions formation, the different fuels reported in Table 2 were tested by keeping the thermal energy supply constant in order to achieve the same temperature at the combustor outlet of almost 900 °C (1173 K). The oxidation of all hydrocarbons was described with a seven-step kinetic mechanism via the Finite Rate—Eddy Dissipation (FRED) approach. The results displayed in Table 3 demonstrate that fuels containing H_2 (BIOM O and SW) presented the highest temperature and nitrogen oxides. On the other hand, the high temperatures led to better combustion efficiency, thus allowing for the complete oxidation of hydrocarbons in CO_2, despite the significant percentages of this species in the fuel compositions.

Table 3. Computational results [35].

	NG	BIOM O	BIOM AD	SW
T_{out} [K]	1176	1140	1165	1183
T_{max} [K]	2557	2753	2511	2689
NO [%]	- -	50.6	−31.6	39.8
CO [%]	- -	−8.3	14.7	−1.8
CH_4 [%]	- -	−27.7	44.2	−93.0

The RQL concept was investigated by Liu et al. [48] as well. The syngas-fueled combustor was simulated with a chemical reactor network model in Chemkin-Pro; it was shown that unlike the lean-premixed combustor, the NO_x emissions of the RQL firstly decreased in the reach-stage and then increased in the lean-stage with the combustor outlet temperature. Moreover, it was affirmed that such a configuration represents a promising solution to prevent flashback of high-H_2 syngas fuels.

It is evident that fuel composition can have a huge impact on the performance and flexibility of the micro gas turbine [49]. Therefore, a modification of the original design of the combustor may be necessary. Nicolosi and Renzi [50] proposed a novel design of the injector with an increased section area to better match the higher flow rate of the injected fuel. Liu et al. [51] performed a multi-fidelity design on a 12 kW MGT system to develop a swirl-stabilized, multi-point injector combustor. Moreover, they presented the overall methodology starting from a preliminary design involving the different design parameters; a detailed analysis was carried out to produce the optimum solution; then the combustor was manufactured; finally, experimental tests were conducted to verify the modeling approach. In particular, they explored the flame stabilization issue that determines the optimal number of injector orifices. The results demonstrate that at each location a more uniform mixture could be formed with an increased number of nozzles. Therefore, a larger number but smaller hole size led to higher local uniformity in fuel-air mixing and less hot-spots, thereby reducing NO_x formation. In Ref. [52] an innovative can-type combustor with a rotating case was numerically studied. The enhancement of the rotational speed increased flame stabilization, efficiency, and a uniform outlet temperature distribution. The last parameter is represented by the pattern factor (PF):

$$PF = \frac{\left(T_{max} - T_{e,avg}\right)}{\left(T_{e,avg} - T_{inlet}\right)} \tag{4}$$

where T_{max} is the maximum temperature in the combustor, $T_{e,avg}$ is the average value at the outlet, and $T_{air,inlet}$ is the air temperature at the burner inlet. A smaller value of this parameter indicates low emissions of NO_x.

In [53] a new vortex type combustor equipped with two adjustable nozzles demonstrated its successful operation with a wide range of biofuels without any need for extra equipment turbine parts. This new design proved to guarantee efficient performance with renewable fuels while keeping a lower value of gaseous emissions (CO_2, NO_x and CO) and outlet temperature distribution uniformity.

Finally, Liu et al. [54] performed experimental tests and numerical simulations on fuel/air mixing in a DLN can-type burner. Indeed, for lean combustion mixing, uniformity significantly affects combustion performance. The authors assessed PF and pollutant emissions for different nozzle positions, diameters and jet angles. The position of the nozzles had a negligible effect on the mixing while diameter and angle optimization could provide benefits on uniformity temperatures decreasing nitrogen oxide emissions but at the expense of combustion efficiency causing a rise in CO emissions.

2.2. Hydrogen

The possibility of using hydrogen as a storage medium of electrical energy has attracted the interest of researchers in several fields. Although, still efforts must be devoted to improving conversion efficiency [55], hydrogen can facilitate the control on energy demand by using it as a programmable energy vector for compensating the breaks of renewable sources, and in cases of excess of produced electricity it can be produced from electrolysis of water [56]. In addition, compared to fuel cells, a double advantage can be obtained from the H_2 feeding in thermal conversion energy systems similar to gas turbines: only water is produced in the exhausts, and it is not necessary to supply the power plant with a high purity level gas. However, gas turbine is a well-established technology mainly for NG and, as seen in the previous sections, changing the fuel usually leads to a performance decay. In addition, the operation of pure hydrogen still presents challenges due to material resistance and flame stability. Indeed, the high adiabatic flame temperature and flashback issues could cause high NO_x emissions and unsafe operation. For these reasons, major redesign efforts would be required to burn 100% H_2 fuel [57,58]. For this project scope, it is better to start using hydrogen mixed with other compounds like enriched H_2 biofuels; as already observed in Table 2 some of them can present significant percentages of H_2; therefore, attention must be paid to the combustion behaviour.

Natural gases from gas fields have different compositions and thermo-physical properties can be further affected by the addition of hydrogen. Indeed, it enlarges the flammability limits and increases compressibility factor and the lower heating value, while density decreases. For this reason, for hydrogen concentration higher than 10%, modification to the components is required [59]. Additionally, in [60] it was demonstrated that by injecting the fuel at a constant rate the outlet temperature became more uniform, but it could not satisfy the thermal resistance of the blades. Air distribution can influence combustion characteristics and pollutants [61]. Finally, Bo et al. [62] examined the flexibility of a 100 kW combustor and provided useful information on the operational issues associated with the fuel change. For a safe operation of the MGT they suggested that the start-up and thermal regulation should be performed using natural gas until the reference load was reached.

The author of this review article analysed a hydrogen—natural gas mixture and its combustion behaviour that results from the adoption of different models [63]. CFD simulations were performed on the same geometry of the reverse flow annular combustor illustrated in Figure 2. Starting with an Eddy Dissipation Concept (EDC) approach, two kinetic schemes were tested: the first one made of 8-reaction for hydrogen oxidation by Jachimowski [64] including the Zel'dovich reactions for nitrogen oxides formation and completed with the kinetic scheme for main hydrocarbons oxidation by Novosselov [65]; while the second one consisting of 53 species and 325 reactions is the well-established GRIMECH 3.0 [66]. The last mechanism was also used to create the look-up table in the flamelet-PDF approach. Moreover, two injection configurations of H_2 were compared with the base case (100% NG) using the flamelet model. Indeed, the injector of the Capstone C30, illustrated in Figure 4, can provide fuel from two lines, the pilot tube and the main holes, when operating at partial and full load, respectively.

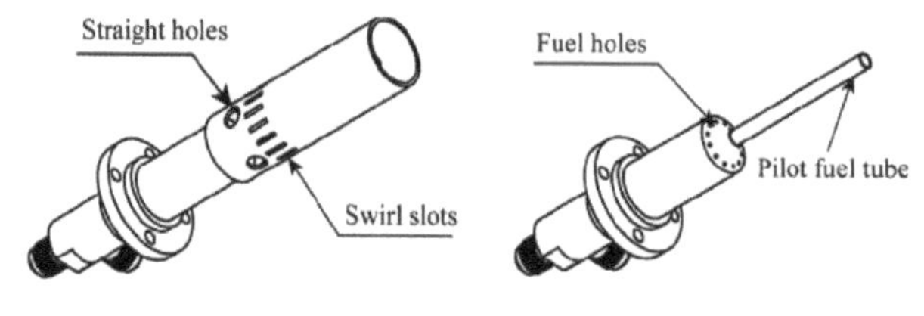

Figure 4. Capstone C30 injector [46].

In the first case 25% of H_2 was introduced in the CC together with NG from the main holes and in the second case, the same percentage of hydrogen was injected from the pilot tube. The results show that in both cases the autoignition of the fuel occurred in the premixed line. This was expected in the first case because, as already mentioned, the straight holes and swirl slots allow air from the external liner to enter into the injector and to be premixed with fuel. On the contrary, in the second case only natural gas was premixed with air since hydrogen was injected directly into the primary region of the combustor. A possible explanation for this is the unsuitability of this specific model to describe the phenomenon.

Based on this experience, the following work of the author [36] used the same strategy of split injection to overcome the limits of hydrogen percentage in the mixture, that was found to be 10% in volume. However, the analysis was conducted on the same burner using only the Finite Rate—Eddy Dissipation and the EDC approaches for the same kinetic mechanisms. Most importantly, the study also involved the operation of MGT under unconventional conditions in terms of load level and fuel supply. The assessment of part load conditions was necessary because, in the real MGT operation, the pilot line is activated only if the load is below 70%. The results in Figure 5 outline different combustion behaviours; indeed, when the fuel was injected from the premixed line the ignition began near the injector. Hence, the splitting of fuel could not only avoid the risk of flashback but could also not be affected by high temperatures thanks to this configuration in which the high reactive zone was located farther from the injector.

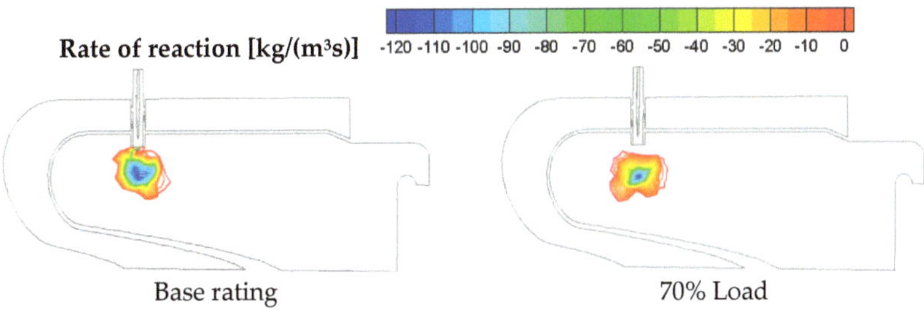

Figure 5. Methane reaction rates at full load and partial load [36].

Two hydrogen contents (10 and 25%) were simulated and compared when introduced from the pilot and from the premixed line, respectively. Results show that the 25% from the premixed line led to higher temperatures and NO emission levels but, more importantly, confirmed the occurrence of unusual autoignition still inside the injector (Figure 6a). On the contrary, Figure 6b demonstrates the effectiveness of the authors' proposal, even with a percentage of hydrogen higher than 10%, the preignition of fuel was prevented and the risk of flashback inside the injector was avoided.

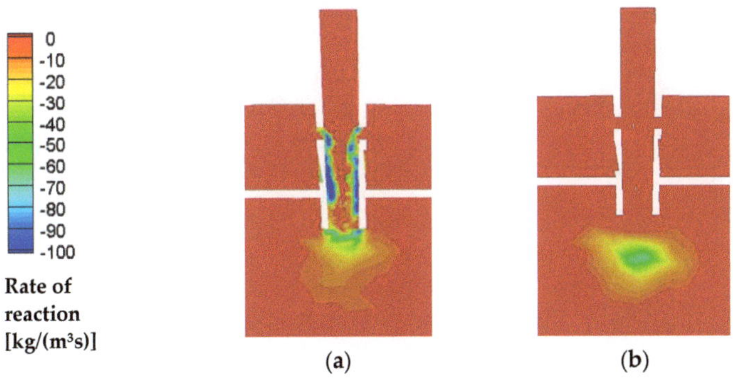

Figure 6. Hydrogen reaction rates when injected from premix (**a**) and from pilot (**b**) lines [36].

Furthermore, Tuccillo et al. [37] compared the response of the C30 combustor that, as already observed, behaves as an RQL type with a lean premixed one. For the second burner, the outputs of CFD simulations were also validated by experimental data. The results clearly illustrate that the addition of hydrogen content higher than 10% from both the pilot and the mixing lines, induces the risk of flashback. Conversely, the split strategy in the RQL type allowed the enhancement of hydrogen content and, at the same time, a better control on nitrogen oxides emissions.

With the aim to prevent flashback, steam injection was proposed as a feasible solution to lower the reactivity of hydrogen [38]. In addition, Reale [67] proved that by introducing steam directly into the combustion chamber it is possible to triple the permissible H_2 contents in the fuel blend; instead of 10%, the percentage could be increased to 30% by volume, leading to a reduction of NO_x emissions and without variating CO emissions.

In spite of the flashback issue, the presence of hydrogen in fuel blends feeding a MGT leads to beneficial effects not only on greenhouse effects [68] and CO [69] emissions, but even on performance. For instance, it was demonstrated that the enrichment of the mixture can compensate the derating of MGT occurring over the summertime due to unfavourable environmental conditions [70].

2.3. Liquid Biofuels

In addition to gaseous fuels, liquid fuels can be obtained from biomass and used to supply micro gas turbines [71]. As for standard liquid fuels, it is necessary to achieve an optimal atomization and vaporization, and usually it is possible to adopt the same solutions [72,73]. However, the different chemical and physical features of vegetable oils require further investigation.

Cavarzere et al. [31] experimentally tested different vegetable oils derived from dedicated crops (rapeseed, sunflower, soybean) and diesel and vegetable oils blends in different concentrations. They found out that oils presented slightly different composition, LHV, specific heat and density. The experimental outputs did not highlight any particular problem such as combustion or noise when the MGT was fed with a mixture of 100% vegetable oil. However, some differences with respect to diesel supply were represented by the reduced rotational speed, due to the reduced LHV at constant fuel flow and by the increased injection pressure consequent to the higher density and viscosity of vegetable oils.

The use of biodiesel obtained from edible plants, e.g., soybean, can have a negative impact on the food chain. Therefore, research is majorly addressed to the use of biodiesel from non-edible raw materials. Tan et al. [74] conducted full properties tests on five types of biodiesel and diesel blends on a 30 kW micro turbine to determine whether they meet international standards for energy production. They found out that physical and chemical properties have a linear correlation with percentage of biodiesel blends. Moisture content (ashes) for biodiesel was higher than standard specification, but this problem could be easily overcome by a heating process. Not only did they prove that the MGT is able to operate on biodiesel but, despite the lower calorific value, a higher thermal efficiency can be achieved.

Cameretti et al. [75] evaluated bioethanol combustion since it can be found in high concentrations in biomass-derived fuels. Results showed that a correct combustion can develop without changing the fuel/air equivalence ratio from the pilot line. Therefore, compared to traditional kerosene a similar efficiency can be attained, but with the benefit of a reduced flame temperature and less NO_x. The same authors [76,77], proposed a different shape and location of the main fuel injector and modified position of the pilot one. As already performed for the syngas, the aim was to reduce thermal and prompt NO by approaching the flameless combustion concept via an internal EGR. Indeed, the pilot injection occurs in poor oxygen but rich inert region. In addition, a different injector, a Delavan type, was adopted and placed at the external wall of the premixed channel, just downstream the air swirling blades. CFD outputs indicated a satisfactory response from both energetic and environmental points of view.

In all the discussed papers, the long-term feasibility of vegetable oils was not demonstrated. Laranci et al. [78] designed and validated, by means of CFD simulation, the design of a new annular RQL combustor of an Elliot TA80 to limit failure caused by overheating and corrosion enhancing the running hours of the MGT. Seljak et al. [79] provided an overview of the main solutions adopted by researchers to face the several challenges encountered when a fuel with different features from the standard ones was supplied to a MGT system. In particular, they explained that the blending of vegetable oils with diesel combined with a moderate preheating of the fuel, represents the most successful approach to limit the changing of the most affecting properties, most commonly viscosity. Enagi et al. [80] designed a new chamber geometry optimized for liquid biofuel combustion.

Finally, in order to detect any damage to the injection system or the combustion chamber in [81], tests on a MGT for power generation were performed to assess the vibrational state of the device. In particular, the chaos analysis methodology was proposed as a possible tool for the real-time characterization of combustion process.

2.4. Gasification

The energy contained in biomasses can be exploited only after thermochemical conversion processes such as gasification. Biomass must be previously pulverised and burned in grate firing systems or fluidised beds. Although in some applications it is possible to directly burn the obtained gas, certain issues such as high temperature corrosion, melting temperature of ashes and agglomeration of particles must be considered, also for logistic reasons, such as transportation and storage. A proper integration of the gasification system with the power plant in which the gas is supplied can represent a viable and economic solution to improve the efficiency of the whole process. Especially, agricultural industries generating huge amount of waste of vegetable origin (wood, leaves, pruning) could benefit from such solutions for CHP production. However, both the LHV of the produced gas and the speed of the gasification process depends on the type of biomass, thus affecting the power of the engine coupled to the gasifier [82].

Gasification can be autothermal and allothermal, depending on the required heat for the partial biomass oxidation that can be provided either by an internal or an external source. Karellas et al. [83] developed an allothermal gasifier called "Biomass Heatpipe Reformer" (BHPR) that is capable of producing hydrogen-rich gases that can supply not only MGT but can also fuel cells. The gasifier consists of a reformer, that is a fluidised bed, and a combustion chamber. The heat for the gasification was introduced into the reformer from the CC with the use of liquid metal heat pipes. The fluidisation of the reformer was obtained via water injection, and the steam was produced from the waste heat of the combustion chamber of BHPR. Therefore, the heat demand was fulfilled by the exhaust gas from the CC of the BHPR and by the sensible heat of the product gas itself after the gasifier, and by the exhaust gas from the microturbine. It was demonstrated that the optimum efficiency was achieved when the gasification took place in high temperature and low steam excess.

Vera et al. [84] modeled and simulated a small-scale CHP plant fueled with olive industry wastes incorporating the gas treatment system, i.e., a fixed bed type downdraft gasifier, gas cleaning and cooling subsystems to the microturbine. To guarantee a complete gasification of the carbon and to get a low content tar syngas, the gasifier operated with a reaction temperature of 800 °C, at atmospheric pressure. Although the obtained gas had a low calorific value (4.8–5.0 MJ/(Nm^3)), the cogeneration system was capable of providing 30 kW$_{el}$ and 60 kW$_{th}$, claiming an overall efficiency of 50%. It was also stated that preheating of the gasification air further increased the gasification efficiency. In [85] a thermoeconomic feasibility analysis of a CCHP system of a real wastewater plant was carried out. The anaerobic digester was optimized to achieve the lower values of total cost and fuel consumption.

Al-attab and Zainal [86] developed a two-stage MGT with a low-speed generator to operate only with the produced gas from the pressurized downdraft gasifier, without any auxiliary fossil fuel for hot air production. Moradi et al. [87] proposed an integrated system configuration with a Heat Recovery Steam Generator (HRSG) that produced steam for both the gasification process and the steam injection in the MGT. The gasifier was a dual-fluidized-bed reactor operating at atmospheric pressure and high temperatures (about 800 °C) which realized the partial oxidation of a hazelnut shell using steam as the oxidant agent. Hazelnut shell is a sustainable biomass because it is a non-edible feedstock from agricultural waste. It is suitable for gasification, allowing for the production of a good quality syngas with a relatively low content of contaminants (sulphur, chlorine, and ash). The authors evaluated the effects of both mass flow rate of the injected steam and the steam/biogas ratio on the performance. In particular, the latter parameter has a significant impact on the tar concentration and the composition of the produced syngas. Results show that the electrical power production increased with the increase in the mass flow rate of steam with a negligible penalization of the electrical efficiency. It was also demonstrated that although passive purification measures were adopted, the use of active purification (catalysts) units was necessary to meet the purity requirements.

Cameretti [88] analysed a hybrid layout in which a solar tower was integrated with a MGT fed by a gasifier which produced syngas from olive pits (Figure 7). The tower solar field provided partial or total replacement of fuel heating to the MGT depending on the month of the year and the hour of the day. For this reason, a parametric study was necessary to assess the behaviour of the system and especially the gasifier, when the working point changed due to the different syngas LHV and, therefore, the fuel flow rate. Results indicated that the power plant was able to satisfy the power demand exploiting exclusively renewable resources but at the expense of the overall efficiency affected by the losses in the gasification process.

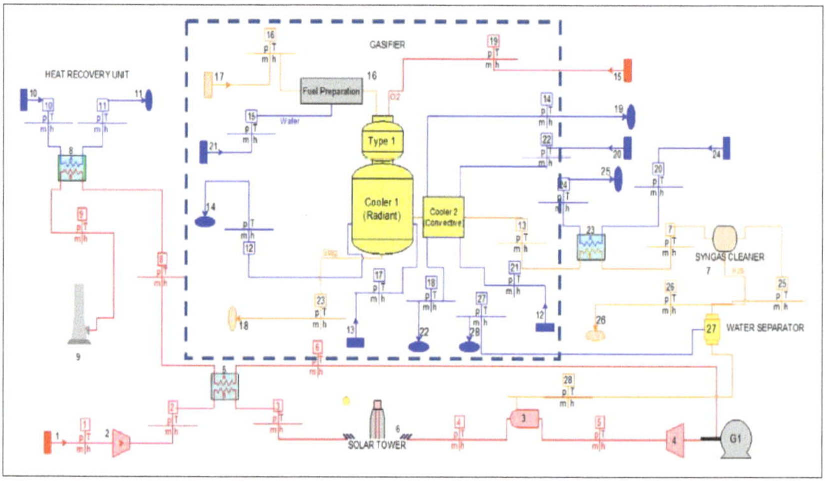

Figure 7. Hybrid MGT-solar power plant with olive pits gasifier [88].

2.5. *Ammonia*

In recent years, ammonia (NH_3) has gained considerable attention because of its carbon-free composition, substantial energy contents, and because it is an option for hydrogen transport and storage in atmospheric conditions. However, there are still challenges to be solved especially regarding high NO_x emissions and low flame stability [89].

Kurata et al. [90] developed a low-NO_x NH_3-air non-premixed combustor based on the rich-lean two-stage combustion. The combustion in the primary region is rich, so to reduce NO formation and to increase the oxidation of the H_2 produced, dilution must occur in two steps. Experimental tests confirmed the possibility of operating gas turbines with ammonia and measurements evidenced a successful reduction of NOx. To enhance flammability stabilization, the same authors [91] studied methane-ammonia mixtures, in particular the flame structure through Particle Image Velocimetry (PIV) and Planar Laser Induced Fluorescence (PLIF) imaging. In addition, they employed the Fourier Transform Infrared (FTIR) gas analysis and Large Eddy Simulation (LES) to assess the flow field and the control of emissions, respectively. Results highlighted the relevance of OH radicals in fuel NOx production. Moreover, rich-lean combustion of methane-ammonia mixtures emitted lower nitrogen oxides than ammonia alone due to the higher flame speed.

Ayaz et al. [92] performed exergy and life cycle-based analysis on a MGT supplied with NG and alternative mixtures of NG-ammonia and NG-methanol (CH_3OH), in different fractions. The mixture composed of 50% natural–50% ammonia showed the lowest CO and CO_2 emissions, whereas NG combustion alone had the lowest NO emissions. Furthermore, NH_3 evidenced a better combustion efficiency with low CO_2 and CO but more nitrogen oxides compared to every same fraction of CH_3OH. However, methanol led to a better life cycle-based environmental performance with respect to other fuel options.

Bonasio and Ravelli [93] explored the potential of direct combustion of ammonia from a thermodynamic point of view. The behaviour of a 100 kW MGT was simulated for loads in the 40–100% range, by increasing NH_3 fraction to predict performance variations of electric, thermal, and total efficiency, and exhaust gas composition. The full replacement of NG with ammonia was found to slightly reduce electric efficiency and power. In addition, CO_2 zero emissions could be achieved only with a proper adjustment of the compressor—turbine matching to compensate the calorific value reduction.

2.6. Externally Fired Gas Turbines

It is worth mentioning that, due to all the above illustrated problems caused by the burning of alternative fuels, some studies attempted to investigate solutions with Externally Fired Gas Turbines (EFGT). Indeed, EFCT can be realized by suppling solid, liquid or gaseous biogenic fuels [94]. Since fuels with a large number of impurities could be burned, the main target should be to optimize the heat exchanger and the recuperator. Whereas the recuperator could use the same technology as in a standard turbine, the heat exchanger needs to be built using special metallic or ceramic materials. Moreover, the economic feasibility could be achieved only when the cost of the biomass combustor and heat exchanger is compensated by the lower price of biomass and waste materials.

Pantaleo at al. [95] evaluated different operational strategies for the CHP with various biomass/natural gas energy input ratios to meet the energy demand required by residential users. Results showed that the energy conversion efficiency decayed, while higher investments are necessary when biomass ratio was increased. However, this solution could facilitate the integration of fossil and renewable fuel systems claiming a higher level of flexibility.

Cordiner and Mulone [96] explained that compared to internal combustion engines, i.e., MGT fuelled with NG, the loss of performance in EFGT was mainly caused by the lower turbine inlet temperature (TIT). This parameter is limited by the high heating value (HHV), by the moisture content, and, therefore, by the maximum temperature in the furnace, and especially by the efficiency of heat exchange between the furnace exhaust gas and the turbine air flow rate.

3. Micro Gas Turbine Integrated with Solar Field

Concentrating Solar Power (CSP) technologies offer a suitable solution in replacing fossil fuel consumption for small-scale energy production, especially in off-grid rural areas with high solar radiation. Through a field of reflectors, solar radiation can be concentrated onto a focal point making possible the achievement of temperatures complying with several end-use applications. Although most of the currently active or under development CSP plants are large-scale plants (>10 MW_{el}) [97,98], due to the modularity feature of such system, the scale-up is a feasible option.

In the last several years, the interest of research has been focused on micro gas turbines presenting a good compromise between flexibility and costs. Indeed, CSP allows a maximum operating temperature of the receiver above 1000 °C, and consequently ensures a satisfying thermodynamic cycle efficiency. Advantages of this integration are also related to the rapid start/stop of the MGT, good compactness, lower level of maintenance and low water consumption, which is particularly favorable for applications in rural and/or arid environments.

Moreover, the hybridization of the energy system consents the independence from the unpredictable fluctuations of solar radiation. In this regard, unlike photovoltaic panels which can be supported by battery units, energy availability is guaranteed during the periods of low or absent solar radiation and by burning a back-up fuel on household demands and achieving a stable operation of the MGT. In addition, the use of expensive batteries significantly increases the cost of produced electricity [99]. On the contrary, the possibility to feed the micro gas turbine with several compatible and renewable fuels, i.e., biogas and methanol, leads to the cost of electrical power becoming competitive with the cost of fossil fuels. In particular, appropriate methodologies of integration allow for efficiency values that exceed the level attainable with a single energy source [100].

Generally, as displayed in Figure 8, a hybrid solar-MGT system consists of two parts: the solar field and the thermal conversion unit. The former, besides reflectors/concentrators and a receiver can be provided with thermal storage, and a hot and a cold tank capable of maintaining temperature levels in a fixed range. The latter is the micro gas turbine with the compressor, the combustion chamber, and a recuperator to compensate the efficiency decay consequent to the reduced scale typical of this kind of plant.

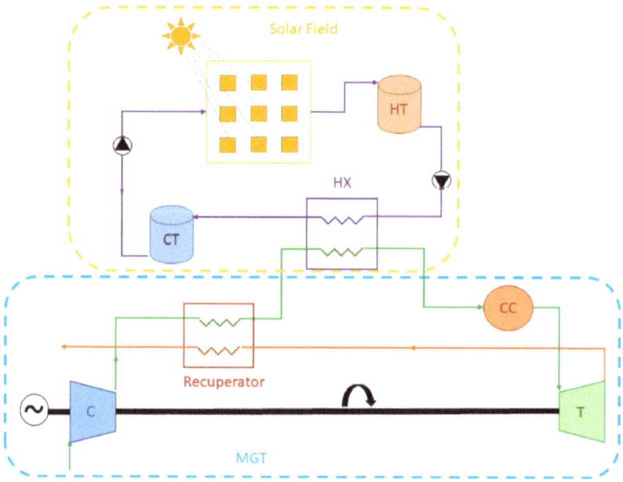

Figure 8. Generic layout of a hybrid MGT-solar power plant.

The connection between the two sections occurs through a heat exchanger (HX), where the solar heat accumulated by a diathermal oil is transferred to the air flow coming from the recuperator or directly from the compressor. If the heat is not sufficient to take the air at the TIT, the remaining increase in temperature is obtained by the combustion inside the CC. It is necessary to highlight that Figure 8 shows a generic layout. Indeed, depending on the solar technology, i.e., tower, parabolic dishes, parabolic trough collectors (PTC) and linear Fresnel reflectors on the MGT fuel and output power, and on the end-use application, the arrangement of the components can present a different asset. For example, parabolic dish and solar tower systems allow for the achievement of temperatures between 800 and 1000 °C [101], values near the typical maximum ones of MGT, in this case the heat exchanger is located between the recuperator and combustion chamber as in Figure 8, or in some projects the CC can be absent or simply bypassed.

Furthermore, as already mentioned, the use of alternative fuels can be coupled with the integration of a solar field for a further pollutant emissions reduction. However, since part of the heat supply is given by the solar plant, it is necessary to evaluate the behaviour of the whole system consequent to a varied fuel mass flow rate that leads to off-design operation. In this regard Cameretti et al. [102] investigated a CHP micro gas turbine/solar field plant (Figure 9) provided with storage (14) to partially cover loading requests during night hours. The 30 kW micro gas turbine was fed with three fuels, namely, NG, hydrogen and syngas obtained by olive pits biomass. Numerical calculations were carried out at a constant heat flow rate in the combustion chamber, hence at constant TIT.

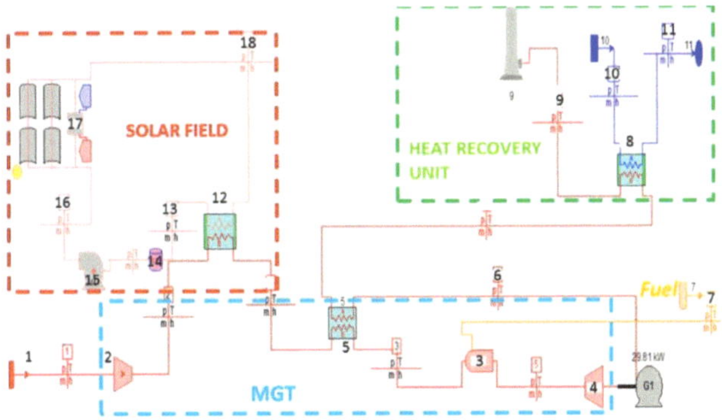

Figure 9. Hybrid MGT-solar power plant [102].

The scheme illustrated in Figure 9 outlines that the solar field solution chosen was the parabolic trough collector's one. In this case the PTC (17) were used to heat the air coming from the compressor (2) to reduce fuel consumption, upstream the recuperator (5). A heat recovery unit (HRU) was included to enhance the performance of the CHP power plant.

In the first phase, a thermodynamic analysis on the daily fuel demand was performed with and without energy storage. In particular, a summer day was chosen, the 5th of July, because the average irradiance and the average temperature are both high and have opposite effects on the systems. The former increases the solar contribution, while the latter causes a decrease of the net power from the MGT due to the enhancement of compressor absorbed work. Secondly, the fuel comparison was conducted relative to the whole year; therefore, with different values of irradiance and temperature. Results showed that thermal storage could provide a substantial contribution during night hours, also allowing for the provision of constant input power to the heat recovery unit. Furthermore, syngas showed the lowest efficiency both on daily and yearly bases, since the reduced LHV required an increase in the fuel flow rate. For the same reason, hydrogen performed better in terms of power and thermal efficiency. However, all the studied parameters presented a minimum during summer, evidencing that the hot temperatures have higher influence than the higher solar radiance. In the final part of the paper, the estimation of pollutants demonstrated that syngas combustion produced the largest amount of CO_2 per kWh, whereas H_2 produced the lowest. Finally, it is important to highlight that the exergy analysis proved a doubled exergy availability at HRU.

In the next two subsections, a review of the main studies and an innovative layout proposal is presented for plants with a parabolic dish and a solar tower.

3.1. Parabolic Dish

Usually, this type of solar collector is coupled with Stirling engines (3–35 kW_{el}) capable of achieving efficiencies of 30–40%. However, the year efficiency is reduced to 10–15% due to the high maintenance costs and reliability issues. Therefore, to prove the economic viability and technical feasibility of MGT integrated with a solar dish, Mazzoni et al. [103] developed a plant simulator for predicting the overall plant performance, and for optimizing the operations under different boundary conditions. The results demonstrate that for a net generated power of 6 $kW_{el,}$ a nominal peak efficiency of 10% could be reached which iscomparable with that of the Stirling engine.

Delavar and Wang [104] studied the effects of cold weather conditions, such as low temperature and variations in solar radiation by referring to the city of Edmonton in a cold remote region of Canada, and the plant layout in Figure 10. They considered altitude, latitude, daylength, irradiance and temperature on the collection and conversion of solar energy for each month and evaluated how these parameters influenced the different components of the cycle and the total electrical output.

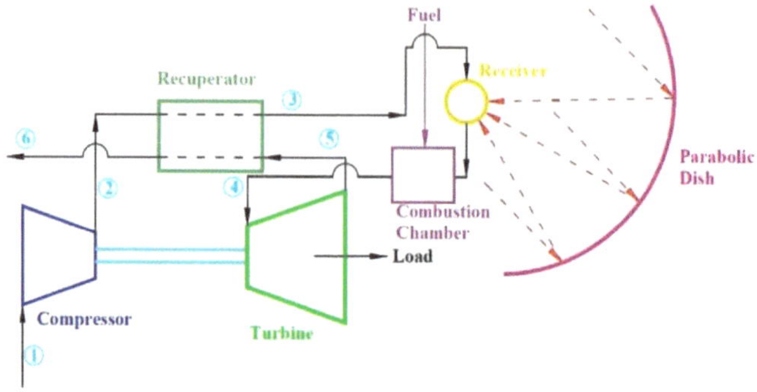

Figure 10. Hybrid MGT-solar dish plant [104].

To account for all the mentioned parameters, the authors selected two other cities, Toronto and Phoenix. It was found out that the higher total daily electrical energy during summer was obtained in Edmonton than the other two cities, while during winter the opposite scenario occurs.

Chahartaghi and Baghaee [105] evaluated the annual performances of a CCHP system for residential buildings, in order to produce electricity, heating and domestic hot water during winter and cooling during summer. Once identified, the building's load profile and solar radiation for three Iranian cities characterized by a specific climatic condition, the best combination of the number of microturbines and the area of the parabolic dishes for each city depended only on the economic evaluation, via the levelized cost of electricity (LCOE), calculated as follows:

$$\text{LCOE}\left[\frac{€}{\text{kWh}}\right] = \frac{\text{Total life cycle cost}}{\text{Total lifetime electricity yeild}} \tag{5}$$

The numerator includes the capital cost of the project (combining power, heating, and cooling), a reduction of the capital costs of the conventional system for cooling production, and the operational costs (maintenance and fuel) and the reduced operational costs of the conventional system. The denominator represents the electricity sold to the grid. It was demonstrated that in most of the cases examined, the LCOE of solar-MGT system was less than the LCOE of the photovoltaics; however, it was still higher than electricity purchase tariffs from grid.

Ghavami et al. [106] carried out a thermo-economic analysis to assess the feasibility of solar dish-MGT plants with respect to Stirling engines and photovoltaic systems. To this end, the authors evaluated different economic indicators that included the capital expenses, operational expenses, annual generated electricity and the LCOE. Such figures were differently affected by the thermodynamic parameters of some of the performances of the technical components. For example, the compressor did not affect the capital costs, but it did substantially influence the overall system performance and, therefore, the LCOE. Hence, various configurations of annual direct normal irradiation (DNI), ambient conditions of the installation site and local economic figures, such as interest rate or incentives, were studied to optimize the design of the setup by finding an effective trade-off between the thermodynamic and economic performances. It was found that a solar dish-MGT system with low-rated power (5 kW_{el}) could generate electricity with costs lower than that of the Stirling engines, but higher than the photovoltaic systems. However, such differences could decrease if the production volume of the dish-MGT units is increased. The same authors in [107] developed and validated a thermodynamic model of a solar MGT coupled with a thermochemical energy storage (TES), and different real-time control strategies for the simultaneous stable operation of the rotational speed and turbine outlet temperature (TOT). Indeed, they analyzed the transient responses of the system under real-word DNI conditions with both short- and long-term disturbances. For the plant scheme illustrated in Figure 11, by combining power and the bypass valve regulations it was possible to cope with real-world weather fluctuations. The proposed strategies allowed the system to safely operate at a constant rotational speed, TOT or both simultaneously, without incurring surging or choking conditions.

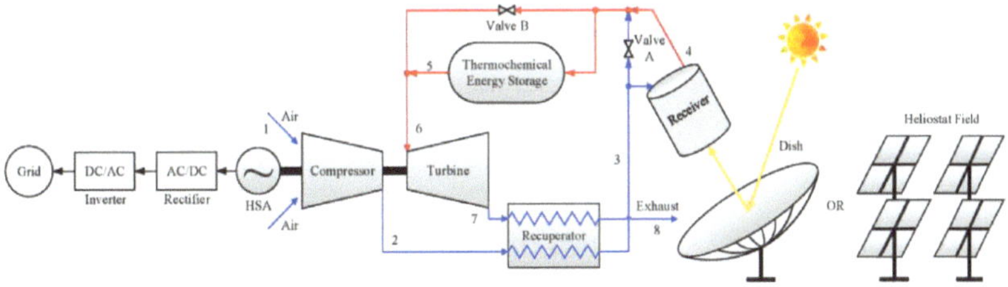

Figure 11. Scheme of the solar-MGT-TCES system [107].

Finally, Bashir et al. [108,109] focused on the improvement of the solar receiver thermal inertia. Indeed, the fluctuations of solar radiation can reduce system's performance and in the worst cases, damage the micro turbine, a thermal energy storage inside the receiver could overcome this issue by ensuring constant TOT in short-term periods. The authors designed an innovative cavity solar receiver integrated with a phase change material (PCM) capable of providing a stable temperature to the working fluid for 15–30 min. The first phase of the design process was based on a 1D thermal model for the heat loss calculation accounting for different geometrical factors, cavity surface temperature, number, and dimensions of the MGT working fluid pipes [108]. The second phase analysis was performed by three-dimensional numerical simulations for the following different PCMs: MgSi, AlSb, NiSi, and Mg_2Si. Latent heat and thermal conductivity proved to be the main factors that affect thermal storage behaviour. However, all PCMs guaranteed a temperature of the working fluid greater than 1000 K, even after 30 min of solar breakdown [109].

3.2. Solar Tower

Solar towers are surrounded by several flat mirrors arranged to reflect the incoming solar radiations in a common point, as displayed in Figure 12, to achieve high thermal radiant energy and temperature [101].

Figure 12. Schemes of the solar towers [110,111].

Cameretti et al. [112] investigated a 30 kW micro-gas turbine plant integrated with a solar tower either to provide only a part, or the total heat necessary to reach the set TIT of

1173 K (Figure 13). Hence, the solar tower is placed right before the combustor. An HRU was included in the layout for the CHP as well. Solar contribution depends on daytime and weather conditions, so that fuel supply to the CC varies over the whole year. To assess the design, the feasibility and the benefits that can be obtained by such integration, a thermo-economic analysis on the whole plant was carried out.

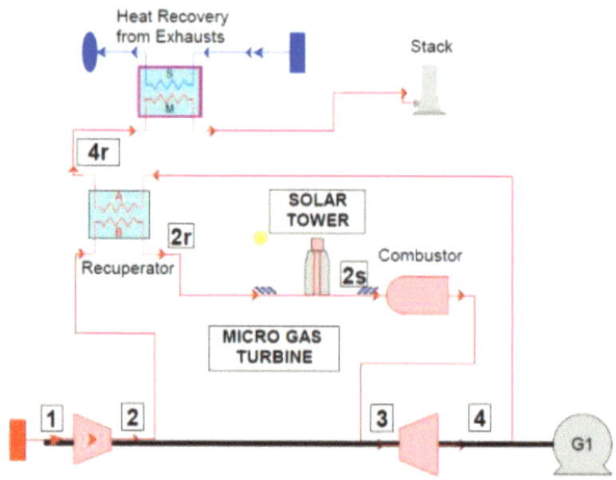

Figure 13. Layout of the MGT—solar tower plant [112].

The plant design consisted of the definition of solar field dimensions: the number of heliostats, their surfaces, and the height of the solar tower by fixing TIT equal to the set value in established hours (8:00, 12:00, 16:00, 19:00) for each month of the year. For each, the total surface of the field was estimated, and the off-design solar-MGT-CHP plant behaviour was simulated over the whole year. Numerical activity was performed with the Thermoflex commercial software that, based on the environmental conditions during the year, calculates the operating conditions of the different components. To simulate the compressor and turbine functioning, the operating maps must be provided for the code. Then, for each plant configuration the overall production of electrical and thermal power, CO_2 emissions and, consequently, investment costs and operative costs, whose sum is the total cost per year, were obtained.

The results demonstrated that two opposite targets could be pursued. Indeed, a larger solar field is able to provide a significant contribution over the year, in terms of fuel cost reduction, but it requires high investment costs (Figure 14).

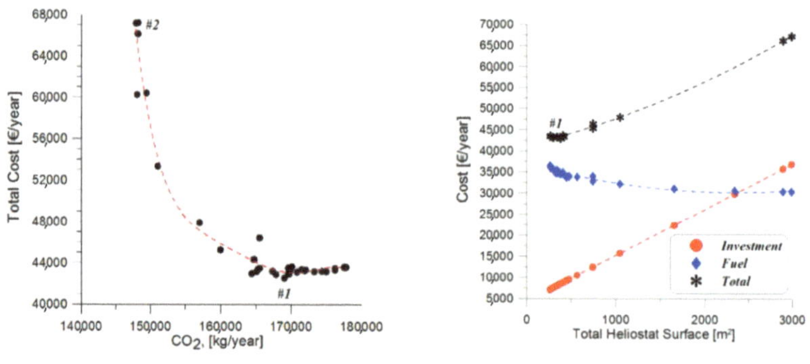

(a) (b)

Figure 14. (**a**) Total cost vs CO_2 emissions (**b**) Total yearly costs [112].

For evaluation of the thermo-economic indicators, i.e., utilization factor, primary energy saving, simple payback, two extreme cases (#1 and #2 in Figure 13) were selected and compared with the "base case" without the solar tower.

Case #1 was aimed at reducing the total cost (Figure 13b) whereas Case #2 corresponded with the hour and day that minimized fuel consumption and CO_2 emission. Both cases led to significant fuel and CO_2 emissions, but the second solution proved especially to be unfeasible due to the excess in heliostats' surfaces required with a simple payback period higher than the solar tower lifetime.

The same authors [113] deepened the study of the combustion development inside the combustor through CFD calculations to provide detailed information on both combustion efficiency and pollutant formation. Indeed, the presence of the solar tower significantly changes the operating conditions of such a component: the fuel amount is reduced, and the inlet air temperature is increased, thus also influencing the equivalence ratio. In this work the solar field was designed with reference to the spring equinox external conditions (temperature and solar irradiance), then the results obtained from thermal cycle analysis were employed as boundary conditions to the 2D and 3D calculations. Moreover, three cases were compared: the standard, and two solar-assisted operations with natural gas and syngas (from anaerobic digestion), respectively. It was outlined by temperature distributions illustrated in Figure 15 that the inlet air temperature enhancement due to solar contribution produced an axial shift on temperatures peaks, which were also reduced with the adoption of a syngas. As reported in Table 4, the inlet air temperature increase did not cause a raise of nitrogen oxides and, at the same time, led to a noticeable decrease of unburned species.

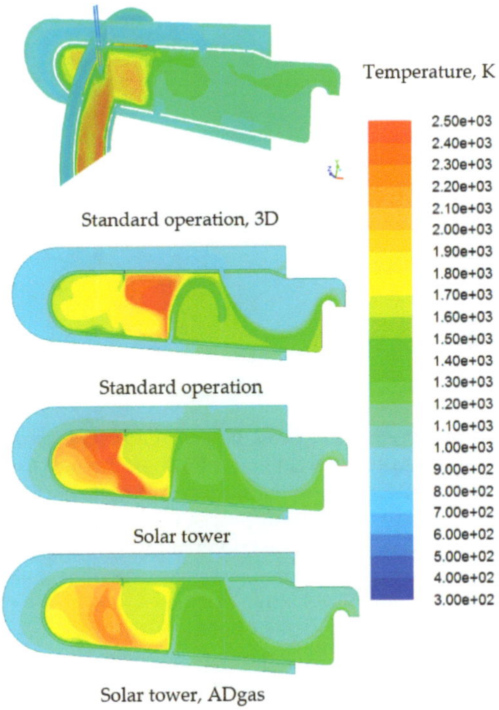

Figure 15. Temperature distributions in the combustor [113].

Table 4. Combustor outputs from CFD based simulations [113].

	Standard Operation 3D	Standard Operation	Solar Tower	Solar Tower ADgas
O_2 [mol%]	11.1	10.2	13.3	12.9
T_{max} [K]	2511	2468	2480	2265
$T_{mean,out}$ [K]	1164	1173	1147	1150
NO [ppm]	226	86	74	7
CO [ppm]	1200	3400	900	1300
UHC [ppm]	2	5	0.5	2

The authors performed off-design simulations at two opposite dates: summer and winter solstices. Although the results showed a satisfactory response of the combustor at full load and when nominal or reduced solar irradiance occurred, during the summer solstice the strong increase in solar radiation caused a further reduction of the equivalence ratio, claiming a weak combustion regime, characterized by a low combustion efficiency. On the contrary, the additional fuel supply required during the winter solstice had as a consequence the increased production of CO.

Rovense et al. [114] presented an MGT/CSP system operating with a constant efficiency level down to 1/5 of the nominal power without using fuel. This result was obtained thanks to an auxiliary compressor (7) and a bleed valve (8) as reported in Figure 16.

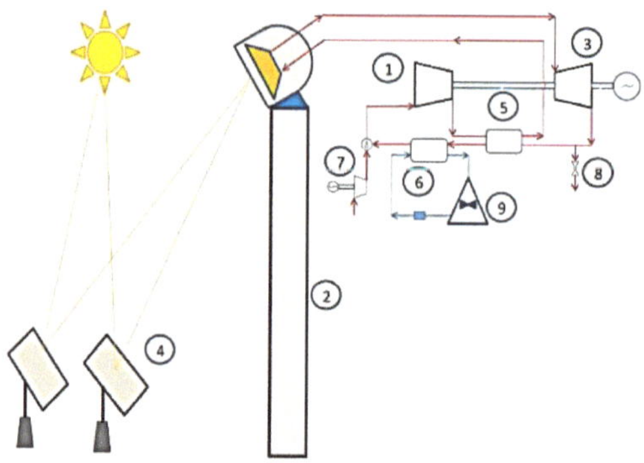

Figure 16. Plant scheme [114].

The proposed system was based on a regenerated 100 kW MGT in a closed loop configuration. After the recuperator (5), a second heat exchanger (6) was included to cool the working fluid. Controlling the maximum temperature of the thermodynamic cycle and, consequently, the output power when the solar radiation changed was achieved with air mass flow rate. In fact, to guarantee a constant net efficiency in all conditions, the auxiliary compressor admits fresh air when the thermal power received by the Sun increases, while the bleed valve discharges the excess air in the atmosphere when the thermal power decreases. In addition, the same authors developed a dynamic model in Matlab/Simulink to evaluate the system behaviour under transient conditions [115].

In Ref. [116] an application of a solar tower-MGT for CHP connected to a micro-grid of a tourist resort consisting of 33 bungalows was addressed. The work was focused on sizing of the components and a control strategy definition. This concept was shown to be flexible and capable of following power and heat demands simultaneously, despite the fluctuations of solar radiance and ambient conditions, thus permitting a fuel savings of 23%.

Ssebabi et al. [117] focused on the prediction of a micro gas turbine performance under a solar-hybrid operation, in order to identify the possible operating range and to propose control strategies considering both steady-state and transient operation. It was found that the addition of solar energy induces pressure losses in the system and, to compensate them, the equilibrium running point shifts to a higher-pressure ratio region of the compressor and it approaches the surge margin. However, an enhancement of both the work output and efficiency were obtained. Moreover, a stable operation was achieved in all seasons except for autumn due to the cloudy weather, and for a solar share between approximately 20 and 80%, depending on the season.

Giostri et al. [110] performed an economic analysis of a small-scale CSP system integrated with an MGT, with a net power in the range of 100–200 kW. In particular, the authors conducted a sensitivity analysis on DNI and turbine size to calculate the LCOE. The results showed that a correct trade-off between the design of the two systems allowed for the reduction of the LCOE to a value that is competitive with large-scale solar towers, although far from that of the photovoltaics.

4. Micro Gas Turbine Integrated with Organic Rankine Cycles

Micro gas turbines usually provide exhaust gases at temperature ranges of 250–300 °C. In addition to CHP, thermal recovery can be exploited by a bottoming ORC for the generation of medium or small electric power. An ORC plant consists of an HRU, a turbine, a condenser, a pump, and, optionally, a recuperator (Figure 17). It was demonstrated that the performances of a 100 kW could be enhanced up to 1/3 and 40% in terms of the net electric power and efficiency, respectively, [118], at the expense of an investment cost increase of 48% [119].

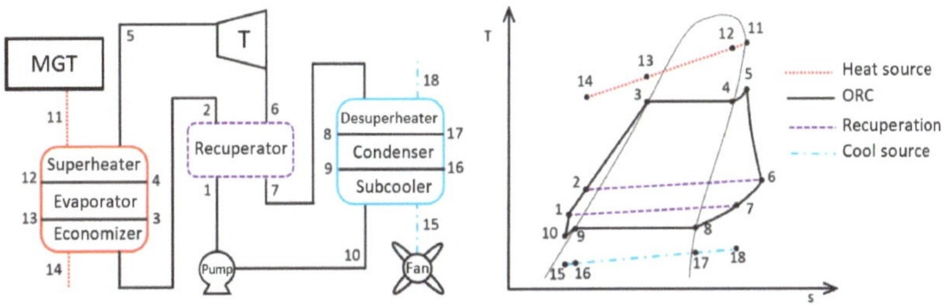

Figure 17. Plant scheme of a generic ORC and temperature vs entropy diagram [120].

In fact, the performance of the ORC and the overall system are mainly affected by the working fluid choice. In Table 5 the properties of water and of some ORC working fluids are reported. Organic fluids feature higher molecular weight and lower latent heat allowing for the reduction of the heat exchanger and machinery dimensions, and a more efficient low-grade heat harnessing [120]. As shown in the temperature-entropy diagram in Figure 16, such fluids can be defined as wet, isentropic, or dry according to the saturation curve slope, in the latter cases problems related to condensation in the expander are avoided.

Table 5. Working fluids properties [120].

	MM [kg/mol]	T_{crit} [°C]	p_{crit} [kPa]	$T_{sat}@p_{amb}$ [°C]
Water	18.01	374	22,064	100
Butane	58.12	151.98	3796	−0.49
Pentane	72.15	196.55	3379	36.06
Heptane	100.2	266.98	2736	98.38
Cyclopentane	70.13	288.87	4571	49.26
Benzene	78.11	288.87	1894	80.07
Toluene	92.14	318.6	4126	110.6

In this regard, Benato et al. [121] presented a plant layout optimization including the fluid selection for a 65 kW MGT. The authors compared the behaviour of the ORC unit in off-design with the following three working fluids: cyclopentane, R141b and cyclohexane. The design optimization was performed with an in-house tool called "ORC-Plant Designer", while the off-design simulations were performed with Aspen Plus. Outcomes demonstrated that the best performances were obtained adopting a management strategy that maintains a TIT constant using cyclopentane.

Yoon et al. [122] compared the performances of an ORC cycle and of a trans-critical CO_2 cycle applied to MGT with the load variation. It was found out that ORC provided higher power when operated at full load. On the contrary, when the micro gas turbine worked at a part load below 75%, the trans-critical CO_2 cycle appeared to be more suitable.

In [123], the ORC was integrated in a HAT cycle to maximize heat recovery at the exhaust of 200 kW MGT and to further increase power output. A saturator was inserted between compressor and CC, an economizer was added to preheat the liquid water before injecting it in the saturator, so that an increasing in fluid flow was obtained due to the substantial presence of vapour in the air entering the combustor. Again, the study was based on a different set of organic fluids and different recovery temperatures reaching optimal combinations using toluene and R245fa for medium and low temperature, respectively, with efficiency levels close to that of the larger power plants. Moreover, despite the high capital costs due to the addition of the saturator and the organic cycle, a 30% reduction in the LCOE with compared to the standalone microturbine was achieved. As a matter of fact, such a solution is potentially applicable to existing MGTs; however, saturator technology cannot be considered mature yet.

Kim et al. [124] proposed an integrated MGT-ORC system fueled by biogas, in which the heat provided by the condenser was used for heating the anaerobic digesters as illustrated in Figure 18. Indeed, the working fluid used was R245fa which is an isentropic fluid; therefore, the low-pressure vapour extracted from the ORC turbine can heat water at suitable temperatures in order to feed the biodigesters.

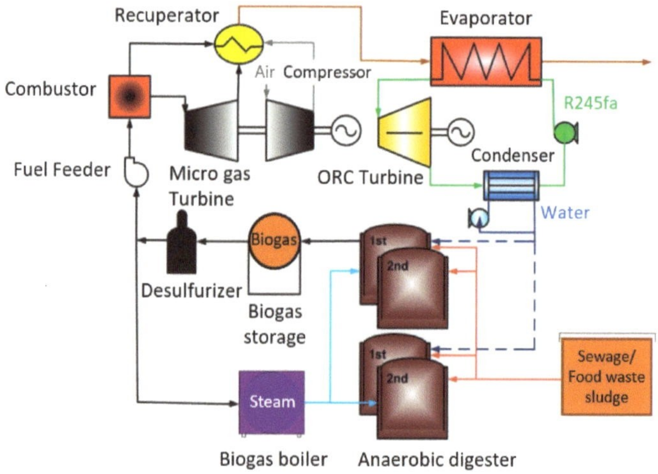

Figure 18. Schematic diagram of a biogas-fuelled MGT with bottoming ORC [124].

The work analyzed thermodynamic performance depending on the system size and the methane concentration in the biogas. The optimal configuration presented a 1000 kW MGT and 150 kW ORC with an anaerobic digester total capacity of 28,000 m^3. However, when operated in CHP configuration, the net power of the ORC dropped to 128.9 kW because of the reduced expansion ratio. Finally, CH_4 concentration of the product gas varied between 55 and 75%.

A further step to reduce fuel consumption was made in paper [113] in which the authors modified the scheme previously presented in Figure 9, that featured a MGT-solar field system by the addition of an ORC plant as illustrated in Figure 19, and compared it with the MGT-solar tower plant as illustrated in Figure 13. The solar array with PTC preheats the air upstream of the MGT recuperator, allowing the exhaust gas temperature to remain at a high level, in order to be particularly suitable to feed a bottoming ORC. The main advantage of this is to improve the recoverable heat availability enabling combined heat and power applications as well. Indeed, the HRU is still present in the layout. In this work, R134a was used as working fluid due to higher output power and energy saving was achievable. The solar field was designed with reference to DNI for a 40.1° latitude at 12:00 of the spring equinox. With respect to the solar tower-MGT system, this plant allowed an increase in the output power of 39%, thus demonstrating major flexibility in terms of the response to variable electricity and heat demands.

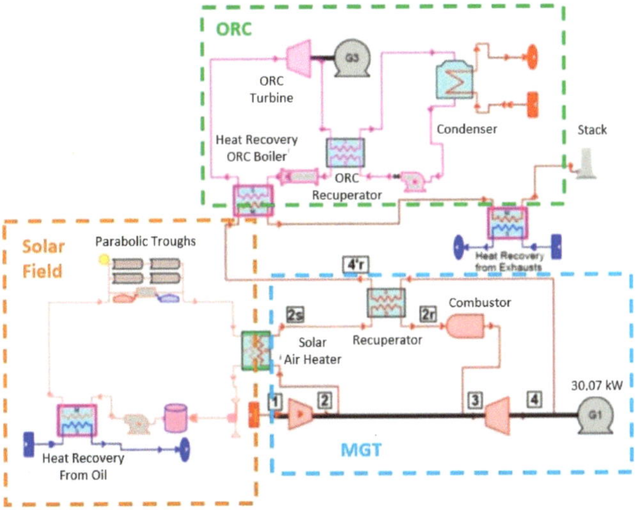

Figure 19. Layout of the combined MGT-ORC power plant with the solar array integration [113].

Javidmehr et al. [125] performed exergy and economic analysis of a hybrid system composed of compressed air energy storage, an MGT, an ORC, a solar dish collector and a multi-effect distillation process for CHP. Exergy results showed that the solar dish collector and combustion chamber are the major contributors to exergy destruction. In addition, the economic performance was improved through a multi-objective optimization of TIT, the pressures of the air before the receiver, the diameter of the dish collector, the inlet pressure in the ORC turbine and the feed water temperature of the desalination process.

Xiao et al. [126] included steam injection in the integrated plant of a 100 kW MGT with solar tower and ORC, as displayed in Figure 20. The peculiarity of this system was represented by the employment of a circle steam receiver placed around the air receiver. The absorbed concentrated solar radiance is used not only for the enhancement of the temperature upstream and the combustor, but to produce the steam that is injected upstream of the recuperator as well. Then the steam in the exhaust is reported in liquid phase by a water-cooled condenser and is pumped again into the steam receiver. Simulations were carried out with reference to four days of the year: spring equinox, summer solstice, autumn equinox and winter solstice. Numerical outcomes evidenced a reduction of fuel consumption for each day of 14.8%, 24.6%, 22.4% and 3.7%, respectively. Finally, with a receiver diameter of at least 1.1 m, steam injection and the addition of ORC, the output power could lead to a total power increase of 37.7%.

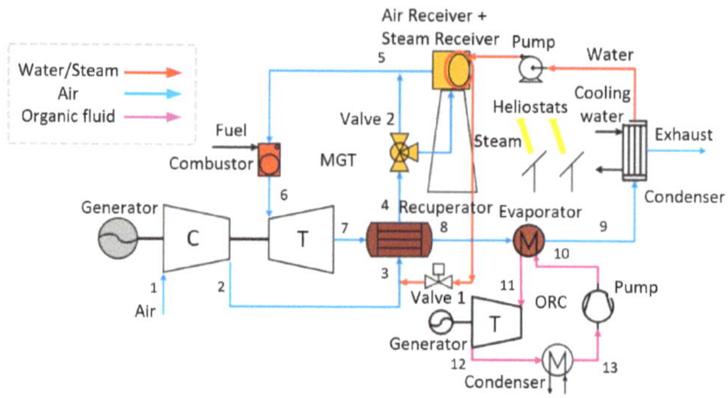

Figure 20. MGT-solar-ORC with steam injection system diagram [126].

5. Micro Gas Turbine Integrated with Fuel Cells

Generally, fuel cells are characterized by higher efficiency levels than thermal engines because they directly convert the chemical potential of the fuel into electrical energy without a thermodynamic cycle. Moreover, they release less pollutant emissions as well.

The combination of FC and MGT proved to be an optimal solution to increase the available power of the whole system while still ensuring good efficiencies and acceptable emissions [127,128]. Limitations related to operating conditions, expensive materials, and in some cases the necessity to use pure fuels, i.e., hydrogen, narrow the possible integration of MGT with the sole high temperatures FCs [129], such as Solid Oxide Fuel Cell (SOFC) and Molten Carbonate Fuel Cell (MCFC), that operate in 800–1000 °C [130] and in 600–700 °C [131] ranges, respectively. Indeed, these high temperatures can ensure a proper expansion of the exhaust gas of the fuel cell generator in the turbine, and in turn the MGT provides compressed air to the FC cathode, following the generic scheme reported in Figure 21 [132]. A pre-reformer (PR) and an indirect internal reformer (IIR) before the anode of the FC, are present to convert fuel (methane or biomass) into H_2 and CO, thanks to a nickel catalyst [133]. Additionally, other devices can be added to the system: an afterburner (AB), in case the MGT requires higher temperatures, or to enhance the flexibility of the power output, a pre-heater to increase efficiency, and components for the recirculation of anode gas from the FC that supplies the steam necessary for reforming.

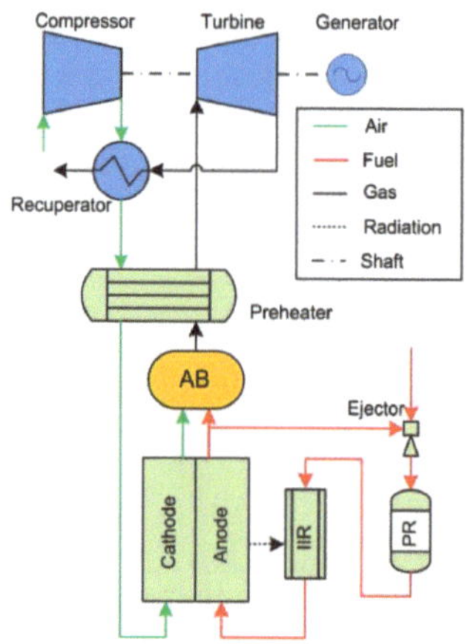

Figure 21. Hybrid SOFC/MGT system [132].

5.1. SOFC

This electrochemical device is based on a solid-state ion-conducting (O^{-2}) electrolyte (yttria-stabilized zirconia) [134]. Because of their highest temperature among fuel cells, in

SOFC systems there is no need to use external heat for fuel reformation to obtain hydrogen (autothermal gasification). Hydrocarbons represent the most suitable fuel; however, mechanical power can be produced directly by the expansion of the exhaust of the FC without the addition of the afterburner [130].

The following principal steam reforming reactions occur in the reformer [130]:

$$\text{Reforming}: \ CH_4 + H_2O \leftrightarrow CO + 3H_2 \tag{6}$$

$$\text{Shift}: \ CO + H_2O \leftrightarrow CO_2 + H_2 \tag{7}$$

The fuel can be methane, or a biogas partially oxidated in CO and H_2; after this process the reformed gas flows in the cell anode where the following electrochemical reactions take place [134]:

$$\text{Anode}: \quad H_2 + O^{-2} \rightarrow H_2O + 2e^- \quad CO + O^{-2} \rightarrow CO_2 + 2e^- \tag{8}$$

$$\text{Cathode}: \quad \frac{1}{2}O_2 + 2e^- \rightarrow O^{-2} \quad \frac{1}{2}O_2 + 2e^- \rightarrow O^{-2} \tag{9}$$

$$\text{Overall} \quad H_2 + \frac{1}{2}O_2 \rightarrow H_2O \quad CO + \frac{1}{2}O_2 \rightarrow CO_2 \tag{10}$$

The voltage of the cell is given by the following correlation:

$$V = V_N - (V_{Ohm} + V_{Act} + V_{Conc}) \tag{11}$$

where, the terms in parenthesis represent ohmic, activation, and concentration losses, respectively; V_N is voltage of the open circuit calculated via the Nernst equation, that beside the standard electrode potential, considers the partial pressures of the components participating in the electrochemical reactions:

$$V_N = -\frac{\Delta G^\circ}{2F} + \frac{RT}{2F}\ln\frac{p_{H_2}p_{O_2}^{0.5}}{p_{H_2O}} \tag{12}$$

As a matter of fact, the stability of the cell operation as well as the whole system, depends on the composition of the gas; indeed, fluctuations were observed when the FC was fed with biogas. For this reason, Kaneko et al. [135] implemented two controllers in a dynamic model to control power by managing the flow rate of biogas, and temperature by adding a bypass valve around the recuperator. Komatsu et al. [130] numerically investigated the performance of a SOFC/MGT hybrid generator demonstrating that with the decrease of the MGT rotational speed, the pressure ratio and the air flow rate decrease, and, consequently, the current density of the SOFC and the total power output were reduced, too. Nevertheless, at the design-point an efficiency level of 60% can be achieved, but in the 60–100% range of power an efficiency above 50% can still be guaranteed. Indeed, SOFC can ensure a safe operation and a high level of efficiency with proper control through a single inverter [136]. Bakalis et al. [132] examined the matching of four micro turbines reported in Table 1 (Capstone C30 and C65, Parallon 75 and Turbec T100) coupled with a fuel cell stack developed by Siemens-Westinghouse. It was found that the smaller the MGT was, the more efficiently the system performed.

Several researchers focused on the effects of mixture composition entering the cell on performances. Di Carlo et al. [137] presented and validated a comprehensive model of biomass gasifier integrated with the FC/MGT system. A sensitivity analysis on enriched air sent to the gasifier and biomass quality proved the feasibility of a wood-gas rich with H_2 and CO used in a solid oxide fuel cell, since almost no tar was produced. As expected, the best result was obtained by using pure oxygen, but the output variation remained in the 2% range, even with a deviation of 10% in the moisture content of the wood chips. Jia et al. [138] developed a zero-dimensional electrochemical and a one-dimensional chemical kinetics model for a SOFC and a downdraft biomass gasifier, respectively, with the aim to evaluate the effects of several parameters, i.e., equivalence ratio, mass flow rate and moisture content in biomass on the overall thermodynamic performance of the power system. It was found that the increase of the equivalence ratio and a decrease of moisture content led to an electrical efficiency up to 56%. The same authors [139] studied the feasibility of co-gasification of wood and animal manure integrated with a SOFC/MGT plant by assessing the effects of air flow rate, moisture content and mass fraction of woody biomass in the fuel blend. Although animal manure is a low-cost product, an increase of woody biomass fraction led to a better conversion of char and improved electrical efficiency (55%).

Fryda et al. [140] proposed an autothermal biogas gasification system expanded with a recovery unit for CHP. Three configurations characterized by different gasification pressures were studied: MGT at 4 bar, SOFC at 1.4 bar and combined SOFC/MGT at 4 bar. This last configuration was demonstrated to be the most efficient, followed by the MGT, whereas the SOFC result was penalized by the operation at a pressure near the atmospheric one. On the other hand, the improved electrical efficiency obtained from the enhanced pressure ratio lowers the available heat for regeneration. A possible solution is to increase the TIT that consequently led to an improvement of the CHP system [141].

The same concept was examined by Perna et al. [142] that estimated plant performances by varying MGT, a Turbec T100, pressure ratio and the steam amount added to the dry syngas (steam to carbon ratio). A cogeneration efficiency of 88% was obtained for a pressure ratio of 4.5, that, as reported in Table 1, is the nominal pressure of the T100, and a steam to carbon value of 0. Hualiang et al. [143] performed an advanced exergy analysis of the plant including CCHP and a multi-effect desalinization system. It was found that the gasifier and the pre-heater are the major sources of exergy destruction [144], and these are the main components to be optimized to enhance efficiency and reduce costs.

Ebrahimi and Moradpoor [145] conducted a parametric study on the thermodynamic behaviour of the cycle and added a bottoming ORC to the already complex hybrid plant. They examined the sensitivity of the system to fuel the flow rate, the steam-to-carbon ratio, the reformer temperature, the fuel utilization factor, the current density, the FC heat loss and pressure, and the maximum and minimum Rankine cycle pressures. The outcomes provided a fuel savings of about 45% and a total efficiency of more than 65%.

Furthermore, Zhimei et al. [146] proposed solar thermochemistry for the decomposition reactions of the fuel (methanol); hence, the H_2-rich fuel was converted in power firstly in SOFC and then in the micro gas turbine, realizing a solar energy cascade utilization. Such a solution allowed us to achieve exergy and an overall efficiency of 59.76% and 80.74%, respectively. In this way it is possible to limit the gasifier exergy destruction as well.

Finally, in ref. [147] a SOFC/MGT system was combined with three-reactors chemical looping hydrogen production. The advantage of using this particular reformer was to stage the conversion of fuel in hydrogen. Indeed, CO_2 was produced only in the first reactor, thus facilitating the sequestration and condensation of this product.

5.2. MCFC

As the name suggests, this type of FC features an electrolyte consisting of molten carbonates (Li_2CO_3) in an inert matrix [131]. MCFC can be fed by both hydrogen and carbon dioxide, making possible the utilization of biogas after a reforming process. Similar to SOFC, it is possible to identify the reactions occurring at the anode and the cathode, and the overall conversion process (Equations (13)–(15)).

$$\text{Anode}: \quad H_2 + CO_3^{2-} \rightarrow CO_2 + H_2O + 2e^- \quad CO + CO_3^{2-} \rightarrow 2CO_2 + 2e^- \tag{13}$$

$$\text{Cathode}: \quad \frac{1}{2}O_2 + CO_2 + 2e^- \rightarrow CO_3^{2-} \quad \frac{1}{2}O_2 + CO_2 + 2e^- \rightarrow CO_3^{2-} \tag{14}$$

$$\text{Overall} \quad H_2 + \frac{1}{2}O_2 \rightarrow H_2O \quad CO + \frac{1}{2}O_2 \rightarrow CO_2 \tag{15}$$

The voltage of the cell is still represented by correlation (8), but the Nernst equation for the open circuit voltage is modified considering the partial pressure of the species at the anode (*a*) and the cathode (*c*), as follows:

$$V_N = -\frac{\Delta G^\circ}{2F} + \frac{RT}{2F}\ln\frac{p_{H_2,a}\, p_{CO_2,c}\, p_{O_2,c}^{0.5}}{p_{H_2O,c}\, p_{CO_2,a}} \tag{16}$$

Leto et al. [131] evaluated the performance characteristics of a MCFC/MGT system for CHP under different operating conditions for both grid-connected systems and stand-alone users. Outputs of the simulations showed in detail the behaviour of the hybrid system by varying the main parameters of the electrical output power, input flow rates, and including the rotation speed of the microturbine. The steady-state simulation of the hybrid system proved that the efficiency of both the electrochemical unit alone and the entire hybrid system, can achieve a value of 60%, or 70% if the contribution of the heat recovery is considered. Huang et al. [148] investigated experimentally and numerically the performance of a MCFC/MGT system bi-fuelled by city gas and biogas. They demonstrated that the system can be able to operate stably with a volume fraction of biogas up to 50% of the total input heat value with efficiency ranging between 39.0% and 42.0%.

Yang et al. [149] developed a dynamic control strategy using Hardware-In-Loop Simulation method. The MCFC was the model subsystem, whereas the MGT (a Capstone C30) was the physical subsystem. To realize the transfer of mass, energy, and momentum between the two subsystems, a Simulation-Stimulation interface model was proposed. Finally, the dynamic models were validated by the prototype system.

6. Micro Gas Turbine in Mobility Applications

In more recent years, micro gas turbines have attracted the attention of the transport sector due to the necessity of reducing fuel consumption and pollutant emissions. In particular, the most efficient way to exploit the characteristics of this power system is to use it as an alternative to ICEs in a series hybrid configuration of a battery electric vehicle (BEV). This solution is called Range Extender Electric Vehicle (REEV) because the vehicle operates as a standard BEV, but once the battery charge is depleted, an Auxiliary Power Unit (APU) is activated to draw energy from a fuel that means to provide electricity to charge the battery and to allow for continuous vehicle operation. Since the propulsion is driven by the electric motor supplied by the battery, the APU delivers only the average power demanded by the vehicle, being decoupled from the load road. For this reason, the APU can be downsized and optimized to work at a fixed nominal point with maximum efficiency. Basically, such features perfectly fit the MGT: it is compact, lightweight, not expensive, requires little maintenance, and produces low noise and vibrations [150]. Moreover, compared to ICEs, micro gas turbines emit less unburned hydrocarbons (HC and CO) and potentially less CO_2.

In Figure 22a scheme of the components' layout is reported. The vehicle controller receives signals by the pedals and gives the traction or brake commands to the electric motor and to the master cylinder, respectively. When the state of charge (SOC) drops down to a minimum value the micro gas turbine range extender (MGTRE) controller starts the MGT to recharge the battery through the power converter; meanwhile, the battery supplies energy to the motor. However, it is possible to charge the battery via grid connection as well [151].

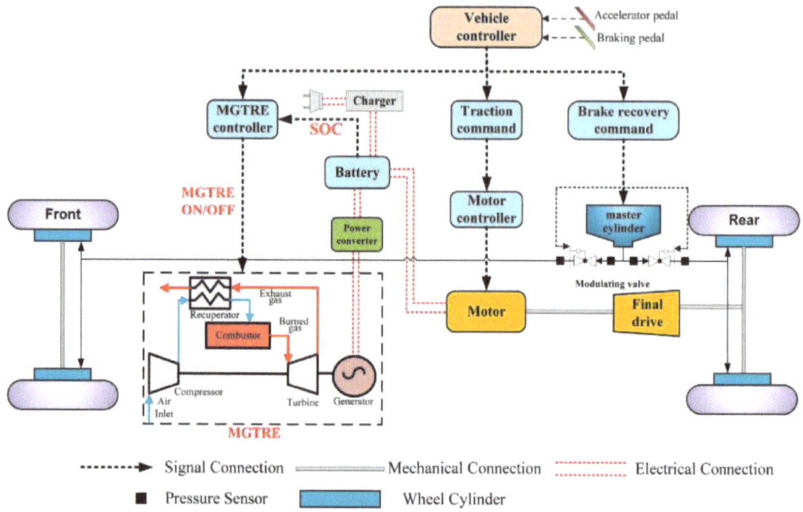

Figure 22. Scheme of a REEV with a MGT [151].

In Ref. [150], a parametric study is presented to examine MGT performance with the aim of meeting the requirements for a REEV and to assess whether this innovative configuration could be competitive in terms of fuel consumption and emission levels, with respect to gasoline or a diesel hybrid power unit, and conventional diesel engines. Calculations were performed with a specific model built in Matlab by varying ambient temperature and MGT fuel supply. The results demonstrated that although the fuel consumption is also strongly influenced by vehicle mass and battery capacity, the use of MGT leads to significant reduction of NO_x and CO emissions. Ji et al. [151] confirmed with an experimental validation that the reduction of pollutants obtainable from the implementation of MGT in REEV compared to ICE. Javed et al. [152] showed that the addition of a MGT allows for a savings in weight battery and charging time in a REEV. Shah et al. [153] also claimed that dynamic strategies can be used for MGT because, unlike internal combustion engines, an almost constant NO_x emission level is emitted with the power demand change.

7. Conclusions

In the latest several decades the energy demand has progressively increased, and this trend is not going to change in the future. In order to satisfy this need, even in remote areas and to ensure environmental sustainability, the way to produce electricity and heat has been changing as well, shifting from a centralized generation to a distributed generation with the exploitation of local renewable resources, creating the so-called "micro grids".

In this review article, the potential or actual role of the micro gas turbine has been explored with reference to the current literature. Micro gas turbine is a mature technology already spread in the market, as demonstrated by the presence of numerous manufacturers that offer power systems of different sizes for the CHP and CCHP. Nevertheless, several researchers are still studying and implementing a lot of improvements to enhance efficiency and to reduce the environmental impact of this power system.

In the first part, the MGT has been analyzed as the protagonist of energy production. Indeed, due to its well-known flexibility, the combustion of different fuels has been investigated through experimental activities and with models for the description of the phenomena that take place inside the combustor. In particular, the greatest attention has been paid to syngas and hydrogen. Such fuels show opposite behaviours, mainly due to different chemical and physical properties. The former consisting of a high percentage of CO_2, cannot always ensure an efficient oxidation process, whereas the latter, due to its high reactivity can cause flashback and increased NO_x emissions. However, in this paper various solutions such as the following have been proposed and illustrated in order to overcome these issues: a high presence of hydrogen in the syngas composition because it can combine the beneficial effects of each fuel on the other one, air staging and fuel staging can be used to control the development of the combustion process and steam injection can mitigate the high production of nitrogen oxides of hydrogen oxidation, just to name a few.

In the second phase, the integration of MGT with other single or multiple power systems has been analyzed and the results are summarized as follows:

(1) Thanks to the higher temperatures achieved by CSP plants, especially with parabolic dishes and solar towers, they can be used as a heat source for the micro gas turbine. In turn, MGT fueling can compensate for solar radiation fluctuations on both a daily and yearly basis. Depending on the overall system design, a significant reduction of fuel consumption can be attained.

(2) The flexibility of MGT reaches the maximum level when combined with an ORC. Indeed, based on the utilizer demand, the high exergy contents in the gas turbine exhausts can be used to produce more electricity.

(3) SOFC and MCFC have proven to perfectly fit with MGTs because these two types of FCs can be fuelled with natural gas and biomass (rich of H_2 and CO), and can operate at high temperatures; hence, the exhaust gases from the FC can properly expand in the turbine of the MGT increasing the produced power, with or without an afterburn process.

Finally, the use of MGT in vehicles for passenger transportation has been illustrated as one of the latest applications in which the MGT is attracting attention.

In all cases, MGT-based energy systems have demonstrated that they are able to achieve high total efficiency and high fuel utilization factors and a significant reduction of pollutant emissions with competitive costs. In particular, the efficiency of a complex micro-grid including a couple, or even all the power systems presented, have shown performances comparable to that of the larger-sized systems.

Such solutions can, therefore, concretely help to achieve a clean production of energy reducing, at the same time, the anthropogenic impact on the environment.

Funding: This research received no external funding.

Data Availability Statement: Not applicable.

Conflicts of Interest: The author declares no conflict of interest.

Abbreviations

AB	Afterburner
APU	Auxiliary Power Unit
BEV	Battery Electric Vehicle
BHPR	Biomass Heatpipe Reformer
CC	Combustion Chamber
CCHP	Combined Cooling Heat and Power
CCS	Carbon Capture and Storage
CFD	Computational Fluid Dynamics
CHP	Combined Heat and Power
CSP	Concentrating Solar Power
CT	Cold Tank
DLN	Dry Low NO_x
DNI	Direct Normal Irradiance
EDC	Eddy Dissipation Concept
EFGT	Externally Fired Gas Turbines
EGR	Exhaust Gas Recirculation
EGT	Exhaust Gas Temperature
FC	Fuel Cell

FLOX	Flameless Oxidation
FRED	Finite Rate—Eddy Dissipation
FTIR	Fourier Transform Infrared
HAT	Humid Air Turbine
HHV	High Heating Value
HRSG	Heat Recovery Steam Generator
HRU	Heat Recovery Unit
HT	Hot Tank
HX	Heat Exchanger
ICE	Internal Combustion Engine
IIR	Indirect Internal Reformer
LCOE	Levelized Cost Of Electricity
LED	Large Eddy Simulation
LHV	Lower Heating Value
MCFC	Molten Carbonate Fuel Cell
MGT	Micro Gas Turbine
MGTRE	Micro Gas Turbine Range Extender
MILD	Moderate and Intense Low-Oxygen Dilution
MM	Molar Mass
NEDC	New European Driving Cycle
NG	Natural Gas
ORC	Organic Rankine Cycle
PCM	Phase Change Material
PF	Pattern Factor
PIV	Particle Image Velocimetry
PLIF	Planar Laser Induced Fluorescence
PR	Pre-reformer
PTC	Parabolic Trough Collectors
RANS	Reynolds Averaged Navier Stokes
REEV	Range Extender Electric Vehicle
RQL	Rich-Quick-Lean
SOC	State Of Charge
SOFC	Solid Oxide Fuel Cell
TCES	Thermochemical Energy Storage
TIT	Turbine Intel Temperature
TOT	Turbine Outlet Temperature

References

1. European Commission. Latest EU Policy Actions on Climate Change. Available online: https://www.consilium.europa.eu/it/policies/climate-change/eu-climate-action/ (accessed on 2 September 2022).
2. European Commission. Recovery Plan for Europe. Available online: https://ec.europa.eu/info/strategy/recovery-plan-europe_en (accessed on 2 September 2022).
3. Guan, J.; Lv, X.; Spataru, C.; Weng, Y. Experimental and numerical study on self-sustaining performance of a 30-kW micro gas turbine generator system during startup process. *Energy* **2021**, *236*, 121468. [CrossRef]
4. Adamou, A.; Turner, J.; Costall, A.; Jones, A.; Copeland, C. Design, simulation, and validation of additively manufactured high-temperature combustion chambers for micro gas turbines. *Energy Convers. Manag.* **2021**, *248*, 114805. [CrossRef]
5. Reale, F.; Sannino, R. Numerical Modeling of Energy Systems Based on Micro Gas Turbine: A Review. *Energies* **2022**, *15*, 900. [CrossRef]
6. Grimm, F.; Enderle, B.; Kislat, O.; Izadi, S.; Zanger, J.; Kutne, P.A.M. A re-design study based on a high pressure cyclonic combustor operated with liquid fuel. In Proceedings of the ASME Turbo Expo 2019, Phoenix, AZ, USA, 17–21 June 2019. GT2019-91528. [CrossRef]
7. De Paepe, W.; Carrero, M.M.; Bram, S.; Parente, A.; Contino, F. Advanced humidified gas turbine cycle concepts applied to micro gas turbine applications for optimal waste heat recovery. *Energy Procedia* **2017**, *105*, 1712–1718. [CrossRef]
8. Xiao, G.; Yang, T.; Liu, H.; Ni, D.; Ferrari, M.; Li, M.; Luo, Z.; Cen, K.; Ni, M. Recuperators for micro gas turbines: A review. *Appl. Energy* **2017**, *197*, 83–99. [CrossRef]
9. Kosowski, K.; Piwowarski, M. Design Analysis of Micro Gas Turbines in Closed Cycles. *Energies* **2020**, *13*, 5790. [CrossRef]

10. Zong, C.; Ji, C.; Cheng, J.; Zhu, T.; Guo, D.; Li, C. Toward off-design loads: Investigations on combustion and emissions characteristics of a micro gas turbine combustor by external combustion-air adjustments. *Energy* **2022**, *253*, 124194. [CrossRef]
11. Lin, P.; Du, X.; Shi, Y.; Sun, X.-M. Modeling and controller design of a micro gas turbine for power generation. *ISA Trans.* **2022**, *124*, 411–426. [CrossRef]
12. Shen, W.; Xing, C.; Liu, H.; Liu, L.; Hu, Q.; Wu, G.; Yang, Y.; Wu, S.; Qiu, P. Exhaust gas recirculation effects on flame heat release rate distribution and dynamic characteristics in a micro gas turbine. *Energy* **2022**, *249*, 123680. [CrossRef]
13. Best, T.; Finney, K.; Ingham, D.; Pourkashanian, M. CO_2-enhanced and humidified operation of a micro-gas turbine for carbon capture. *J. Clean. Prod.* **2018**, *176*, 370–381. [CrossRef]
14. Bellas, J.-M.; Finney, K.; Diego, M.; Ingham, D.; Pourkashanian, M. Experimental investigation of the impacts of selective exhaust gas recirculation on a micro gas turbine. *Int. J. Greenh. Gas Control* **2019**, *90*, 102809. [CrossRef]
15. De Paepe, W.; Coppitters, D.; Abraham, S.; Tsirikoglou, P.; Ghorbaniasl, G.; Contino, F. Robust Operational Optimization of a Typical micro Gas Turbine. *Energy Procedia* **2018**, *158*, 5795–5803. [CrossRef]
16. Pilavachi, P. Mini- and micro-gas turbines for combined heat and power. *Appl. Therm. Eng.* **2002**, *22*, 2003–2014. [CrossRef]
17. Carrero, M.M.; De Paepe, W.; Bram, S.; Parente, A.; Contino, F. Does humidification improve the micro Gas Turbine cycle? Thermodynamic assessment based on Sankey and Grassmann diagrams. *Appl. Energy* **2017**, *204*, 1163–1171. [CrossRef]
18. De Paepe, W.; Pappa, A.; Carrero, M.M.; Bricteux, L.; Contino, F. Reducing waste heat to the minimum: Thermodynamic assessment of the M-power cycle concept applied to micro Gas Turbines. *Appl. Energy* **2020**, *279*, 115898. [CrossRef]
19. Pratt, R.; Balducci, P.; Gerkensmeyer, C.; Katipamula, S.; Kintner-Meyer, M.; Sanquist, T.; Schneider, K.; Secrest, T. *The Smart Grid: An Estimation of the Energy and CO_2 Benefits*; Pacific Northwest National Laboratory: Richland, WA, USA, 2010.
20. Balli, O.; Caliskan, H. Various thermoeconomic assessments of a heat and power system with a micro gas turbine engine used for industry. *Energy Convers. Manag.* **2022**, *252*, 114984. [CrossRef]
21. Maghsoudia, P.; Sadeghi, S. A novel economic analysis and multi-objective optimization of a 200-kW recuperated micro gas turbine considering cycle thermal efficiency and discounted payback period. *Appl. Therm. Eng.* **2020**, *166*, 114644. [CrossRef]
22. Akorede, M.F.; Hizam, H.; Pouresmaeil, E. Distributed energy resources and benefits to the environment. *Renew. Sustain. Energy Rev.* **2010**, *14*, 724–734. [CrossRef]
23. Mirzaeea, M.; Zareb, R.; Sadeghzadehc, M.; Maddaha, H.; Ahmadid, M.; Acıkkalpe, E.; Chen, L. Thermodynamic analyses of different scenarios in a CCHP system with micro turbine—Absorption chiller, and heat exchanger. *Energy Convers. Manag.* **2019**, *198*, 111919. [CrossRef]
24. Nascimento, M.R.; de Oliveira Rodrigues, L.; Santos, E.C.; Gomes, E.E.B.; Dias, F.L.G.; Velásques, E.G.; Carrillo, R.M. Micro Gas Turbine Engine: A Review. In *Progress in Gas Turbine Performance*; BoD–Books on Demand: Norderstedt, Germany, 2014; Chapter 5. [CrossRef]
25. Malmquist, A.; Aglén, O.; Keller, E.; Suter, M.; Wickström, J. Microturbines: Speeding the Shift to Distributed Heat and Power. *ABB Rev.* **2000**, *3*, 22–30.
26. Capstone. Available online: https://www.capstonegreenenergy.com/ (accessed on 15 November 2022).
27. ETGroup. Available online: https://www.etgroup.nl/en/micro-gas-turbine/ (accessed on 10 October 2022).
28. Yinger, R.J. *Behavior of Capstone and Honeywell Microturbine Generators During Load Changes*; Lawrence Berkeley National Lab. (LBNL): Berkeley, CA, USA, 2000. [CrossRef]
29. Visser, W.; Shakariyants, S.; Oostveen, M. Development of a 3 kW Microturbine for CHP Applications. *J. Eng. Gas Turbines Power* **2011**, *133*, 042301-1. [CrossRef]
30. Available online: https://www.ansaldoenergia.com/business-lines/new-units/microturbines/ (accessed on 12 October 2022).
31. Cavarzere, A.; Morini, M.; Pinelli, M.; Spina, P.; Vaccari, A.; Venturini, M. Experimental Analysis of a Micro Gas Turbine Fuelled with Vegetable Oils from Energy Crops. *Energy Procedia* **2014**, *45*, 91–100. [CrossRef]
32. Kun-Balong, A.; Sztankó, K. Reduction of pollutant emissions from a rapeseed oil fired micro gas turbine burner. *Fuel Process. Technol.* **2015**, *134*, 352–359. [CrossRef]
33. Cameretti, M.C.; Tuccillo, R. Combustion features of a bio-fuelled micro-gas turbine. *Appl. Therm. Eng.* **2015**, *89*, 280–290. [CrossRef]
34. Delattin, F.; Di Lorenzo, G.; Rizzo, S.; Bram, S.; De Ruyck, J. Combustion of syngas in a pressurized microturbine-like combustor: Experimental results. *Appl. Energy* **2010**, *87*, 1441–1452. [CrossRef]
35. Abagnale, C.; Cameretti, M.; De Robbio, R.; Tuccillo, R. CFD Study of a MGT Combustor supplied with Syngas. *Energy Procedia* **2016**, *101*, 933–940. [CrossRef]
36. Cameretti, M.C.; De Robbio, R.; Tuccillo, R. CFD Study of a Micro—Combustor Under Variable Operating Conditions. In Proceedings of the ASME Turbo Expo 2017, Charlotte, NC, USA, 26–30 June 2017. GT2017-63661. [CrossRef]
37. Tuccillo, R.; Cameretti, M.; De Robbio, R.; Reale, F.; Chiariello, F. Methane—Hydrogen Blends in Micro Gas Turbines: Comparison of Different Combustor Concepts. In Proceedings of the ASME Turbo Expo 2019, Phoenix, AZ, USA, 17–21 June 2019. GT2019-90229. [CrossRef]
38. Pappa, A.; Bricteux, L.; Benard, P.; De Paepe, W. Can water dilution avoid flashback on a hydrogen enriched micro gas turbine combustion?—A large eddy simulation study. *J. Eng. Gas Turbines Power* **2021**, *143*, 041008-1. [CrossRef]

39. Adinolfi, G.; Cigolotti, V.; Graditi, G.; Ferruzzi, G. Grid integration of Distributed Energy Resources: Technologies, potentials contributions and future prospects. In Proceedings of the International Conference on Clean Electrical Power, Alghero, Italy, 11–13 June 2013. [CrossRef]
40. Sharf, M.; Romm, I.; Palman, M.; Zelazo, D.; Cukurel, B. Economic dispatch of a single micro gas turbine under CHP operation with uncertain demands. *Appl. Energy* **2022**, *309*, 118391. [CrossRef]
41. Directive, E.U. Directive 2001/77/ec of the european parliament and of the council of 27 September 2001 on the promotion of electricity produced from renewable energy sources in the internal electricity market. *Off. J. Eur. Communities* **2001**, *283*.
42. Available online: https://www.ipcc.ch/sr15/chapter/glossary/ (accessed on 20 October 2022).
43. Gómez, M.Á.; Martín, R.; Collazo, J.; Porteiro, J. CFD Steady Model Applied to a Biomass Boiler Operating in Air Enrichment Conditions. *Energies* **2018**, *11*, 2513. [CrossRef]
44. Zornek, T.; Monz, T.; Aigner, M. Performance analysis of the micro gas turbine Turbec T100 with a new FLOX-combustion system for low calorific fuels. *Appl. Energy* **2015**, *159*, 276–284. [CrossRef]
45. Fortunato, V.; Giraldo, A.; Rouabah, M.; Nacereddine, R.; Delanaye, M.; Parente, A. Experimental and Numerical Investigation of a MILD Combustion Chamber for Micro Gas Turbine Applications. *Energies* **2018**, *11*, 3363. [CrossRef]
46. Corrêa, P.S.J.P.; Zhang, J.; Lora, E.S.; Andrade, R.V.; de Mello e Pinto, L.; Ratner, A. Experimental study on applying biomass-derived syngas in a microturbine. *Appl. Therm. Eng.* **2019**, *146*, 328–337. [CrossRef]
47. Liu, A.; Yang, Y.; Chen, L.; Zeng, W.; Wang, C. Experimental study of biogas combustion and emissions for a micro gas turbine. *Fuel* **2020**, *267*, 117312. [CrossRef]
48. Liu, H.; Qian, W.; Zhu, M.; Li, S. Kinetics modeling on NOx emissions of a syngas turbine combustor using Rich-Burn, Quick-Mix, Lean-Burn combustion method. *J. Eng. Gas Turbines Power* **2020**, *142*, 021005, GTP-19-1654. [CrossRef]
49. Di Nardo, A.; Bo, A.; Calchetti, G.; Giacomazzi, E.; Messina, G. Study on the Fuel Flexibility of a Microgas Turbine Combustor Burning Different Mixtures of H2, CH4, and CO_2. *J. Eng. Gas Turbines Power* **2020**, *142*, 061001, GTP-19-1277. [CrossRef]
50. Nicolosi, F.F.; Renzi, M. Design and CFD simulation of a micro gas turbine combustor fuelled with low LHV producer gas. In Proceedings of the ASME Turbo Expo 2020, Virtual, 21–25 September 2020. GT2020-14978. [CrossRef]
51. Liu, Y.; Nikolaidis, T.; Madani, S.H.; Sarkandi, M.; Gamil, A.; Sainal, M.F.; Hosseini, S.V. Multi-Fidelity Combustor Design and Experimental Test for. *Energies* **2022**, *15*, 2342. [CrossRef]
52. Ajvad, M.; Shih, H.-Y. Modeling syngas combustion performance of a can combustor with rotating casing for an innovative micro gas turbine. *Int. J. Hydrogen Energy* **2020**, *45*, 31188–31201. [CrossRef]
53. Bazooyar, B.; Darabkhani, H.G. Design, manufacture and test of a micro-turbine renewable energy. *Energy Convers. Manag.* **2020**, *213*, 112782. [CrossRef]
54. Liu, A.; Fan, R.; Liu, Q.; Xi, L.; Zeng, W. Numerical and Experimental Study on Combustion Characteristics of Micro-Gas Turbine Biogas Combustor. *Energies* **2022**, *15*, 8302. [CrossRef]
55. Escamilla, A.; Sánchez, D.; García-Rodríguez, L. Assessment of power-to-power renewable energy storage based on the smart integration of hydrogen and micro gas turbine technologies. *Int. J. Hydrogen Energy* **2022**, *47*, 17505–17525. [CrossRef]
56. di Gaeta, A.; Reale, F.; Chiariello, F.; Massoli, P. A dynamic model of a 100 kW micro gas turbine fuelled with natural gas and hydrogen blends and its application in a hybrid energy grid. *Energy* **2017**, *129*, 299–320. [CrossRef]
57. Cappelletti, A.; Martelli, F.; Bianchi, E.; Trifoni, E. Numerical redesign of 100kw MGT combustor for 100% H2 fueling. *Energy Procedia* **2014**, *45*, 1412–1421. [CrossRef]
58. García-Conde, A.; Martínez, C.; Prada, G.; Vélez-Bracho, V.; Benito, J. Modifications of an existing microturbine to use hydrogen as fuel. In Proceedings of the 20th Worlds Hydrogen Energy Conference 2014, Gwangju, Republic of Korea, 15–20 June 2014.
59. Deymi-Dashtebayaz, M.; Ebrahimi-Moghadam, A.; Pishbin, S.; Pourramezan, M. Investigating the effect of hydrogen injection on natural gas thermo-physical properties with various compositions. *Energy* **2019**, *167*, 235–245. [CrossRef]
60. Meziane, S.; Bentebbiche, A. Numerical study of blended fuel natural gas-hydrogen combustion in rich/quench/lean combustor of a micro gas turbine. *Int. J. Hydrogen Energy* **2019**, *44*, 15610–15621. [CrossRef]
61. Lu, J.; Fu, Z.; Liu, J.; Pan, W. Influence of air distribution on combustion characteristics of a micro gas turbine fuelled by hydrogen-doped methane. *Energy Rep.* **2022**, *8*, 207–216. [CrossRef]
62. Bo, A.; Giacomazzi, E.; Messina, G.; Di Nardo, A. Analysis of a Fuel Flexible Micro Gas Turbine Combustor Through Numerical Simulations. *J. Eng. Gas Turbines Power* **2018**, *140*, 121504-1. [CrossRef]
63. De Robbio, R. Innovative combustion analysis of a micro-gas turbine burner supplied with hydrogen-natural gas mixtures. *Energy Procedia* **2017**, *126*, 858–866. [CrossRef]
64. Jachimowski, C. *An Analytical Study of Hydrogen-Air Reaction Mechanism with Application to Scramjet Combustion*; NASA-TP-2791; National Aeronautics and Space Administration: Washington, DC, USA, 1988.
65. Novosselov, I.; Malte, P. Development and Application of an Eight-Step Global Mechanism for CFD and CRN Simulations of lean-premixed combustors. In Proceedings of the ASME Turbo Expo 2007, Montreal, QC, Canada, 1 January 2007. GT2007-27990. [CrossRef]
66. Smith, G. Available online: http://www.me.berkeley.edu/gri_mech/ (accessed on 11 November 2022).
67. Reale, F. Effects of Steam Injection on the Permissible Hydrogen Content and Gaseous Emissions in a Micro Gas Turbine Supplied by a Mixture of CH4 and H2: A CFD Analysis. *Energies* **2022**, *15*, 2914. [CrossRef]

68. Nazari, M.A.; Alavi, M.; Salem, M.; El Haj Assad, M. Utilization of hydrogen in gas turbines: A comprehensive review. *Int. J. Low-Carbon Technol.* **2022**, *17*, 513–519. [CrossRef]
69. Toit, M.; Engelbrecht, N.; Oelofse, S.; Bessarabov, D. Performance evaluation and emissions reduction of a micro gas turbine via the co-combustion of H2/CH4/CO2 fuel blends. *Sustain. Energy Technol. Assess.* **2020**, *39*, 100718. [CrossRef]
70. de Santoli, L.; Basso, G.L.; Barati, S.; D'Ambra, S.; Fasolilli, C. Seasonal energy and environmental characterization of a micro gas turbine fueled with H2NG blends. *Energy* **2020**, *193*, 116678. [CrossRef]
71. Chiaramonti, D.; Rizzo, A.; Spadi, A.; Prussi, M.; Riccio, G.; Martelli, F. Exhaust emissions from liquid fuel micro gas turbine fed with diesel oil, biodiesel and vegetable oil. *Appl. Energy* **2013**, *101*, 349–356. [CrossRef]
72. Annunziata, G.; Cameretti, M.; De Robbio, R.; Tuccillo, R. Numerical Investigation of Spray Development in a Micro Gas Turbine LPP Combustor with Airblast Atomizar. In Proceedings of the ASME Turbo Expo 2018, Oslo, Norway, 11–15 June 2018. GT2018-75789. [CrossRef]
73. Adamou, A.; Kennedy, I.; Farmer, B.; Hussein, A.; Copeland, C. Experimental and Computational Analysis of an Additive Manufactured Vaporization Injector for a Micro-Gas Turbine. In Proceedings of the ASME Turbo Expo 2019, Phoenix, AZ, USA, 17–21 June 2019. GT2019-90245. [CrossRef]
74. Tan, E.; Kuruman, P.; Indra, T.; Yoshikawa, K. Effect of Non-Edible Biodiesel Physical and Chemical Properties as Microturbine Fuel. *Energy Procedia* **2017**, *142*, 413–418. [CrossRef]
75. Cameretti, M.C.; Tuccillo, R.; Reale, F.; Piazzesi, R. Liquid Bio-Fuels in an EGR Equipped Micro Gas Turbine. In Proceedings of the ASME 2011 Turbo Expo, Vancouver, BC, Canada, 6–10 June 2011. GT2011-46341. [CrossRef]
76. Cameretti, M.C.; Tuccillo, R.; Piazzesi, R. Study of an exhaust gas recirculation equipped micro gas turbine supplied with bio-fuels. *Appl. Therm. Eng.* **2013**, *59*, 162–173. [CrossRef]
77. Cameretti, M.; Tuccillo, R.; Reale, F.; Piazzesi, R. Numerical Simulation of Liquid Bio-Fuel Combustion in an EGR Equipped Micro Gas Turbine. *Am. J. Appl. Sci.* **2015**, *12*, 463–478. [CrossRef]
78. Laranci, P.; Zampilli, M.; D'Amico, M.; Bartocci, P.; Bidini, G.; Fantozzi, F. Geometry optimization of a commercial annular RQL combustor of a micro gas turbine for use with natural gas and vegetable oils. *Energy Procedia* **2017**, *126*, 875–882. [CrossRef]
79. Seljak, T.; Pavalec, K.; Buffi, M.; Valera-Medina, A.; Chiaramonti, D.; Katrasnik, T. Challenges and Solutions for Utilization of Bioliquids in Microturbines. *J. Eng. Gas Turbines Power* **2019**, *141*, 031401. [CrossRef]
80. Enagi, I.I.; Al-attab, K.; Zainal, Z.; Teoh, Y. Palm biodiesel spray and combustion characteristics in a new micro gas turbine combustion chamber design. *Energy* **2022**, *254*, 124335. [CrossRef]
81. Allouis, C.; Amoresano, A.; Capasso, R.; Langella, G.; Niola, V.; Quaremba, G. The impact of biofuel properties on emissions and performances of a micro gas turbine using combustion vibrations detection. *Fuel Process. Technol.* **2018**, *179*, 10–16. [CrossRef]
82. Roy, P.; Datta, A.; Chakraborty, N. An assessment of different biomass feedstocks in a downdraft gasifier for engine application. *Fuel* **2013**, *106*, 864–868. [CrossRef]
83. Karellas, S.; Karl, J.; Karakas, E. An innovative biomass gasification process and its coupling with microturbine and fuel cell systems. *Energy* **2008**, *33*, 284–291. [CrossRef]
84. Vera, D.; Jurado, F.; Panopoulos, K.; Grammelis, P. Modelling of biomass gasifier and microturbine for the olive oil industry. *Int. J. Energy Res.* **2012**, *36*, 355–367. [CrossRef]
85. Movahed, P.; Avami, A. Techno-economic optimization of biogas-fueled micro gas turbine cogeneration systems in sewage treatment plant. *Energy Convers. Manag.* **2020**, *218*, 11296. [CrossRef]
86. Al-attab, K.; Zainal, Z. Performance of a biomass fueled two-stage micro gas turbine (MGT) system with hot air production heat recovery unit. *Appl. Therm. Eng.* **2014**, *70*, 61–70. [CrossRef]
87. Moradi, R.; Marcantonio, V.; Cioccolanti, L.; Bocci, E. Integrating biomass gasification with a steam-injected micro gas turbine and an Organic Rankine Cycle unit for combined heat and power production. *Energy Convers. Manag.* **2020**, *205*, 112464. [CrossRef]
88. Cameretti, M.C. Modelling of a Hybrid Solar Micro-Gas Turbine fuelled by biomass from agriculture product. *Energy Rep.* **2020**, *6*, 105–116. [CrossRef]
89. Lee, H.; Lee, M. Recent advances in ammonia combustion technology in thermal power generation system for carbon emission reduction. *Energies* **2021**, *14*, 5604. [CrossRef]
90. Kurata, O.; Iki, N.; Inoue, T.; Matsunuma, T.; Tsujimura, T.; Furutani, H.; Kawano, M.; Arai, K.; Okafor, E.; Hayakawa, A.; et al. Development of a wide range-operable, rich-lean low-NOx combustor for NH3 fuel gas-turbine power generation. *Proc. Combust. Inst.* **2019**, *37*, 4587–4595. [CrossRef]
91. Okafor, E.C.; Kunkuma, K.; Somarathne, A.; Ratthanan, R.; Kawano, M.; Kudo, T.; Kurata, O.; Iki, N.; Tsujimura, T.; Furutani, H.; et al. Control of NOx and other emissions in micro gas turbine combustors fuelled with mixtures of methane and ammonia. *Combust. Flame* **2020**, *211*, 406–416. [CrossRef]
92. Ayaz, S.; Altuntas, O.; Caliskan, H. Enhanced life cycle modelling of a micro gas turbine fuelled with various fuels for sustainable electricity production. *Renew. Sustain. Energy Rev.* **2021**, *149*, 111323. [CrossRef]
93. Bonasio, V.; Ravelli, S. Performance Analysis of an Ammonia-Fueled Micro Gas Turbine. *Energies* **2022**, *15*, 3874. [CrossRef]
94. Kautz, M.; Hansen, U. The externally-fired gas-turbine (EFGT-Cycle) for decentralized use of biomass. *Appl. Energy* **2007**, *84*, 795–805. [CrossRef]
95. Pantaleo, A.M.; Camporeale, S.; Shah, N. Natural gas–biomass dual fuelled microturbines: Comparison of operating strategies in the Italian residential sector. *Appl. Therm. Eng.* **2014**, *71*, 686–696. [CrossRef]

96. Cordiner, S.; Mulone, V. Experimental–numerical analysis of a biomass fueled microgeneration power-plant based on microturbine. *Appl. Therm. Eng.* **2014**, *71*, 905–912. [CrossRef]
97. The National Renewable Energy Laboratory. Concentrating Solar Power Projects. Available online: https://solarpaces.nrel.gov (accessed on 16 November 2022).
98. Llamas, J.M.; Bullejos, D.; de Adana, M.R. Optimal Operation Strategies into Deregulated Markets for 50 MWe Parabolic Trough Solar Thermal Power Plants with Thermal Storage. *Energies* **2019**, *12*, 935. [CrossRef]
99. Aichmayer, L.; Spelling, J.; Laumert, B. Preliminary design and analysis of a novel solar receiver for a micro gas-turbine based solar dish system. *Sol. Energy* **2015**, *114*, 378–396. [CrossRef]
100. Mingaleeva, G.; Afanaseva, O.; Nguen, D.T.; Pham, D.N.; Zunino, P. The Integration of Hybrid Mini Thermal Power Plants into the Energy Complex of the Republic of Vietnam. *Energies* **2020**, *13*, 5848. [CrossRef]
101. Kumar, L.; Ahmed, J.; El Haj Assad, M.; Hasanuzzaman, M. Prospects and Challenges of Solar Thermal for Process Heating: A Comprehensive Review. *Energies* **2022**, *15*, 8501. [CrossRef]
102. Cameretti, M.C.; Cappiello, A.; De Robbio, R.; Tuccillo, R. Comparison between Hydrogen and Syngas fuels in an integrated Mi-cro Gas Turbine/Solar field with Storage. *Energies* **2020**, *13*, 4764. [CrossRef]
103. Mazzoni, S.; Cerria, G.; Chennaoui, L. A simulation tool for concentrated solar power based on micro gas turbine engines. *Energy Convers. Manag.* **2018**, *174*, 844–854. [CrossRef]
104. Delavar, M.A.; Wang, J. Simulation of a hybrid system of solar-microturbines in cold climate regions. *Appl. Therm. Eng.* **2021**, *182*, 1359–4311. [CrossRef]
105. Chahartaghi, M.; Baghaee, A. Technical and economic analyses of a combined cooling, heating and power system based on a hybrid microturbine (solar-gas) for a residential building. *Energy Build.* **2020**, *217*, 110005. [CrossRef]
106. Ghavami, M.; Al-Zaili, J.; Sayma, A.I. A methodology for techno-economic and operation strategy optimisation of micro gas turbine-based solar powered dish-engine systems. *Energy* **2022**, *251*, 123873. [CrossRef]
107. Yang, J.; Xiao, G.; Ghavami, M.; Al-Zaili, J.; Yang, T.A.; Ni, D. Thermodynamic modelling and real-time control strategies of solar microgas turbine system with thermochemical energy storage. *J. Clean. Prod.* **2021**, *304*, 127010. [CrossRef]
108. Bashir, M.A.; Giovannelli, A.; Ali, H.M. Design of high-temperature solar receiver integrated with short-term thermal storage for Dish-Micro Gas Turbine systems. *Sol. Energy* **2019**, *190*, 156–166. [CrossRef]
109. Bashir, M.A.; Daabo, A.; Amber, K.; Khan, M.; Arshad, A.; Elahi, H. Effect of phase change materials on the short-term thermal storage in the solar receiver of dish-micro gas turbine systems: A numerical analysis. *Appl. Therm. Eng.* **2021**, *195*, 117179. [CrossRef]
110. Giostri, A.; Binotti, M.; Sterpos, G.; Lozza, G. Small scale solar tower coupled with micro gas turbine. *Renew. Energy* **2020**, *147*, 570–583. [CrossRef]
111. Available online: https://www.energy.gov/eere/articles/concentrating-solar-power-could-provide-flexibility-and-reliability-us-electric-grid (accessed on 15 November 2022).
112. Cameretti, M.; De Robbio, R.; Pirone, E.; Tuccillo, R. Thermo-Economic Analysis of a Hybrid Solar Micro Gas Turbine Power Plant. *Energy Procedia* **2017**, *126*, 667–674. [CrossRef]
113. Abagnale, C.; Cameretti, M.; De Robbio, R.; Tuccillo, R. Thermal cycle and Combustion Analysis of a Solar-assisted Micro Gas Turbine. *Energies* **2017**, *10*, 773. [CrossRef]
114. Rovense, F.; Amelio, M.; Scornaienchi, N.; Ferraro, V. Performance analysis of a solar-only gas micro turbine, with mass flow control. *Energy Procedia* **2017**, *126*, 675–682. [CrossRef]
115. Amelio, M.; Pèrez, M.S.; Ferraro, V.; Rovense, F.; Bova, S. Dynamic simulation of the temperature inlet turbine control system for an unfired micro gas turbine in a concentrating solar tower. *Energy Procedia* **2018**, *148*, 712–719. [CrossRef]
116. Rovense, F.; Reyes-Belmonte, M.; Romero, M.; González-Aguilar, J. Combined Heat/Cooling and Power Generation Using Hybrid Micro Gas Turbine in a CST Plant for a Residential Off-grid Application. *AIP Conf. Proc.* **2020**, *2303*, 080006. [CrossRef]
117. Ssebabi, B.; Dinter, F.; van der Spuy, J.; Schatz, M. Predicting the performance of a micro gas turbine under solar-hybrid operation. *Energy* **2019**, *177*, 121–135. [CrossRef]
118. Invernizzi, C.; Iora, P.; Silva, P. Bottoming micro-Rankine cycles for micro-gas turbines. *Appl. Therm. Eng.* **2007**, *27*, 100–110. [CrossRef]
119. de Campos, G.B.; Bringhenti, C.; Traverso, A.; Tomita, J.T. A Review on Combining Micro Gas Turbines with Organic Rankine Cycles. *E3S Web Conf.* **2019**, *113*, 03007. [CrossRef]
120. de Campos, G.B.; Bringhenti, C.; Traverso, A.; Tomita, J.T. Thermoeconomic optimization of organic Rankine bottoming cycles for micro gas turbines. *Appl. Therm. Eng.* **2020**, *164*, 114477. [CrossRef]
121. Benato, A.; Stoppato, A.; Mirandola, A.; Del Medico, M. Design and Off-Design Analysis of an ORC Coupled with a Micro-Gas Turbine. *Energy Procedia* **2017**, *129*, 551–558. [CrossRef]
122. Yoon, S.; Kim, M.; Kim, I.; Kim, T. Comparison of micro gas turbine heat recovery systems using ORC and trans-critical CO_2 cycle focusing on off-design performance. *Energy Procedia* **2017**, *129*, 987–994. [CrossRef]
123. Chacartegui, R.; Becerra, J.; Blanco, M.; Muñoz-Escalona, J.M. A Humid Air Turbine–Organic Rankine Cycle combined cycle for distributed microgeneration. *Energy Convers. Manag.* **2015**, *104*, 115–126. [CrossRef]
124. Kim, S.; Sung, T.; Kim, K.C. Thermodynamic Performance Analysis of a Biogas-Fuelled Micro-Gas Turbine with a Bottoming Organic Rankine Cycle for Sewage Sludge and Food Waste Treatment Plants. *Energies* **2017**, *10*, 275. [CrossRef]

125. Javidmehr, M.; Joda, F.; Mohammadi, A. Thermodynamic and economic analyses and optimization of a multi-generation system composed by a compressed air storage, solar dish collector, micro gas turbine, organic Rankine cycle, and desalination system. *Energy Convers. Manag.* **2018**, *168*, 467–481. [CrossRef]
126. Xiao, G.; Chen, J.; Ni, M.; Cen, K. A solar micro gas turbine system combined with steam injection and ORC bottoming cycle. *Energy Convers. Manag.* **2021**, *243*, 114032. [CrossRef]
127. Zhu, Y.; Tomsovic, K. Development of models for analyzing the load-following performance of microturbines and fuel cells. *Electr. Power Syst. Res.* **2002**, *62*, 1–11. [CrossRef]
128. Damo, U.; Ferrari, M.; Turan; Massardo, A. Solid oxide fuel cell hybrid system: A detailed review of an environmentally clean and efficient source of energy. *Energy* **2019**, *168*, 235–246. [CrossRef]
129. Traverso, A.; Magistri, L.; Massardo, A. Turbomachinery for the air management and energy recovery in fuel cell gas turbine hybrid systems. *Energy* **2010**, *35*, 764–777. [CrossRef]
130. Komatsu, Y.; Kimijima, S.; Szmyd, J. Performance analysis for the part-load operation of a solid oxide fuel cell–micro gas turbine hybrid system. *Energy* **2010**, *35*, 982–988. [CrossRef]
131. Leto, L.; Dispenza, C.; Moreno, A.; Calabrò, A. Simulation model of a molten carbonate fuel cellemicroturbine hybrid system. *Appl. Therm. Eng.* **2011**, *31*, 1263–1271. [CrossRef]
132. Bakalis, D.P.; Stamatis, A.G. Incorporating available micro gas turbines and fuel cell: Matching considerations and performance evaluation. *Appl. Energy* **2013**, *103*, 607–617. [CrossRef]
133. Bang-Møller, C.; Rokni, M. Thermodynamic performance study of biomass gasification, solid oxide fuel cell and micro gas turbine hybrid systems. *Energy Convers. Manag.* **2010**, *51*, 2330–2339. [CrossRef]
134. Calise, F.; Dentice d'Accadia, M.; Palombo, A.; Vanoli, L. Simulation and exergy analysis of a hybrid Solid Oxide Fuel Cell (SOFC)–Gas Turbine System. *Energy* **2006**, *31*, 3278–3299. [CrossRef]
135. Kaneko, T.; Brouwer, J.; Samuelsen, G. Power and temperature control of fluctuating biomass gas fueled solid oxide fuel cell and micro gas turbine hybrid system. *J. Power Sources* **2006**, *160*, 316–325. [CrossRef]
136. Baudoin, S.; Vechiu, I.; Camblong, H.J.-M.; Barelli, L. Sizing and control of a Solid Oxide Fuel Cell/Gas microTurbine hybrid power system using a unique inverter for rural microgrid integration. *Appl. Energy* **2016**, *176*, 272–281. [CrossRef]
137. Di Carlo, A.; Borello, D.; Bocci, E. Process simulation of a hybrid SOFC/mGT and enriched air/steam fluidized bed gasifier power plant. *Int. J. Hydrogen Energy* **2013**, *38*, 5857–5874. [CrossRef]
138. Jia, J.; Abudula, A.; Wei, L.; Sun, B.; Shi, Y. Effect of operating parameters on performance of an integrated biomass gasifier, solid oxide fuel cells and micro gas turbine system. *Biomass Bioenergy* **2015**, *75*, 35–45. [CrossRef]
139. Jia, J.; Shu, L.; Zang, G.; Xu, L.; Abudula, A.; Ge, K. Energy analysis and techno-economic assessment of a co-gasification of woody biomass and animal manure, solid oxide fuel cells and micro gas turbine hybrid system. *Energy* **2018**, *149*, 750–761. [CrossRef]
140. Fryda, L.; Panopoulos, K.; Kakaras, E. Integrated CHP with autothermal biomass gasification and SOFC–MGT. *Energy Convers. Manag.* **2008**, *49*, 281–290. [CrossRef]
141. Wongchanapai, S.; Iwai, H.; Saito, M.; Yoshida, H. Performance evaluation of a direct-biogas solid oxide fuel cell-micro gas turbine (SOFC-MGT) hybrid combined heat and power (CHP) system. *J. Power Sources* **2013**, *223*, 9–17. [CrossRef]
142. Perna, A.; Minutillo, M.; Jannelli, E.; Cigolotti, V.; Nam, S.; Yoon, K.J. Performance assessment of a hybrid SOFC/MGT cogeneration power plant fed by syngas from a biomass down-draft gasifier. *Appl. Energy* **2018**, *227*, 80–91. [CrossRef]
143. You, H.; Han, J.; Liu, Y. Conventional and advanced exergoeconomic assessments of a CCHP and MED system based on solid oxide fuel cell and micro gas turbine. *Int. J. Hydrogen Energy* **2020**, *45*, 12143–12160. [CrossRef]
144. Karimi, M.H.; Chitgar, N.; Emadi, M.; Ahmadi, P.; Rosen, M.A. Performance assessment and optimization of a biomass-based solid oxide fuel cell and micro gas turbine system integrated with an organic Rankine cycle. *Int. J. Hydrogen Energy* **2020**, *45*, 6262–6277. [CrossRef]
145. Ebrahimi, M.; Moradpoor, I. Combined solid oxide fuel cell, micro-gas turbine and organic Rankine cycle for power generation (SOFC–MGT–ORC). *Energy Convers. Manag.* **2016**, *116*, 120–133. [CrossRef]
146. Zheng, Z.; Liu, T.; Liu, Q.; Lei, J.; Fang, J. A distributed energy system integrating SOFC-MGT with mid-and-low temperature solar thermochemical hydrogen fuel production. *Int. J. Hydrogen Energy* **2021**, *46*, 19846–19860. [CrossRef]
147. Isfahani, S.N.R.; Sedaghat, A. A hybrid micro gas turbine and solid state fuel cell power plant with hydrogen production and CO_2 capture. *Int. J. Hydrogen Energy* **2016**, *41*, 9490–9499. [CrossRef]
148. Huang, H.; Li, J.; He, Z.; Zeng, T.; Kobayashi, N.; Kubota, M. Performance Analysis of a MCFC/MGT Hybrid Power System Bi-Fueled by City Gas and Biogas. *Energies* **2015**, *8*, 5661–5677. [CrossRef]
149. Yang, C.; Deng, K.; He, H.; Wu, H.; Yao, K.; Fan, Y. Real-Time Interface Model Investigation for MCFC-MGT HILS Hybrid Power System. *Energies* **2019**, *12*, 2192. [CrossRef]
150. Karvountzis-Kontakiotis, A.; Andwari, A.; Pesyridis, A.; Russo, S.; Tuccillo, R.; Esfahanian, V. Application of Micro Gas Turbine in Range-Extended Electric Vehicles Application of Micro Gas Turbine in Range-Extended Electric Vehicles. *Energy* **2018**, *147*, 351–361. [CrossRef]
151. Ji, F.; Zhang, X.; Du, F.; Ding, S.; Zhao, Y.; Xu, Z.; Wang, Y.; Zhou, Y. Experimental and numerical investigation on micro gas turbine as a range extender for electric vehicle. *Appl. Therm. Eng.* **2020**, *173*, 115236. [CrossRef]

152. Javed, A.; Khalid, H.; Arif, S.U.; Imran, M.; Rezk, A.; Khan, Z.A. Micro Gas Turbine Small-Scale Effects in Range Extended Electric Vehicles. *J. Energy Resour. Technol.* **2021**, *143*, 120906. [CrossRef]
153. Shah, R.M.R.A.; McGordon, A.; Rahman, M.; Amor-Segan, M.; Jennings, P. Characterisation of micro turbine generator as a range extender using an automotive drive cycle for series hybrid electric vehicle application. *Appl. Therm. Eng.* **2021**, *184*, 116302. [CrossRef]

www.ingramcontent.com/pod-product-compliance
Lightning Source LLC
Chambersburg PA
CBHW041400210926
51850CB00013B/3138
9783725825738

MDPI AG
Grosspeteranlage 5
4052 Basel
Switzerland
Tel.: +41 61 683 77 34

Energies Editorial Office
E-mail: energies@mdpi.com
www.mdpi.com/journal/energies